高等学校土建类专业英文系列教材

Civil Engineering Construction Technology and Organization
土木工程施工技术与组织

赵延辉　张淑朝　编著

中国建筑工业出版社

图书在版编目（CIP）数据

土木工程施工技术与组织 = Civil Engineering Construction Technology and Organization / 赵延辉，张淑朝编著. — 北京：中国建筑工业出版社，2022.5
高等学校土建类专业英文系列教材
ISBN 978-7-112-28166-4

Ⅰ.①土… Ⅱ.①赵…②张… Ⅲ.①土木工程-工程施工-高等学校-教材-英文 Ⅳ.①TU7

中国版本图书馆CIP数据核字（2022）第217580号

The book introduces the basic concepts, theory, technological process and construction methods related to Civil Engineering Construction Technology and Organization. The book contents comply with normalization and focus on requirements of practical operation; include necessary figures, tables, formulas and examples in every chapter of this book.

This book can be used as a textbook for undergraduate students majoring in civil engineering, civil project management, civil engineering materials and civil engineering cost, also can be used as training materials and reference books for engineers and technicians of architectural engineering design and construction fields.

责任编辑：李笑然　吉万旺
责任校对：姜小莲

高等学校土建类专业英文系列教材
Civil Engineering Construction Technology and Organization
土木工程施工技术与组织
赵延辉　张淑朝　编著

*

中国建筑工业出版社出版、发行（北京海淀三里河路9号）
各地新华书店、建筑书店经销
北京鸿文瀚海文化传媒有限公司制版
建工社（河北）印刷有限公司印刷

*

开本：787毫米×1092毫米　1/16　印张：22¼　字数：551千字
2023年7月第一版　2023年7月第一次印刷
定价：**59.00**元
ISBN 978-7-112-28166-4
（38958）

版权所有　翻印必究
如有内容及印装质量问题，请联系本社读者服务中心退换
电话：（010）58337283　QQ：2885381756
（地址：北京海淀三里河路9号中国建筑工业出版社604室　邮政编码：100037）

Foreword

Civil engineering construction technology and organization is one of the compulsory courses for students majoring in civil engineering and engineering management, a course with characteristics of strong practicality, a wide covering range, and rapid developing for studying construction technology and construction organization of civil engineering.

This textbook is compiled according to current national codes, technical specifications, related standards and "Course Syllabus of Civil Engineering Construction Technology" issued by national authority departments. It focuses on training students to master the basic theory and skills of civil engineering construction, and have a certain ability to solve practical engineering problems.

Scientific and advanced features are reflected on the content arrangement. The book has abandoned the outdated contents which have been less used in present construction, as well as those that are not in line with the development direction; while the new and commonly used theory, technologies and methods are retained and updated.

Practicability is another feature at the contents introductions. The book takes construction technology and processes as its main clue, focuses on the introduction of process principle and methods, which have certain theoretical depth and are easy to be applied to practices. As for the contents of construction organization, it emphasizes on the introduction of organization principle and scientific organization method, which has certain operability.

The contents need to be mastered, familiar with and understood, as well as the key points and difficult points, are listed before each chapter; questions and exercises are listed at end of chapter. It is helpful for students to read and learn basic concept, theories and applications clearly and systematically.

This text book is edited by Zhao Yanhui as chief editor, Zhang Shuchao as associate editor. The authors of each chapter are as follows: Zhao Yanhui: Chapter $4 \sim 9$. Zhang Shuchao: Chapter $1 \sim 3$.

In the process of writing this book, many experts and scholars have gave guidance and constructive suggestions. I convey my gratitude to all those who have helped me in the way of preparing the book.

I have made every effort to make this book as complete and quality as possible, due to its short writing period and more writing contents, there might be inadequacies or even mistakes in the book. I will feel grateful to the teachers and users of this book for pointing out any mistakes and for suggestions for further improvement to make the book more useful.

Directory

Chapter 1 Earthwork and Foundation Pit Engineering ················ 1
  1.1 Introduction ················ 1
  1.2 Engineering classification of soil ················ 2
  1.3 The engineering properties of soil ················ 2
  1.4 Foundation pit (trench) dewatering ················ 5
    1.4.1 Natural drainage method ················ 6
    1.4.2 Artificial dewatering method ················ 7
    1.4.3 Dewatering hazard and prevention method ················ 18
  1.5 Earthwork calculation ················ 20
    1.5.1 Earthwork calculation for foundation pit ················ 20
    1.5.2 Earthwork calculation for groove, pipe trench ················ 21
    1.5.3 Earthwork calculation for site leveling ················ 22
  1.6 Earthwork construction ················ 28
    1.6.1 Preparations before earthwork construction ················ 28
    1.6.2 Earthwork construction machinery and methods ················ 29
    1.6.3 Check of foundation subsoil ················ 35
  1.7 Foundation pit engineering ················ 37
  1.8 Earth filling and compaction ················ 40
    1.8.1 The selection of filling material ················ 40
    1.8.2 Compaction methods ················ 40
    1.8.3 The influence factors of filling soil compaction ················ 42
    1.8.4 Filling construction requirements ················ 45
    1.8.5 Quality control of compaction ················ 46
  Questions ················ 47

Chapter 2 Foundation Engineering ················ 49
  2.1 Introduction ················ 49
  2.2 Construction of shallow foundation ················ 50
    2.2.1 Construction of non-reinforced foundation ················ 50
    2.2.2 Construction of reinforced concrete foundation ················ 52
  2.3 Construction of pile foundation ················ 56
    2.3.1 Construction of precast concrete pile ················ 56
    2.3.2 Construction of steel piles ················ 73
  2.4 Construction of cast-in-place pile ················ 73

2.5　Post grouting of cast-in-place pile ······ 87
2.6　Construction of pile cap ······ 88
Questions ······ 89

Chapter 3　Masonry and Construction Facilities Engineering ······ 91
3.1　Introduction ······ 91
3.2　Masonry materials ······ 91
　　3.2.1　Masonry mortar ······ 92
　　3.2.2　Masonry blocks ······ 95
3.3　Brick masonry work ······ 99
　　3.3.1　Bonds in brick masonry work ······ 100
　　3.3.2　Preparation for masonry work ······ 104
　　3.3.3　Brick masonry construction technology ······ 106
　　3.3.4　Quality standard of brick masonry ······ 109
　　3.3.5　Construction of wall with reinforced concrete constructional column ······ 112
3.4　Concrete block masonry construction ······ 113
　　3.4.1　Preparation work ······ 113
　　3.4.2　Key construction points ······ 114
　　3.4.3　Construction of reinforced concrete core column ······ 116
3.5　Stone masonry ······ 116
　　3.5.1　Rubble masonry ······ 116
　　3.5.2　Ashlar masonry ······ 118
3.6　Scaffold ······ 120
　　3.6.1　The outdoor scaffold ······ 121
　　3.6.2　The indoor scaffold ······ 130
3.7　Vertical transportation facilities ······ 131
Questions ······ 134

Chapter 4　Concrete Structure Engineering ······ 135
4.1　Formwork engineering ······ 135
　　4.1.1　Overview of formwork ······ 136
　　4.1.2　Detailing of formwork ······ 137
　　4.1.3　Design of formwork ······ 148
　　4.1.4　Installation of formwork ······ 152
　　4.1.5　Quality requirements of formwork ······ 164
　　4.1.6　Removal of formwork ······ 165
4.2　Reinforcement engineering ······ 166
　　4.2.1　Classification categories and acceptance of reinforcement ······ 167
　　4.2.2　Reinforcement blanking ······ 170
　　4.2.3　Reinforcement substitution ······ 177
　　4.2.4　Reinforcement Fabrication ······ 178

4.2.5	Reinforcement connection	179
4.2.6	Reinforcement binding and installation	185
4.2.7	Reinforcement splice and fixing	186
4.2.8	Quality control	190

4.3 Concrete engineering ... 191
  4.3.1 Concrete preparation ... 191
  4.3.2 Concrete transportation ... 197
  4.3.3 Concrete conveying ... 199
  4.3.4 Concrete placing ... 201
  4.3.5 Concrete compacting ... 212
  4.3.6 Concrete curing ... 216
  4.3.7 Concrete quality control ... 219
  4.3.8 Crack control of mass concrete ... 220
  4.3.9 Repair of concrete defects ... 222
Questions ... 224
Exercise ... 225

Chapter 5 Prestressed Concrete Structure Engineering ... 226
5.1 Overview of prestressed concrete ... 226
5.2 Pre-tensioning ... 227
  5.2.1 Tendons, abutment, clamp, equipment and tools ... 228
  5.2.2 Construction technology of pre-tensioning pre-stressed concrete ... 231
5.3 Construction technology of bonded post-tensioning prestressed concrete ... 234
  5.3.1 Materials, equipment and tool in bonded post-tensioning process ... 234
  5.3.2 Construction technology of bonded post-tensioning prestressed concrete ... 236
5.4 Construction technology of un-bonded post-tensioning prestressed concrete ... 241
  5.4.1 Un-bonded prestressed tendons, equipment and tools ... 241
  5.4.2 Construction technology of un-bonded post-tensioning prestressed concrete ... 242
Questions ... 243

Chapter 6 Structure Installation Engineering ... 245
6.1 Rigging equipment ... 245
  6.1.1 Pulley blocks ... 245
  6.1.2 Winch ... 246
  6.1.3 Steel wire rope ... 246
6.2 Lifting machinery and equipment ... 248
  6.2.1 Derrick crane ... 248
  6.2.2 Self-propelled derrick crane ... 249
  6.2.3 Tower crane ... 254
6.3 Structure installation engineering of reinforced concrete single-story industrial building ... 257

  6.3.1 Preparation for structural installation ··· 257
  6.3.2 Component installation process of the components ······················· 260
  6.3.3 Structural installation scheme ··· 269
 Questions ·· 273

Chapter 7 Waterproof Engineering ·· 274
 7.1 Roof waterproof engineering ·· 274
  7.1.1 Construction details of coiled sheet waterproof roof ······················· 277
  7.1.2 Construction of coating waterproof roof ···································· 280
  7.1.3 Construction of rigid waterproof roof ······································ 281
 7.2 Underground waterproof engineering ·· 283
  7.2.1 Construction of concrete itself waterproof ·································· 284
  7.2.2 Construction of additional waterproof layer ······························· 287
  7.2.3 Construction of prevention and drainage method ·························· 289
 7.3 Interior waterproof engineering ··· 290
  7.3.1 Selection of waterproof materials ··· 291
  7.3.2 Construction of interior waterproof works ································· 292
 Questions ·· 295

Chapter 8 Principle of Flow Construction ··· 297
 8.1 Overview flow construction ··· 297
  8.1.1 Concept of flow construction ·· 297
  8.1.2 Expression of flow construction ··· 300
  8.1.3 Categories of flow construction ·· 300
 8.2 Parameters of flow construction ··· 301
  8.2.1 Technology parameter ·· 301
  8.2.2 Space parameter ·· 302
  8.2.3 Time parameter ··· 304
 8.3 Organization of flow construction ··· 307
  8.3.1 Fixed rhythm Flow Construction ··· 307
  8.3.2 Multiple rhythm flow construction ·· 309
  8.3.3 No rhythm flow construction ··· 312
 Questions ·· 315
 Exercise ··· 315

Chapter 9 Network Plan Technology ·· 317
 9.1 Basic concepts of network plan ·· 317
  9.1.1 Implication of network plan technology ···································· 317
  9.1.2 The principle of network plan ··· 317
  9.1.3 Characteristics of network plan ·· 318
  9.1.4 Categories of network plan ··· 318
 9.2 Activity on line network plan (double codes network diagram) ················ 319

  9.2.1 Content and basic symbols of activity on line diagram ·································· 319
  9.2.2 Drawing of activity on line diagram ··· 322
  9.2.3 Calculation of time parameters in activity on line network diagram ················ 326
 9.3 Activity on node network plan (Single code network diagram) ···················· 332
  9.3.1 Content and basic symbols of activity on line network diagram ···················· 332
  9.3.2 Drawing of activity on node network diagram ·· 333
  9.3.3 Calculation of time parameters in activity on node network diagram ·············· 335
 9.4 Activity on line network plan with time scaled ·· 338
  9.4.1 Concept of activity on line network plan with time scaled ·························· 338
  9.4.2 Drawing activity on line network plan with time scaled ····························· 339
  9.4.3 Calculation of time parameters in activity on node network diagram with time scaled ······ 340
Questions ·· 343
Exercise ··· 344
Reference ··· 345

Chapter 1 Earthwork and Foundation Pit Engineering

Mastery of Contents: the looseness properties of soil; earthwork calculation of foundation pit (groove); earthwork excavation methods of foudation pit; earthwork compaction method and factors affecting compaction.

Familiarity of Contents: soil moisture content, soil permeability and the concept of earth slope; dewatering methods for foundation pit and their applicability, dewatering hazards and prevention methods; retaining structures for foundation pit and their applicability; foundation pit (groove) checking contents and methods; quality requirements for selection of filling soil and compaction.

Understanding of Contents: soil engineering classification; design of light well point and tube well dewatering; earthwork calculation for site leveling.

Key Points: the looseness coefficients; dewatering hazards and prevention methods.

Difficult points: calculation of earthwork.

1.1 Introduction

In the construction of civil engineering, earthwork is the first step of construction. Earthwork includes site leveling, excavation of foundation pit (trench) or pipe trench, excavation of underground engineering, foundation pit back-filling and ground, road or dam filling, etc.

The difficulty degree of earthwork construction is related to soil types, engineering properties of soil, the volume of earthwork, depth and method of excavation, as well as geological conditions and topography of the construction site. The characteristics of earthworks are large quantities of works, wide construction scope, wide variety of soil, the construction greatly influenced by the regional climate, geological and topographical features, and the construction conditions are complex. Therefore, it is necessary to do a thorough investigation and experimental research before construction in order to make a reasonable construction plan.

During the excavation of earthwork, when the excavation depth is big, soil maybe collapse, some measures should be taken to prevent the collapse of soil, which can guarantee the safety of foundation construction, surrounding roads, building structures and underground pipelines. If the excavation is under groundwater table, water-proof structures can be built to prevent the in-flowing of water to the foundation pit and some dewatering methods should be adopted to lower the groundwater level to below the excavation surface, which can provide a relative dry working surface for the machinery and workers in the construction of earthwork and foundation. Therefore, as the auxiliary works of earthwork, dewatering and foundation pit supporting engineering are first introduced in this chapter.

1.2 Engineering classification of soil

In earthwork construction and engineering budget quota, soil is divided into 8 types (16 grades), as list in Table 1-1, according to the difficulty degree of excavation.

The first four types are general soils, i.e. grade Ⅰ soil is soft soil, grade Ⅱ soil is common soil, grade Ⅲ soil is hard soil, grade Ⅳ soil is gravel hard soil. The last four types are rocks, i.e. grade Ⅴ-Ⅵ soil are soft rocks, grade Ⅶ-Ⅸ soil are secondary hard rocks, grade Ⅹ-ⅩⅢ soil are hard rocks, grade ⅩⅣ-ⅩⅥ soil are very hard rocks. The higher the soil grade is, the harder it is and more difficult to excavate, however, the stabler the soil structure is, the less easy to loosen and collapse after excavation.

1.3 The engineering properties of soil

The engineering properties of soil includes looseness, water content, permeability, earth slope, etc. which determine the construction method, machinery selection, foundation pit (trench) dewatering method and the cost of earthworks etc.

1. The looseness of soil

After excavation, the original structure of soil is destroyed and the volume of natural soil increases due to loosening, in spite of vibrating and compaction, it still can not be fully recovered to original volume, this nature of the soil is called looseness, which can be expressed by the looseness factors.

Initial looseness factor:

$$K_s = \frac{V_2}{V_1} \quad (1\text{-}1)$$

Eventual looseness factor:

$$K'_s = \frac{V_3}{V_1} \quad (1\text{-}2)$$

Where V_1——Soil volume in natural state (m^3);
V_2——Soil volume after excavation (m^3);
V_3——Soil volume after compaction (m^3).

In the digging and filling earthwork, the looseness of soil is an important parameter for calculation of earthwork machinery productivity, number of earth-moving machines, back-filling earthwork volume, elevation design and earthwork balance allocation of site leveling.

Chapter 1 Earthwork and Foundation Pit Engineering

Engineering classification and excavation methods of soil

Table 1-1

Types of soil	Grade	Name of soil	Looseness factors K_s	Looseness factors K'_s	Excavation methods and tools
1st type (soft soil)	I	Sand silt, alluvial sandy soil	1.08~1.17	1.01~1.03	dug by spade, hoe, partly need kicking
		Loose planting soil, mud(peat)	1.20~1.30	1.03~1.04	
2nd type (ordinary soil)	II	Silty clay, humid loess, sand containing gravels and pebbles, silt mixing with pebbles(gravel), planting soil, filling	1.14~1.28	1.02~1.05	dug by spade, hoe, sometimes should be loosened by pickax
3rd type (hard soil)	III	Soft and medium dense clay, heavy silty clay, gravel soil, dry loess, silty clay, compacted filling soil	1.24~1.30	1.04~1.07	mainly dug by pickax, a few use spade, hoe, partly use crowbar
4th type (sandy gravel soil)	IV	Heavy clay, clay containing gravels and pebbles, coarse pebbles, dense loess, natural grading sand and gravel	1.26~1.32	1.06~1.09	use pickax, crowbar first, and dug by spade, partly use wedge and hammer
		Soft marlstone and opal	1.33~1.37	1.11~1.15	
5th type (soft rock)	V-VI	Hard clay, medium dense shale, marl, chalk, slight cemented conglomerate, soft limestone, shell limestone	1.30~1.45	1.10~1.20	dug by pickax or crowbar and hammer, partly by explosion
6th type (secondary hard rock)	VII-IX	Mudstone, sandstone, conglomerate, hard shale, marl, dense limestone, weathering granite, gneiss and syenite	1.30~1.45	1.10~1.20	by explosion, partly by pneumatic pick
7th type (hard rock)	X-XIII	Marble, diabase, porphyrite, coarse, medium grained granite, solid dolomite, sandstone, conglomerate, gneiss, limestone, micrweathering andesite and basalt	1.30~1.45	1.10~1.20	by explosion
8th type (very hard rock)	XIV-XVI	Andesite, basalt, granite gneiss, solid fine grained granite, diorite, quartzite, gabbro, diabase, porphyrite	1.45~1.50	1.20~1.30	by explosion

2. Water content of soil

The water content is the mass of water in soil divided by the mass of dry soil solids and multiplied by 100%. That is:

$$w = \frac{m_w}{m_s} \times 100\% \tag{1-3}$$

Where m_w——The mass of water (g);

m_s——The mass of dry soil solids (g).

The water content of soil is influenced by climatic conditions, rainfall and groundwater level change, and has a direct influence on the excavation, the slope stability and the compaction effect of earthwork fill.

3. Permeability of soil

The property of soil can be permeated by water is referred to permeability, which can be expressed by the permeability coefficient. The seepage of groundwater in soil is resisted by soil particles, and the seepage quantity Q per unit time is related to soil permeability coefficient, water head difference and seepage distance.

The permeability coefficient of soil is relate to the particle gradation, degree of density and so on, which can be determined by lab or field experiments, referred as Table 1-2. It is the main parameter for foundation pit (trench) dewatering calculation, selection of dewatering method, etc..

$$Q = AK\frac{\Delta h}{l} \rightarrow v = \frac{Q}{A} = Ki \tag{1-4}$$

Where Q——Volume of seepage water in unit time (m^3/d or cm^3/s);

A——Seepage area (m^2);

K——Permeability coefficient (m/d or cm/s);

Δh——Water head difference (m);

l——Seepage length (m);

i——Hydraulic gradient, $i = \Delta h / l$.

The reference permeability coefficients of different soils Table 1-2

Types of soil	Permeability coefficient (m/d)	Types of soil	Permeability coefficient (m/d)
Clay	<0.005	Medium sand	5~20
Clayey soil	0.005~0.10	Uniform medium sand	35~50
Silty soil	0.10~0.50	Coarse sand	20~50
Loess	0.25~0.50	Uniform coarse sand	60~75
Silt sand	0.50~1.00	Round gravel	50~100
Fine sand	1.00~5.00	Pebble	100~500

4. Earth slope

The ability that soil can keep the stability of inclined surface in a certain state can be expressed by slope gradient and slope coefficient. Slope gradient is the ratio of slope height H to slope width B, as shown in Figure 1-1. In civil engineering, $1:m$ is commonly used to represent the size of the slope, where m is called slope coefficient, i.e.

$$\tan \alpha = \frac{H}{B} = \frac{1}{B/H} = \frac{1}{m} = 1:m \quad (1-5)$$

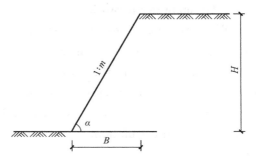

Figure 1-1 Slope diagram

Soil properties, excavation depth, slope excavation method, existing time of slope can all affect the slope gradient of soil. The coefficient of earthwork slope directly affects the calculation of earthwork volume. Some slop coefficients of different grade soils can be referred as Table 1-3.

Start depth and slope coefficient for different soil and excavation method   Table 1-3

Soil	Artificial digging	Mechanical digging		Start depth (m)
		Digging on the bottom of groove and pit	Digging above the groove and pit	
I and II grade	1:0.50	1:0.33	1:0.75	1.2
III grade	1:0.33	1:0.25	1:0.67	1.5
IV grade	1:0.25	1:0.10	1:0.33	2.0

1.4 Foundation pit (trench) dewatering

When the foundation pit or trench is excavated in the area with high groundwater table, the aquifers of soil will be cut off and the groundwater will flow into the foundation pit continually. During construction in rainy season, rainwater will also fall into the foundation pit. In order to provide a relative dry working place for the excavation and following construction and prevent the occurrence of quicksand, bottom soil heaving, slope instability and decline of foundation bearing capacity, dewatering and drainage must be done before or during the excavation of foundation pit (trench). The dewatering methods of foundation pit (trench) include natural drainage method and artificial dewatering method.

1. Groundwater

Groundwater is the water below the ground surface, which can be divided into three types: upper stagnant water, phreatic water and confined water, as shown in Figure 1-2. Foundation pit dewatering is mainly to reduce the phreatic water table, sometimes it is also necessary to reduce the confined water table.

2. Groundwater network

When the water seepage is stable in the soil, the water flow does not change with time, the void ratio and saturation of the soil are also unchanged. The water flowing into any soil unit is equal to the water flowing out of that unit to maintain the balance. A flow network can be used to represent stable seepage, it is composed of a group of seepage lines and equi potential lines, as shown in Figure 1-3, which is the theory basis of the dewatering well calculation.

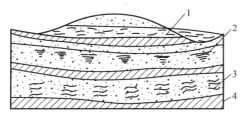

Figure 1-2 Groundwater diagram
1—stagnant water; 2—phreatic water;
3—confined water; 4—impervious stratum

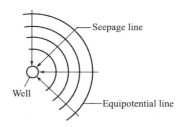

Figure 1-3 Flow network

If the corresponding flow network is drawn according to the dewatering scheme, the seepage path of water in the soil can be visually inspected. More importantly, the flow network can be used to calculate the seepage volume (water yield) of the foundation pit (trench) and determine the water head and hydraulic gradient of each point in the soil.

1.4.1 Natural drainage method

Natural drainage method, also known as open drainage method or water collecting well method, belongs to gravity drainage. This method uses interception, dredging and pumping for water drainage, that is, during the excavation of foundation pit, the drainage ditches are excavated along the perimeter or the center of the foundation pit bottom, and a certain number of water collecting wells are set up which can collect water when groundwater flows into the ditch, then the water is pumped out, as shown in Figure 1-4.

During the excavation, the cross section size of the drainage ditch and the number of water collecting wells should be determined according to the water yield at the bottom of the foundation pit (trench), the shape of the foundation and the pumping capacity of water pump. The drainage ditch and water collecting well shall be set more than 0.4m away from the bottom edge of the foundation. When the foundation pit (trench) bottom is sandy soil, the edge of the drainage ditch should be more than 0.3m away from the slope toe to avoid influencing the stability of the slope. The width and depth of the drainage ditch are generally 0.3~0.4m and 0.4~0.5m respectively, and the longitudinal gradient to the water collecting well is about 0.2%~0.3%. The spacing of water collecting well is generally 20~40m. Its diameter or width is 0.6~0.8m. The depth should be always 0.7~

Chapter 1 Earthwork and Foundation Pit Engineering

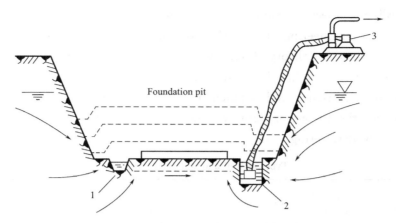

Figure 1-4 Natural drainage diagram of foundation pit
1—drainage ditch, (0.3~0.4) m× (0.4~0.5) m, $i=2~3‰$; 2—water-collecting well,
width 0.6~0.8mm, depth 0.7~1.0m, @20~40m; 3—water pump

1.0m lower than the excavation surface. When the water in the collecting well reaches a certain depth, the water can be pumped out of the foundation pit by diving pump or centrifugal pump in time. After the foundation pit (trench) is excavated to the design elevation, the bottom of the water collecting well shall be 0.5m lower than the bottom of the drainage ditch, and some gravels shall be laid on the bottom of the wells as filter layer. In order to prevent the well wall from collapsing due to the agitation of the mud and sand at the bottom of the well during long pumping time, the well wall can be simply reinforced with bamboo, wood, brick, cement pipe, etc.

Natural drainage method is suitable for large area drainage of coarse grained soil, and can also be used to cohesive soil with small seepage speed, but it is not suitable for fine sandy soil and silty soil layers.

Natural drainage method has some advantages as simple equipment, convenient drainage, economy, so it is widely used for the shallow foundation pit dewatering.

1.4.2 Artificial dewatering method

Artificial dewatering method needs to bury a certain number of wells around or inside the foundation pit (trench) before the excavation, and continuously pump out the groundwater by pumping equipment, so that the groundwater table can be lowered below the foundation pit bottom quickly. In this way, the excavation surface can be kept dry all the time and the construction conditions can be improved. At the same time, it also makes the hydrodynamic pressure direction downward, fundamentally prevents the occurrence of quicksand and heaving, increases the effective stress in the soil, and improves the strength and effective density of the soil.

Artificial dewatering method includes light well point, jet well point, electro-osmosis well point and tube well point (large opening well). During construction, it can be selected according to specific conditions such as permeability coefficient of soil, required dewatering depth, engineering characteristics, equipment conditions and cost, refer to Table 1-4. At present, the theory of light well point is the most perfect, and large opening well method is widely used in a lot of deep foundation pits, whose dewatering design is mainly based on experience and supplemented by theoretical calculation. This section mainly introduces the theory of light well point and the successful experience of large opening well point dewatering.

Types of well and suitable conditions   Table 1-4

Types of well	Permeability coefficient (m/d)	Dewatering depth (m)	Applicable soil layer	Hydrogeological conditions
Light well point	0.1~20.0	Single level <6	Fill, silty soil, clayey soil, sandy soil	Small amount of phreatic water
		Multiple level <20		
Jet well point	0.1~20.0	<20		
Electro-osmosis well point	<0.1	Used with former methods	Clay	
Tube well	1.0~200.0	>5	Silty, sandy, clayey soil, gravelly soil, loose rock, broken rock	Phreatic water, confined water and fissure water with abundant volume

1. Light well point

Light well point as shown in Figure 1-5 embeds well pipes (the bottom is filter tube, $\phi 38 \sim 57$mm steel tube with many $\phi 12 \sim 18$mm holes) in the aquifer around the foundation pit or on one side at a certain span. The upper ends of well pipes are connected to the main pipe, the underground water will be pumped out through filter tube, well pipe and main pipe by pumping equipment, thereby the groundwater table drops below the bottom of foundation pit.

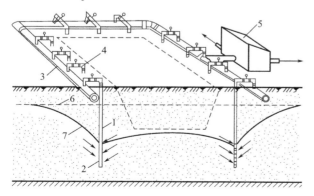

Figure 1-5 Light well system
1—well pipe; 2—filter pipe; 3—main pipe; 4—bend pipe; 5—pump equipment;
6—original groundwater table; 7—groundwater table after dewatering

Light well point device composed of piping system and pumping equipment. Piping system includes filter tubes, well pipe, bend pipe and main pipe, etc., see Figure 1-6.

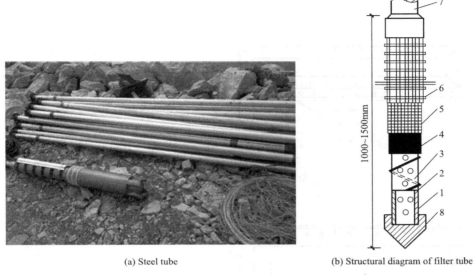

(a) Steel tube (b) Structural diagram of filter tube

Figure 1-6　Steel tube and filter tube

1—steel pipe; 2—small holes on pipe wall; 3—winding iron wire; 4—fine mesh; 5—coarse mesh; 6—coarse wire mesh; 7—well pipe; 8—cast iron head

1) The plane layout of light well points

The plane layout of light well points shall be determined according to the foundation pit size and depth, soil properties, groundwater table, flow direction and dewatering depth. Whether the well point layout is appropriate or not has a great influence on dewatering effect and construction speed.

(1) Single line layout

When the width of foundation pit or trench is less than 6m, dewatering depth is not more than 6m, the wells can be arranged in a single line on the upstream side of the groundwater flow, see Figure 1-7. The extend length B on both sides should be no less than the width of trench.

(2) Double rows or circular layout

When the width is more than 6m or the permeability coefficient is big, the well points should be arranged in two rows along the longer sides of the foundation pit. When the area of the foundation pit is big ($L/B \leqslant 5$, dewatering depth $S \leqslant 5$m, foundation pit width B less than 2 times of pumping influence radius R), the well point can be arranged around the pit, see Figure 1-8. When the foundation pit area is too large or $L/B > 5$, it can be arranged in sections.

No matter which layout scheme, the distance between well point pipe and foundation pit (trench) bottom edge is generally not less than $0.7 \sim 1.0$m to prevent vacuum

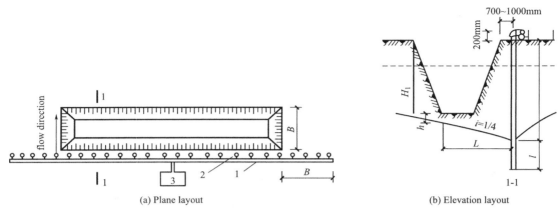

Figure 1-7 Single line layout of light well points
1—main pipe; 2—well pipe; 3—pump

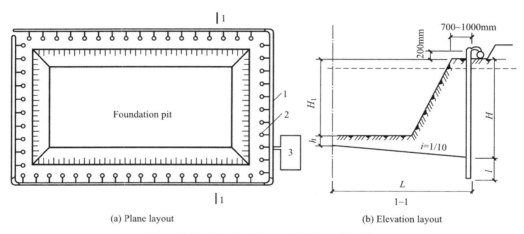

Figure 1-8 Circular layout of light well points
1—main pipe; 2—well pipe; 3—pump

leakage. The well point pipe spacing shall be determined according to the soil properties, dewatering depth, engineering properties etc., generally it is 0.8~1.6m.

2) The elevation layout of well points

Considering the head loss of pumping equipment, well point dewatering depth is generally not more than 6 m (not including the filter pipes).

The length of solid pipe:
$$H \geqslant H_1 + h + iL \qquad (1\text{-}6)$$

The embedded depth of well:
$$H_0 = H_1 + h + iL + l \qquad (1\text{-}7)$$

Where H_1——The distance from buried face to the bottom of foundation pit (m).

h——The distance from the bottom center of foundation pit (bottom level of far side for single line well) to the highest water level after dewatering, generally it is 0.5~1.0m.

i——Draw-down slope, circular line well point: 1/10, single line well point: 1/4.

L——The horizontal distance from the well point to highest water level (far side for single line well) (m).

l——The length of filter pipe (m).

When single level well point system can not meet the dewatering depth, double level system can be used after the groundwater dewatered to a certain depth by the first level well points, then the second well points are built to increase dewatering depth, see Figure 1-9.

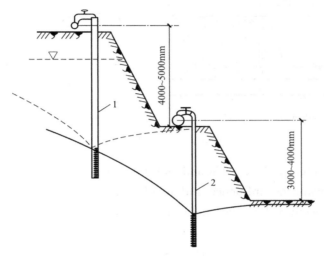

Figure 1-9 Double level well point
1—first level well; 2—second level well

3) Design of light well points

The design of light well points includes the calculation of water yield, number of well points, well spacing, and the selection of pumping equipment.

(1) Water yield calculation

The calculation of water yield of light well point system is based on the well theory. According to the geological conditions of groundwater, which aquifer the filter pipe is in, wells are divided into phreatic water well and confined water well. When the bottom of the well reaches the impermeable layer, it is called complete well, otherwise it is called incomplete well, see Figure 1-10. The calculation formulas of water yield are different for different types of wells.

① Complete well in phreatic aquifer

$$Q = 1.336 K \frac{(2H-S)S}{\lg(1+R/r_0)} \tag{1-8}$$

Where Q——Water yield of foundation pit (m³/d).

K——Permeability coefficient of soil (m/d).

H——Thickness of phreatic aquifer (m).

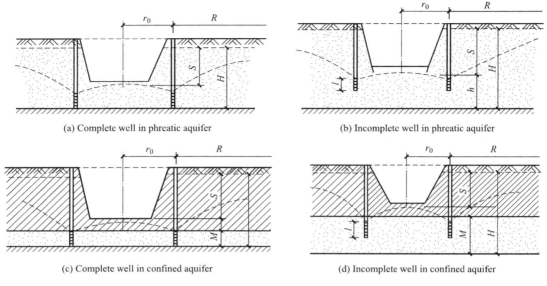

(a) Complete well in phreatic aquifer
(b) Incomplete well in phreatic aquifer
(c) Complete well in confined aquifer
(d) Incomplete well in confined aquifer

Figure 1-10 Types of light well point

S——Dewatering depth of foundation pit (m).

R——Dewatering influence radius of a single well (m), which should be determined by experiment or local experience.

The phreatic aquifer is calculated according to the following formula:

$$R = 2S\sqrt{KH} \tag{1-9}$$

The confined aquifer is calculated according to the following formula:

$$R = 10S\sqrt{K} \tag{1-10}$$

r_0——Equivalent radius of the foundation pit (m), when the foundation pit is circular, use its radius, and the equivalent radius of the rectangular foundation pit is calculated according to the following formula:

$$r_0 = 0.29(L + B) \tag{1-11}$$

Where L and B are the length and width of foundation pit (m)

For irregular foundation pit, its equivalent radius is calculated according to the following formula:

$$r_0 = \sqrt{\frac{A}{\pi}} \tag{1-12}$$

Where A is the area of the foundation pit (m²).

② Incomplete well in phreatic aquifer

$$Q = 1.366K \frac{H^2 - h_m^2}{\lg\left(1 + \frac{R}{r_0}\right) + \frac{h_m - l}{l}\lg\left(1 + 0.2\frac{h_m}{r_0}\right)} \tag{1-13}$$

Where h——Distance from dewatering table to aquifer bottom (m), $h = H - S$.

h_m——Average value of H and h (m), $h_m = (h + H)/2$.

l——Length of filter pipe (m).

③ Complete well in confined aquifer

$$Q = 2.73K \frac{MS}{\lg(1+R/r_0)} \quad (1\text{-}14)$$

Where M——Thickness of confined aquifer.

④ Incomplete well in confined aquifer

$$Q = 2.73K \frac{MS}{\lg\left(1+\frac{R}{r_0}\right) + \frac{M-l}{l}\lg\left(1+0.2\frac{M}{r_0}\right)} \quad (1\text{-}15)$$

(2) Number calculation of well points

The water can be discharged by a single well point g (m³/d):

$$g = 65\pi dl \sqrt{K} \quad (1\text{-}16)$$

Where d——Inner diameter of filter pipe (m).

l——Length of filter pipe (m).

K——Permeability coefficient of soil (m/d).

So the minimum number of well points n is calculated according to the following formula:

$$n = 1.1 \frac{Q}{g} \quad (1\text{-}17)$$

(3) Well spacing D (m)

$$D = \frac{L}{n} \quad (1\text{-}18)$$

Where L——Length of main pipe (m).

(4) Selection of pump equipment

The commonly used pumping equipment are vacuum pump, centrifugal pump and jet pump, which can be selected according to the water yield, suction water head and total water head. Pump flow should be 10%~20% larger than water yield of foundation pit (trench). In general, one vacuum pump corresponds to one centrifugal pump, but two centrifugal pumps can be equipped when the permeability coefficient and water yield of soil are large.

4) Construction technology of light well point

Construction preparation→well point layout→main pipe layout→pipe embedment→bending pipe connection→pumping equipment installation→pump running→removing.

5) The installation of light well point

According to the dewatering scheme, the installation steps include line alignment, trench digging, main pipe layout, punching holes, well point pipe installation, sand filter burying, clay sealing, connection of well point pipe with main pipe by bending pipe, pumping equipment installation, test running.

As shown in Figure 1-11, well point pipe can be buried by flushing method, which includes two process, punching and pipe burying.

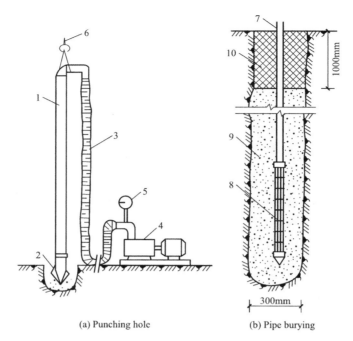

(a) Punching hole   (b) Pipe burying

Figure 1-11 Installation of light well point

1—punching pipe; 2—spray nozzle; 3—rubber pipe; 4—high pressure water pump; 5—pressure gauge;
6—hanger of crane; 7—well pipe; 8—filter pipe; 9—sand filling; 10—clay sealing

6) The running and checking of light well

Once the light wells begin running, pumping should be continuous. During the running, silting check should be carried out generally by listening to water sound in pipe, touching pipe to feel vibration and temperature (warm in winter and cool in summer).

After under-structure and back-filling are finished, stop pumping and remove the well point system.

2. Jet well point

When the excavation of the foundation pit (trench) is deep, the groundwater table is high and the dewatering depth is more than 6m, one-level light well point can not meet the requirements, so it is necessary to use the multi-level light well point to achieve the expected effect, but this method will increase the number of equipment, the excavation and backfilling earthwork of the foundation pit (trench), the construction time will be increased, so it is uneconomical. Then, jet well point can be used, and the dewatering depth can reach 8~20m. It is most effective in the sandy soil with $K=3\sim50$ m/d. For the silt soil and mucky soil with $K=0.1\sim0.3$ m/d, the effect is also very significant.

According to the jetting media, jet well point can be divided into water jetting well point and air jetting well point. The main equipment consists of jet well point, high pressure water pump (or air compressor), and pipeline, see Figure 1-12.

The layout of jet well is the same with light well: single side (width of pit $\leqslant$10m),

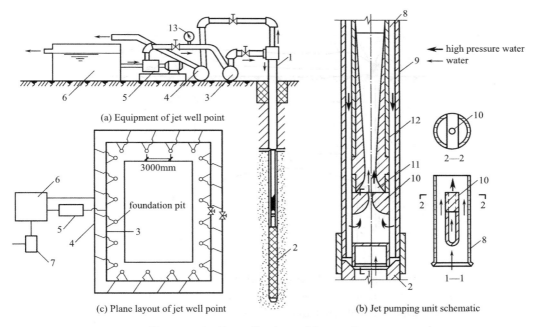

Figure 1-12 Jet well point and layout diagram

1—injection well pipe; 2—filter pipe; 3—inlet main pipe; 4—outlet main pipe; 5—high-pressure water pump; 6—collecting basin; 7—water pump; 8—inner pipe; 9—outer pipe; 10—spry nozzle; 11—mixing chamber; 12—diffusion pipe; 13—pressure gauge

double sides (width of pit ≥10m) and circular layout. Each jet well point system can control about 30 wells, the space is 2~3m, the water yield calculation and embedding method are similar to the light well point.

3. Electro-osmosis well point

When $K < 0.1$m/d, the dewatering effect of light well and jet well is not good, electro-osmosis well point can be adopted to speed the seepage of groundwater.

Electro-osmosis well point uses light well point or jet well point as cathode, steel pipe (diameter $\phi 50 \sim 70$mm) or steel bar ($> \phi 25$mm) is buried on the inside of the well point pipe as anode, see Figure 1-13. The anode shall be buried vertically and shall not contact with the adjacent cathode. The anode shall be 200~400mm long above the ground, and its penetration depth shall be 500mm deeper than the well point pipe to ensure that the groundwater can be reduced to the required depth. The spacing between the anode and cathode is generally 0.8~1.0m for light well point or 1.2~1.5m for jet well point, and arranged in parallel and staggered way. The number of anode and cathode should be equal, but the number of anode can be more than that of cathode if necessary.

Due to the positive charge of pore water and the negative charge of soil particles, the electro-osmosis well point can accelerate the flow of pore water to the well. Electro-osmosis well point is a supplement of light well point and jet well point. It is suitable to clay, silty clay and mud.

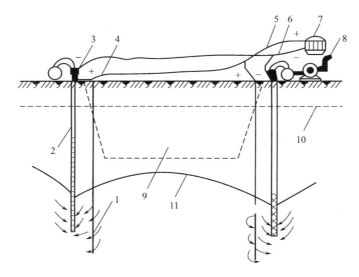

Figure 1-13 Electro-osmosis well point diagram

1—anode; 2—cathode; 3—cathode connected with the flat steel, bolt, or wires;
4—anode connected with the steel bar or wire; 5—anode and engine connected by wires;
6—cathode and engine connected by wires; 7—Dc generator (or dc arc welding machine);
8— water pump; 9— foundation pit; 10—original water level; 11—water level after pumping

4. Tube well

Tube well is composed of filter pipe, solid wall pipe, suction pipe and water pump. When sand-free concrete pipe is used as filter pipe, solid concrete pipe can be used above it to protect the upper part of well. When steel tube with holes on the wall is used as filter pipe, steel tube without hole can be used above it. Both concrete and steel tube can be made in sections and joint during the sinking. The suction pipe inserted into the well can be 50~100mm diameter steel pipe, rubber pipe or plastic pipe, as shown in Figure 1-14. The suction pipe is connected with the water pump. Depending on the amount of water yield and the pumping capacity of the water pump, there can be one well with one pump, or one well with multiple pumps. After the groundwater seeps into the filter well pipe, it is pumped out by the water pump through the suction pipe.

The layout of tube well is mainly based on practical experience, supplemented by theoretical calculation. Tube well can be set inside or outside the foundation pit. The well spacing is 8~25m, the well depth is 8~30m, the well diameter (inner diameter) is 300~800mm, mostly 400mm and 500mm, as shown in Figure 1-15. The hole forming diameter is 500~900mm, and coarse sand filter layer needs to be filled around the well.

The construction technology of tube well is: layout of wells→forming hole in soil→ filter tube sinking → connection of tube → tube sinking → alignment → sand filter filling between well and hole → well cleaning → pumping → well sealing or backfilling after construction.

The tube well point has wide applicability. During pumping, the groundwater table in

Chapter 1　Earthwork and Foundation Pit Engineering

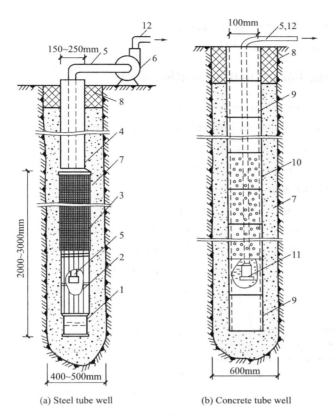

(a) Steel tube well　　　(b) Concrete tube well

Figure 1-14　Tube well diagram

1—sand sinking pipe; 2—steel ring skeleton; 3—suction pipe; 4—mesh; 5—steel pipe;
6—centrifugal pump; 7—small stone filtration layer; 8—clay sealing; 9—concrete pipe;
10—concrete filter tube; 11—diving pump; 12—outlet pipe

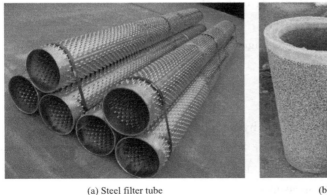

(a) Steel filter tube　　　(b) Concrete filter tube

Figure 1-15　Filter tube for tube well

well and the influence range of pumping can be adjusted by controlling the pumping capacity of the pump. It can even control the dewatering process by the methods such as stopping pumping, sealing the well and reducing the pumping frequency, so the success rate of tube well dewatering is quite high. It is suitable for dewatering of various soils and foundation pits or trenches with different shapes and sizes.

1.4.3 Dewatering hazard and prevention method

During the dewatering, because the groundwater table decreases inside the foundation pit, the water pressure inside and outside the foundation pit loses its balance, which will generate some hazards to the stability and safety of the foundation pit. The hazards commonly include quicksand and heaving.

1. Hydrodynamic pressure

Hydrodynamic pressure refers to the pressure of groundwater acting on soil particles when water flows in soil, which is consistent with the direction of water flow. The properties of hydrodynamic pressure can be explained by the tests in Figure 1-16.

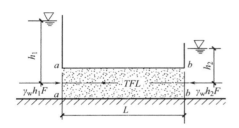

Figure 1-16 Seepage force

In Figure 1-16, due to the waterhead difference between the left side of high water level (head h_1) and the right side of low water level (head h_2), water flows from the left side to the right side, which creates pressure on the soil, while the soil particle skeleton produces resistance to water flow. According to the principle of acting force equals reaction force, the formula can listed as

$$\gamma_w h_1 F - \gamma_w h_2 F - TFL = 0 \Rightarrow T = \frac{h_1 - h_2}{L}\gamma_w = \frac{\Delta h}{L}\gamma_w = i\gamma_w \quad (1-19)$$

$$G_D = T = i\gamma_w \quad (1-20)$$

Where T——Seepage force resistance of unit soil.

G_D——Hydrodynamic pressure.

F——Cross section area of seepage.

L——Length of seepage.

h——Difference of water head.

i——Hydraulic gradient.

The formula (1-20) shows that the hydrodynamic pressure G_D is proportional to the hydraulic gradient, the greater the water level difference $\Delta h = h_1 - h_2$, the greater the G_D, and the longer the seepage distance, the smaller the G_D.

2. Quicksand and prevention measures

Under the action of groundwater table difference, dynamic water pressure will act on the unit soil (soil particle), see Figure1-17, and the direction of the dynamic water pressure is the same as direction of the water flow. For unit soil at position 1, the flow direction is downward, that is, the hydrodynamic pressure is consistent with the gravity direction, and the soil tends to stabilize. For unit soil at position 2, the flow direction is upward, that is, the hydrodynamic pressure is upward. When the hydrodynamic pressure G_D is equal to or greater than the effective specific gravity γ' of the soil, the soil particles become suspended in the water, and flow into the foundation pit with the seepage water, that is, the phenomenon of quicksand occurs.

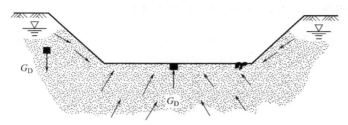

Figure 1-17 Hydrodynamic pressure direction

There are three main ways to prevent quicksand, the first is to reduce or balance the hydrodynamic pressure, the second is to change the direction of hydrodynamic pressure, the third is to improve soil quality. The specific measures are as follows:

(1) Digging when in low groundwater level period to reduce the hydrodynamic pressure.

(2) Digging in water to balance the hydrodynamic pressure.

(3) Build water-prove structure around foundation pit to increase seepage path and reduce the hydrodynamic pressure.

(4) Dewatering outside foundation pit to reduce groundwater level and change the direction of hydrodynamic pressure.

(5) Steam method or freezing method toimprove soil quality.

(6) Grouting in the soil with water-glass or other materials to improve the soil quality.

3. Heaving and prevention measures

When the bottom of the foundation pit (groove) is located in the impermeable soil layer and there is a confined water layer just below the impermeable layer, if the thickness of the impermeable layer is not enough, the weight of the impermeable layer at the bottom of the foundation pit (groove) is less than the upward pressure of the confined water, the impermeable layer will be raised up, which is called heaving. When the heaving deformation is too big, the soil layer will be broken and piping occurs at the bottom of the foundation pit (trench), see Figure 1-18.

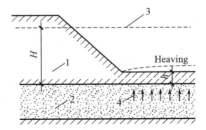

Figure 1-18 Heaving phenomenon
1—clayey soil; 2—sandy soil; 3—confined water table; 4—confined water pressure

Since the excavation depth of the foundation pit is unchangeable, so the only way to reduce the occurrence opportunity of heaving is to reduce the water pressure of the confined water layer bydewatering well or water retaining structure.

1.5 Earthwork calculation

1.5.1 Earthwork calculation for foundation pit

Earthwork quantity of foundation pit can be approximately calculated as cylinder (made up of two parallel plane for the bottom of a polyhedron, see Figure 1-19) volume formula.

$$V = \frac{H}{6}(A_1 + 4A_0 + A_2) \qquad (1\text{-}21)$$

Where H——Depth of a foundation (m).

A_1, A_2——The bottom and top area of the foundation pit (m^2).

A_0——The area of the middle cross-section of the foundation pit (m^2).

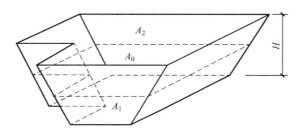

Figure 1-19 Earthwork calculation diagram of foundation pit

In addition to the cushion size, the width of construction working face and drainage ditch shall also be considered for the bottom size of foundation pit (trench) or pipe trench excavation, as shown in Figure 1-20. The width of the construction working face a is determined according to the foundation form, generally no more than 0.8m. The width of the drainage ditch b is determined according to the amount of groundwater inflow, generally no more than 0.5m. The width for the protection of slope foot c is generally 0.3m.

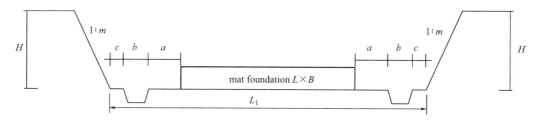

Figure 1-20 Foundation pit size diagram

So the area of different section of foundation pit can be calculated by following formulas:

$$A_1 = L_1 \times B_1 \qquad L_1 = L + 2(a+b+c) \qquad B_1 = B + 2(a+b+c) \qquad (1\text{-}22)$$

$$A_2 = L_2 \times B_2 \quad L_2 = L_1 + 2mH \quad B_1 = B_1 + 2mH \qquad (1\text{-}23)$$
$$A_0 = L_0 \times B_0 \quad L_0 = L_1 + mH \quad B_0 = B_1 + mH \qquad (1\text{-}24)$$

1.5.2 Earthwork calculation for groove, pipe trench

The length and width of foundation trench and pipe trench are longer and smaller than that of foundation pits. In order to ensure the accuracy of calculation, earthwork can be calculated in sections along the length direction, see Figure 1-21.

$$V_i = \frac{L_i}{6}(A_1 + 4A_0 + A_2) \qquad (1\text{-}25)$$

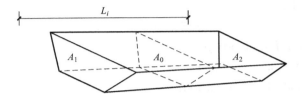

Figure 1-21 Earthwork calculation diagram of trench

The first and last section can be calculated as 3-slop foundation pit when there are slops at 2 ends of the trench.

The total volume of earthwork is the sum of each section's.

$$V = \sum_{i=1}^{n} V_i \qquad (1\text{-}26)$$

Example 1-1

The size of a cushion for a mat foundation is 40m×20m, natural ground surface level is −0.5m, the bottom elevation of the cushion is −3.5m, the ground soil is 2nd grade ($K_s = 1.20$, $K'_s = 1.02$). Excavating slope on 4 sides, the slope gradient is 1 : 0.65, the width of foundation working face is 0.8m.

Try to calculate the following items (results keep two decimal):

(1) How many m³ of the foundation pit and earthwork after excavation?

(2) If the foundation volume under the natural groundsurface is 1150m³, the remaining space will be back-filled by the excavated soil. How many m³ excavated soil needed?

(3) How many m³ excavated soil should be abandoned?

(4) If the abandoned soil will be transported outside by 10m³ dump truck, how many truck×time needed?

Solution:

(1) The volume of foundation pit and excavated soil

The depth of foundation pit:
$$H = -0.5 - (-3.5) = 3.0\text{m}$$

The length and width of foundation pit bottom:

$$L_1 = 40 + 0.8 \times 2 = 41.6\text{m}; \quad B_1 = 20 + 0.8 \times 2 = 21.6\text{m}$$

The area of foundation pit bottom:
$$A_{\text{bottom}} = 41.6 \times 21.6 = 898.56\text{m}^2$$

The length and width of top surface:
$$L_2 = 41.6 + 0.65 \times 3.0 \times 2 = 45.5\text{m}; \quad B_2 = 21.6 + 0.65 \times 3.0 \times 2 = 25.5\text{m}$$

The area of top surface:
$$A_{\text{top}} = 45.5 \times 25.5 = 1160.25\text{m}^2$$

The length and width of middle section:
$$L_2 = 41.6 + 0.65 \times 3.0 = 43.55\text{m}; \quad B_2 = 21.6 + 0.65 \times 3.0 = 23.55\text{m}$$

The area of middle section:
$$A_{\text{middle}} = 43.55 \times 23.55 = 1025.60\text{m}^2$$

The volume of the foundation pit:
$$V_1 = (1/6) \times 3.0 \times (898.56 + 4 \times 1025.60 + 1160.25) = 3080.61\text{m}^3$$

The volume of soil after excavation:
$$V_2 = V_1 \times K_s = 3080.61 \times 1.20 = 3696.73\text{m}^3$$

(2) The excavated soil needed for back-filling
$$V_2' = V_3 \times K_s' = (3080.61 - 1150) \div 1.02 \times 1.2 = 2271.31\text{m}^3$$

(3) Excavated soil abandoned outside
$$V_2'' = V_2 - V_2' = 3696.73 - 2271.31 = 1425.42\text{m}^3$$

(4) Trucks and times needed
$$1425.42/10 = 142.54 \approx 143 \text{ truck} \cdot \text{time}$$

1.5.3 Earthwork calculation for site leveling

Site leveling is to transform the natural ground within the scope of the building into the design plane required by the project. The difference between the height of the design plane and the natural ground elevation gives the construction height of each point of the site, from which the amount of site leveling earthwork can be calculated.

There are 2 methods in earthwork calculation of site leveling: cross-section method and square grid method. Cross section method is to divide the calculated site into several cross sections, calculate the earthwork by cross section calculation formula section by section. The cross-section method has low calculation accuracy and can be used in areas with large topographic relief. The square grid method is often adopted for areas with relatively flat terrain. The steps of calculating site leveling earthwork by grid method are as follows:

1. Grid division and natural elevation calculation of grid points

1) Setting up a rectangular coordinate system on a given contour map through the square grid with the edge length $a = 10 \sim 40\text{m}$.

2) Calculate the natural elevation of grid points by interpolation method between 2 adjacent contour lines. Mark the natural elevation for the corner points of every square

grid.

Interpolation calculation of the natural elevation is shown in Figure 1-22 and Formula (1-27). When there is no contour map, the grid can be divided by timber piles on site, and then the elevation of each grid point can be measured with instruments.

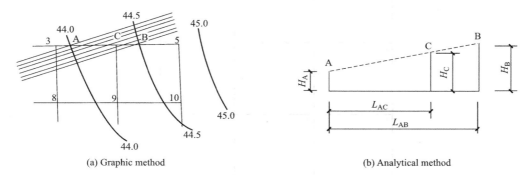

(a) Graphic method (b) Analytical method

Figure 1-22 Calculation of nature elevation

$$H_C = H_A + L_{AC} \times \frac{H_B - H_A}{L_{AB}} \tag{1-27}$$

There in:

H_C——Elevation of point C which is at a corner of a certain square grid (m).

H_A——Elevation of point A which lie on a certain contour line (m).

H_B——Elevation of point B which lie on another certain contour line (m).

L_{AC}——Distance from point A to point C (m).

L_{AB}——Distance from point A to point B (m).

3) Determination of the site design elevation

Principles for determining the design elevation:

(1) Meet the requirements of building planning, leveling process and transportation.

(2) Make full use of terrain to reduce the amount of excavating and filling earthwork.

(3) Balance the excavating and filling of earthwork to minimize transportation cost.

(4) Make a certain drainage slope (≥0.2%) to meet the ground drainage requirements.

(5) Consider the highest flood level requirements.

If there is no clear provisions and special requirements on the site design elevation in design document, design elevation can be determined by the "best plane method" or the "excavating and filling balance method". The "best plane method" can balance the excavating and filling earthwork and minimize the total earthwork at the same time, but the calculation method is complex. The "excavating and filling balance method" is simple, and the accuracy meets the construction requirements, so it is often used in actual construction projects though it can not guarantee the minimum total earthwork. The calculation steps are as follows:

(1) Calculation of initial design elevation (based on excavating and filling balance method)

Illustration and formula for the calculation of initial design elevation are shown in Figure 1-23 and Formula (1-28).

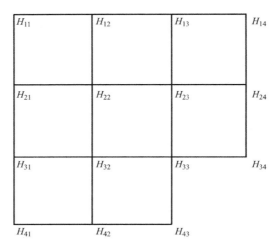

Figure 1-23　Calculation of design elevation

$$H_0 = \frac{1}{4N} \times [(H_{11}+H_{12}+H_{22}+H_{21})+(H_{12}+H_{13}+H_{23}+H_{22})+\cdots] \quad (1\text{-}28)$$

There in:

H_0——Initial design elevation (m).

N——Number of the square grids.

H_{ij}——Nature elevation of square grid corner points (m).

(2) Calculation of final design elevation H_d

The initial design elevation calculated by the former formula is a theoretical value. In fact, the initial value need to be adjusted considering the following factors to get final elevation.

① Because looseness of soil, the final elevation will be increased.

As shown in Figure 1-24, in theoretical calculation:

$$V_e = V_f \quad (1\text{-}29)$$

There in:

V_e——Excavation volume (m³).

V_f——Filling volume (m³).

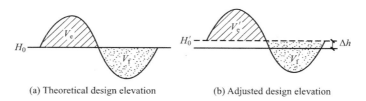

(a) Theoretical design elevation　　(b) Adjusted design elevation

Figure 1-24　Adjustment of design elevation considering looseness of soil

Δh is the supposed value more than the H_0 due to the looseness of soil. Then the total excavation volume will be reduced to:

$$V'_e = V_e - A_e \Delta h \qquad (1\text{-}30)$$

There in:

V'_e——Excavation volume after adjustment (m^3).

A_e——Excavation area before adjustment (m^2).

After the adjustment of elevation, the filling volume changes to:

$$V'_f = V_f + A_f \Delta h \qquad (1\text{-}31)$$

There in:

V'_f——Filling volume after adjustment (m^3).

A_f——Filling area before adjustment (m^2).

Considering the looseness of soil, after the filling work, the excavation volume totally changes to filling volume:

$$V'_f = V'_e K'_s = (V_e - A_e \Delta h) K'_s \qquad (1\text{-}32)$$

There in:

K'_s——Final looseness factor.

Combining formula (1-29), (1-31) and (1-32)

$$\Delta h = \frac{V_e(K'_s - 1)}{A_f + A_e K'_s} \qquad (1\text{-}33)$$

② The influences that the filling earthwork (as roadbed) above H_0 can reduce the final elevation and the excavation earthwork (as pool) below H_0 can raise the final elevation.

③ Unequal of excavating and filling earthwork due to slope construction.

④ Borrowing or discarding soil nearby outside the site after economic comparison.

(3) Calculation of final design elevation when there is drainage slope

After adjustment of the design elevation, the whole site surface is in a same level. When there is drainage requirement, the design elevation for construction should be calculated according to the drainage slop in one-way drainage or two-way drainage.

① One-way drainage

As shown in Figure 1-25 (a), use H_0 as the center line perpendicular to the drainage direction, the design elevation of any points in the site can be calculated as

$$H_d = H_0 \pm li \qquad (1\text{-}34)$$

There in:

H_d——Design elevation of any points on the site (m).

l——Distance for that point to the center line (m).

i——Drainage slope (%).

② Two-way drainage

As shown in Figure 1-25 (b), use H_0 as the elevation of center point, the design elevation of any points on the site can be calculated as

$$H_d = H_0 \pm l_x i_x \pm l_y i_y \qquad (1\text{-}35)$$

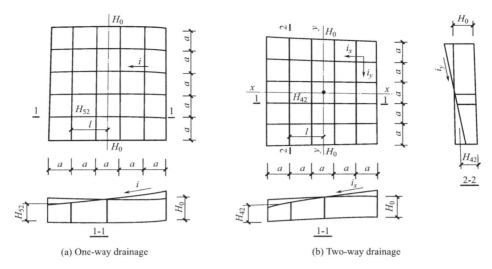

(a) One-way drainage   (b) Two-way drainage

Figure1-25 Ground drainage diagram

There in:

l_x, l_y——Distance for that point to the center point along x, y direction (m).

i_x, i_y——Drainage slope along x, y direction (%).

2. Earthwork calculation of site leveling

1) Calculation of excavating or filling height

Calculate the excavating or filling height of each corner point of every square grid by Formula (1-36).

$$h_0 = H_d - H_n \tag{1-36}$$

There in:

h_0——Excavating or filling height (m).

H_d——Design elevation (m).

H_n——Nature elevation (m).

According to the result of the calculation, positive value means filling work, negative value means excavating work.

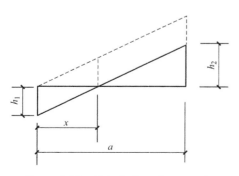

Figure 1-26 Calculation of zero point

2) Calculation of zero point and zero line

For any square grid unit, if the signs are different at the two points of an edge, there must be a point where the height of filling and excavation is zero which means there is no filling or excavation at this point. It is called zero point. Connecting the zero points will form zero line across through edges of the square grid units. Zero line is the boundary of excavating and filling area.

The process of calculation zero point is shown in Figure 1-26 and Formula (1-37).

Chapter 1 Earthwork and Foundation Pit Engineering

$$x = |h_1| \times \frac{a}{|h_1| + h_2} \tag{1-37}$$

There in:

x——Distance of the zero point to the excavation point (m).

$|h_1|$——Absolute value of the excavation height (m).

h_2——Height of the filling point (m).

a——Length of the square grid side (m).

3) Earthwork quantity of filling and excavating

The formula adopted in earthwork quantity of filling and excavating is shown in Formula (1-38), (1-39), (1-40), all of the h_i (height of excavation) in these formulas are taken as an only absolute value.

Case 1: Excavating at all corner points of square grid unit, as shown in Figure 1-27.

$$V_e = \frac{a^2}{4}(h_1 + h_2 + h_3 + h_4) \tag{1-38}$$

There in:

a——Length of the square grid side (m).

h_1, h_2, h_3, h_4——Height of excavating (m).

Case 2: Filling at two points and excavating at the other two points of the square grid unit, as shown in Figure 1-28.

$$V_e = \frac{a^2}{4}\left(\frac{h_1^2}{h_1 + h_4} + \frac{h_2^2}{h_2 + h_3}\right) \tag{1-39a}$$

$$V_f = \frac{a^2}{4}\left(\frac{h_3^2}{h_2 + h_3} + \frac{h_4^2}{h_1 + h_4}\right) \tag{1-39b}$$

There in:

V_e——Volume of excavating earthwork (m³).

V_f——Volume of filling earthwork (m³).

h_1, h_2——Excavating height (m).

h_3, h_4——Filling height (m).

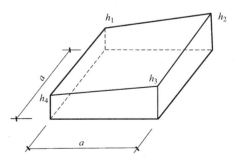

Figure 1-27 Excavating at 4 points

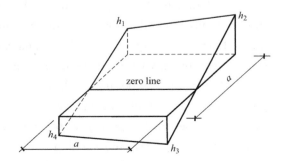

Figure 1-28 Excavating at 2 points and filling at another 2 points

Case 3: Filling at one point and excavating at the other three points of square grid unit, as shown in Figure 1-29.

$$V_f = \frac{a^2}{6} \cdot \frac{h_4^3}{(h_1+h_4)(h_3+h_4)} \qquad (1\text{-}40a)$$

$$V_e = \frac{a^2}{6}(2h_1+h_2+h_3-2h_4)+V_f \qquad (1\text{-}40b)$$

There in:

h_1, h_2, h_3 ——Excavating height (m).

h_4 ——Filling height (m).

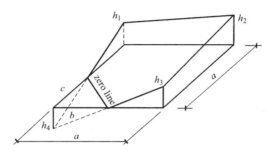

Figure 1-29 Excavating at 3 points and filling at another 1 point

In the case of all points filling, one point excavation and the other three points filling, the formulas are still related to those above, but the filling and excavating are opposite in sign.

1.6 Earthwork construction

1.6.1 Preparations before earthwork construction

Before earthwork construction, the following preparations shall be made.

(1) Site clearance: including demolition of buildings and underground obstacles in the construction area, clearing of arable soil and river silt, etc.

(2) Drainage of surface water: the accumulated water in the site will affect the construction, and the drainage ditch, intercepting ditch, retaining bank and other methods are generally used for the drainage of surface water. Temporary drainage facilities shall be combined with permanent drainage facilities as far as possible.

(3) Set up temporary facilities: set up necessary temporary buildings, such as processing shed, tool warehouse, material warehouse, temporary office and living rooms, etc. Set up temporary water supply, power supply and compressed air pipelines (for rock excavation), and running test for water, power and gas.

(4) Construction of transportation road: the road for mechanical operation on the site should be built (it should be built in combination with the permanent road). The road surface should be two lanes with a width of no less than 6m, and the roadside should be

provided with drainage ditch.

(5) Arrange equipment operation: carry out maintenance, inspection and running test of earthwork machinery, transportation vehicles and various auxiliary equipment to be used in the construction, and transport them to the site.

(6) Preparation of earthwork construction organization design: it is mainly to determine the dewatering scheme of foundation pit (trench), the methods of excavation and filling earthwork and slope treatment, the selection and organization of earthwork excavation machinery, the selection of filling earth and backfilling methods.

1.6.2 Earthwork construction machinery and methods

Generally, earthwork can be excavated by machine, manpower or combination method according to the foundation pit size, shape, cost and local construction conditions. Nowadays, the keys to speed up the construction progress, ensure the construction quality and reduce the project cost are to select the earth moving machinery reasonably, make all kinds of machinery coordinate in the construction and give full play to the mechanical efficiency. The commonly used earthwork machines are bulldozer, scraper, excavator, etc.

1. Main earthwork machines and construction characteristics

1) Bulldozer

The bulldozer is composed of a tractor and ablade which can be lifted and lowered. According to the moving way, bulldozers can be divided into caterpillar bulldozer and tyre bulldozer. Caterpillar bulldozers have strong adhesion with ground, good climbing performance and adaptability. Tyre bulldozers have high driving speed, good flexibility.

Bulldozer can be used to shoveling, transporting, spreading, compacting and scarifying soil, it is suitable for the excavation of 1st to 3rd grade soil, soil more than 4th grade should be loosened before working.

Bulldozers are mainly used in the site leveling, sub-grade construction, foundation pit excavation within 1.5m depth, filling trenches and cooperate the works of scraper, excavators and so on. Ground leveling within 100m distance (40~60m) is most efficient. Behind the bulldozer, the scarifier can also be installed to loosen soil, or the sheep foot roller can be hung for earth compaction.

The commonly working methods of bulldozer are downhill pushing, parallel pushing, groove digging, multi-shoveling and transporting, blade with side plate, etc.

2) Scraper

According to the moving way, scraper can be divided into tractor-moving scraper and self-moving scraper. According to the bucket control system, it can be divided into two kinds: hydraulic control and mechanical control.

As shown in Figure 1-30, the working method of scraper is downhill shoveling and scraping soil pushed by bulldozer.

Figure 1-30　Working method of scraper

Scraper can independently complete the shoveling, transporting, unloading, filling and leveling work. It is most suitable for excavation of 1st, 2nd grade soil with water content less than 27%.

According to the distribution of the filling and excavating area, combined with the specific local conditions, reasonably chose running route and promote production efficiency. Circular route and "8" glyph route are two common types, as shown in Figure 1-31.

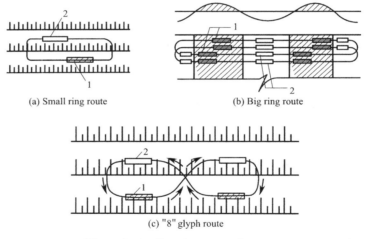

Figure 1-31　Running route of scraper
1—scrape soil; 2—dump soil

3) Single-scoop excavator

Single-scoop excavator is the most commonly used earthwork machinery in the excavation of large foundation pit (trench), pipe trench etc.

According to different working device, single-scoop excavator can be classified as forward shovel, back-acting shovel, pull shovel and clam-shell shovel (Figure 1-32).

According to different operating mechanism, shovels can be classified as mechanical and hydraulic shovels.

(1) Forward shovel

The working characteristics of forward shovel are forward and upward moving, cutting soil by force. It can excavate 1st to 4th grade soil, rock and frozen soil after

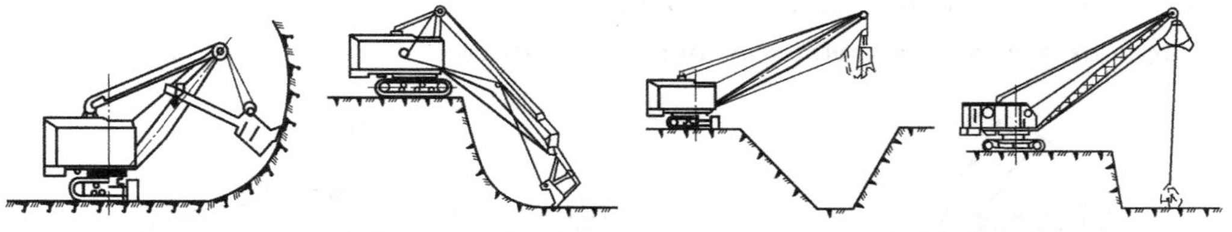

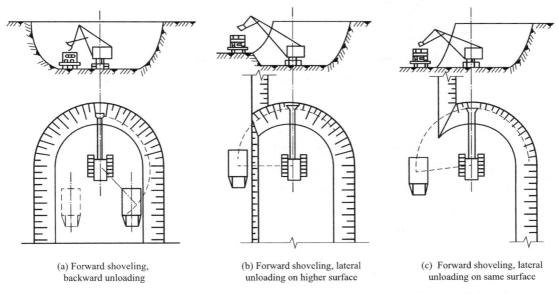

(a) Forward shovel   (b) Back-acting shovel   (c) Pull shovel   (d) Clam shell shovel

Figure 1-32 Types of single-scoop excavator

blasting. In earthwork construction, the forward shovel is often used to excavate large dry foundation pit and earth mound, etc. It can complete the excavation and transportation task with the cooperation of trucks.

The excavation methods of the forward shovel include forward shoveling, backward unloading and forward shoveling, lateral dumping, as shown in Figure 1-33. The common excavation steps are layered excavation, multi-layered excavation, central excavation, up and down rotation excavation, shovel excavation and interval excavation.

(a) Forward shoveling, backward unloading   (b) Forward shoveling, lateral unloading on higher surface   (c) Forward shoveling, lateral unloading on same surface

Figure 1-33 Excavation methods of forward shovel

(2) Back-acting shovel

The working characteristics of back-acting shovel are backward and downward moving, cutting soil by force. It can excavate 1st to 3rd grade soil below the stop surface, as well as soil with large water content or high groundwater table. In earthwork construction, back-acting shovel is often used to excavate foundation pit, foundation trench or pipe trench, etc. It can also complete the excavation and transportation tasks with the cooperation of trucks.

Back-acting shovel can excavate soil in two ways, digging at one end or one side of the trench, as shown in Figure 1-34. The common excavation methods are strip excavation, ditch angle excavation, layered excavation and multi-layer relay excavation.

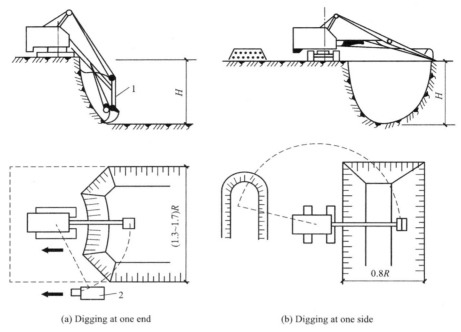

(a) Digging at one end (b) Digging at one side

Figure 1-34 Excavation methods of back-acting shovel
1—Back-acting shovel; 2—Truck

(3) Pull shovel

The working characteristics of pull shovel are backward and downward moving, cutting soil by self weight. Using inertia, pull shovel throws off shovel, digging and unloading soil by tightening or loosing wire rope, the shovel cuts soil downwards by self weight. It can excavate 1st and 2nd class soils below the stop surface, and is especially suitable for the soft loose soil under water, soil with high water content and ordinary soil. But the accuracy of the excavation is poor. In earthwork construction, the pull shovel is often used to excavate deep and large foundation pit (groove), ditch, sub-grade and embankment filling, etc.

Pull shovel can excavate soil in two ways, digging at one end or one side of the trench, as shown in Figure 1-35. The common excavation methods are triangle excavation, section excavation, layered excavation, sequential excavation, circle excavation and fan-shaped excavation, etc.

(4) Clam-shell shovel

The working characteristics of clam-shell shovel are downward and upward moving, cutting soil by self weight. It can excavate 1st and 2nd class soils below the stop surface, and is suitable for underwater excavation, loading and unloading gravels, slags and other loose materials. Clam-shell shovel is often used to excavate deep foundation pit, trench

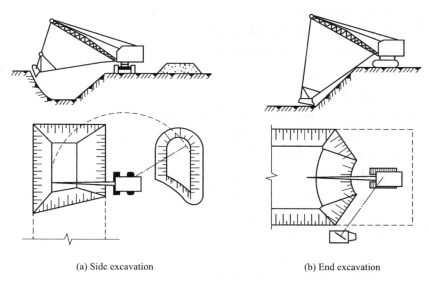

(a) Side excavation   (b) End excavation

Figure 1-35 Excavation methods of pull shovel

with narrow workspace, caisson, pile hole, diaphragm wall in soft loose soil. The excavation method is shown in Figure 1-36.

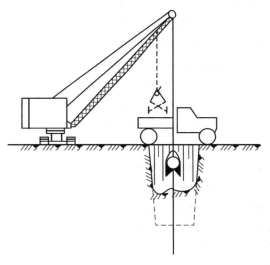

Figure 1-36 Excavation method of clam-shell shovel

2. Earthwork excavation construction

Before excavation, the excavation plan of foundation pit and the construction plan of groundwater control should be determined according to the data of engineering structure form, depth of foundation pit, geological condition, surrounding environments, construction method, construction period and ground load.

1) Construction keys for shallow foundation pit excavation

(1) Excavation procedure: outline surveying→layered excavation→water drainage→

slope repair→leveling→proper soil layer reserved for base protection, etc.

(2) The width of each side of foundation pit bottom should be 15~30cm wider than that of foundation to facilitate construction operation.

(3) The disturbance to base soil should be prevented as far as possible during excavation of foundation pit. In order to avoid destroying the base soil, a layer of soil should be reserved above the base elevation for manual excavation and repair. When machinery are used for excavation, the thickness of the retaining soil layer is 15~20cm for scraper or bulldozer excavation, and 20~30cm for single-scoop excavator. When manual excavation is adopted, the next working procedure can not be carried out immediately after the foundation pit is excavated, 15~30cm thick soil should be left, and then excavated to the design elevation just before the next working procedure begins.

(4) Earthwork excavation should be carried out continuously and completed as soon as possible. During the construction, the ground surface around the foundation pit should be waterproof and drainage treatment, and the surface water such as rainwater should be strictly prevented from seeping into the soil around the foundation pit and flowing into the foundation pit, so as to avoid the collapse of slope or the destruction of base soil. During the rainy season construction, the foundation pit should be excavated in sections and cushion should be cast subsequently. Water drainage or dewatering measures should be taken to reduce the water table to 50cm below the bottom of foundation pit to facilitate excavation. Water drainage or dewatering should continue until foundation construction (including backfill below groundwater level) is completed.

2) Construction keys for deep foundation pit excavation

Before the excavation of deep foundation pit, the excavation scheme and construction organization should be determined in details, the necessary monitoring and protection of supporting structures, groundwater level and surrounding environment should be carried out, and information construction technology can be adopted during excavation. In addition to following the key points of shallow foundation pit excavation, the following key points should also be paid attention to deep foundation pit excavation.

(1) The main excavation methods are center island, basin-shaped (see Figure 1-37), slope excavation and "cover and excavation method". The slop method is suitable for excavation without supporting structure, the latter three are suitable for excavation with supporting structure.

(2) The order and method of earthwork excavation in deep foundation pit must be consistent with the requirements of construction organization design, and follow the principle of "slotted excavation for supporting structure construction, bracing before digging, layered excavation, strict prohibition of over-excavation".

(3) It is necessary to prevent the rebound deformation of the base soil after excavation of deep foundation pit. The effective measure to reduce the rebound deformation is to reduce the exposure time and prevent the foundation soil from water immersion.

Therefore, during and after excavation, the dewatering work should be carried out continuously, and the cushion and foundation should be cast as soon as possible after excavating to the design elevation. If necessary, the soil layer under foundation should be strengthened.

(4) The excavation of deep foundation pit is usually carried out after the construction of pile foundation. Reasonable construction sequence and technical measures should be worked out to prevent the displacement and tilt of piles during earthwork excavation. Therefore, it is advisable to stay for a certain time after the foundation pile construction, and reduce the groundwater level first. After soil stress and pore water pressure accumulated in the soil due to pile construction is released, and the disturbed soil is reconsolidated, then earthwork excavation can be carried out. The excavation should be uniform, layered to reduce the earth pressure difference around piles as far as possible.

(5) The excavation should be combined with construction of bracing structure. Because excavation method directly affects the load of earth pressure to the supporting structures, it is necessary to make the load of the supporting structure uniform and reduce the deformation as far as possible. Therefore, the excavation should be carried out in a layered, block, balanced and symmetrical way to ensure the stability of the supporting structures and the safety of construction.

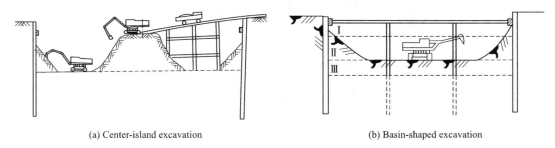

(a) Center-island excavation (b) Basin-shaped excavation

Figure 1-37　Excavation methods of deep foundation pit

1.6.3　Check of foundation subsoil

After the foundation pit (groove) is excavated to the design elevation and cleaned up, all building foundation pits (grooves) should be inspected. The construction party must work with the prospecting, design and owner (or supervision) parties to carry out the inspection of the foundation pit (groove), and only after the foundation pit qualified, can the foundation engineering construction be carried out.

The observation method is the main method, supplemented by drilling rod method.

1. Rod detection

Steel bars with a diameter of 22~25mm and a length of 2.1~2.6m are used to make rods, the rod tip is in the tapered shape of 60° angle. The hammer of 8~10kg is lifted 50cm high by manpower or machinery and freely dropped.

The rod is penetrated into the soil layer, and the required number of hammer hits for a unit depth is recorded. According to the number of hammer hits, the soft and hard conditions of the soil and whether there are tombs, dead wells, caves, weak substratum etc. can be evaluated. It can provide acceptance basis for quality indexes such as bearing capacity of foundation design, geological exploration results and evenness of basal soil layer.

Before drilling, according to the plan of foundation pit (groove), draw the layout plan of drill points and number them in turn. The vertical and horizontal distance of drill points are all 1.5m in quincunx shape, see Figure 1-38. During the detection, the same project should use same diameter rods, hammer weight, dropping distance, record number of hammer hits for every 30cm drilling depth and the final drilling depth is 2.1m. After finishing the drilling, the drilling record from top to bottom of one point and between horizontal drill points should be analyzed, drill points with too many or too few hammer hits should be checked. The hole should be filled with sand after drilling.

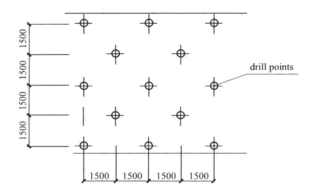

Figure 1-38　Quincunx layout of drilling rod detection

2. Observation

The observation method is to observe the foundation pit (groove) according to the construction experience. The main contents of the observation are as follows:

(1) According to the design drawing, check whether the plane position, dimension and bottom elevation of the excavation meet the design requirements.

(2) Carefully observe the soil type and uniformity of the soil wall and basal soil, whether there is abnormal soil quality, verify whether the soil quality at the bottom of the foundation pit is consistent with the prospecting report, whether the water content of the soil is too dry or too wet, whether the soil structure at the bottom of the groove is artificially damaged.

(3) Check whether there are old foundations, tombs, caves, dead wells and any other underground structures under the bottom of the foundation pit. If the above situation exists, it should be tracked along its direction to find out its range, extension direction, length, depth and width under the bottom of the foundation pit.

(4) To inspect, verify and analyze the drilling data, and recheck abnormal drilling points.

(5) Check whether the foundation pit slope is stable or not, and the distance between the outer edge of the foundation pit slope and the nearby buildings, and analyze whether the excavation of the foundation pit has an effect on the stability of the building.

(6) The focus positions of the inspection should be on the base of the column, the corner of the wall, the load-bearing wall, or other parts with greater pressure.

After the inspection of foundation pit, if geological conditions are found to be inconsistent with the design data, the treatment plan should be made with the relevant parties such as prospecting and design.

1.7 Foundation pit engineering

Before earthwork excavation, in order to ensure the safety of earthwork excavation and foundation construction and surrounding environment, slope or supporting structures should be determined when compiling the construction organization design of earthwork.

Foundation pit supporting structures are retaining, reinforcement and protection measures for soil wall to meet the requirements of the construction of underground structure, and to protect the safety of foundation pit surrounding environment.

According to the construction orders of excavation and bracing structure, foundation pit construction can be classified into 2 types:

Excavation and cover method: build vertical retaining structure→excavate soil and build horizontal bracing structures alternately from top to bottom→build foundation→build underground structure from bottom to top and remove bracing structure orderly.

Cover and excavation method: build vertical structure as soil retaining structure and permanent external wall of basement→build horizontal beam and slab of underground structure as bracing structure, and excavate soil alternately from top to bottom→build foundation.

According to the surrounding environment, excavation depth and soil quality, the following supporting structures can be selected.

1. Slop

When the excavation depth is small, the soil properties are good and there is enough space on site, slope excavation can be used around outline of foundation. The slope surface can be natural surface or strengthened by soil nail, steel bar net and sprayed concrete.

1) Slope shape

As shown in Figure 1-39, the shape of slope can be straight line, polyline, stepped and multilevel, which can be determined according to soil type, depth of excavation, surrounding environment, construction season, construction organization mode and rationality of technology and economy.

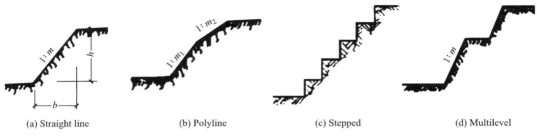

(a) Straight line   (b) Polyline   (c) Stepped   (d) Multilevel

Figure 1-39 Slope diagram

2) Slop setting

Some soils with good properties have self-stability which can be excavated vertically to a certain depth without bracing. Some reference limit vertical excavation depth of different soils are list in Table 1-5.

Limited vertical excavation depth   Table 1-5

Soil type	Limit vertical excavation depth(m)
Dense, middle dense sand and gravel soil	1.0
Hard plasticity, plasticity silt clay and silt	1.25
Hard plasticity, plasticity clay and gravel soil(filling is cohesive soil)	1.5
Hard clay	2.0

2. Vertical earth-retaining structures

For the foundation pit (groove) and pipe ditch with poor and unstable soil quality or deeper excavation, earth-retaining structures should be adopted for earthwork excavation when slope excavation can not guarantee the construction safety, or there is no slope space on site.

Water-retaining measures should also be adopted when the groundwater level is higher than the bottom surface of the foundation pit and the dewatering is carried out inside foundation pit.

As shown in Figure 1-40, steel sheet pile, concrete pile, diaphragm wall, soil-cement mixing wall, soil-cement mixing wall with steel H-section pile and various composite structures are common earth-retaining structures, in which diaphragm wall and soil-cement mixing wall can be used both as earth and water retaining structures. Different structures can be used separately for earth retaining or used in combination for both earth and water retaining.

3. Horizontal bracing structure

The upper mentioned earth-retaining structures can be used as cantilever when the depth of foundation pit is small. But when the excavation depth is big, horizontal bracing structures have to be set up to improve the stiffness and stability of retaining structures. The commonly used bracing structures are steel beam, steel pipe, concrete truss, steel

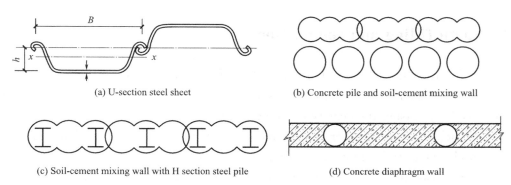

Figure 1-40　Earth-retaining structures

truss, steel beam with prestressed wire rope truss and anchor etc, as shown in Figure 1-41.

Figure 1-41　Bracing structures

1.8 Earth filling and compaction

Earth filling is the process of filling soil for foundation pit (trench) or pipe ditch, indoor floor, outdoor site leveling, road sub-grade, earth dam and so on. Compaction is the process of increasing the density of filling soil to a certain value by outer force to pack the soil particles closer together with a reduction in the volume of air but there is no significant change in the volume of water in the soil. So, after compaction, the filling soil will become denser and has higher strength and bearing capacity. The construction of earth filling and compaction include the selection of filling material, compaction machinery, thickness of compaction layer, compaction efforts and quality checking etc.

1.8.1 The selection of filling material

Filling soil should satisfy the design requirements to ensure the strength and stability of the filled soil. If there is no design requirement, the following regulations shall be complied with.

(1) Soil which contains a lot of organic matter, gypsum and water-soluble sulfate (content more than 5%) and mud, frozen soil, expansive soil, etc. should not be used as filling material.

(2) When clay was selected, water content should be checked whether is in the control range, clay with high water content is unfavorable for the filling soil.

(3) General gravel soil, sandy soil and rock blasting slag can be used below the surface, the biggest size shall not be more than 2/3 of the thickness of each compaction layer.

(4) Silt and silty soil can not be used as filling soil, but in soft soil areas, if the treated water content meets the compaction requirements, it can be used in the secondary parts of the filling work.

1.8.2 Compaction methods

There are three kinds of compaction methods, such as rolling compaction, tamping compaction and vibrating compaction, as shown in Figure 1-42.

1. Rolling method

Rolling method relies on the compaction pressure of rolling drum or wheel on the surface of filling soil to compact soil. It is suitable for large area filling engineering. Rolling machines include flat roller, sheep foot roller, vibration roller and tyre roller, see Figure 1-43. The flat roller, also known as the smooth roller, is divided into three types according to the weight: light grade (3~5t), medium grade (6~10t) and heavy grade (12~15t), respectively. They are suitable for the sandy soil and cohesive soil. Sheep foot roller has small contact area with soil, the pressure per unit area is large and the

Chapter 1　Earthwork and Foundation Pit Engineering

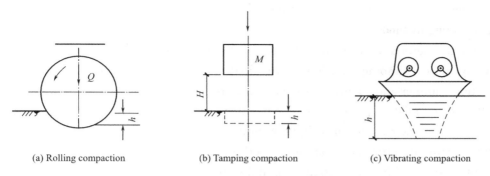

(a) Rolling compaction　　(b) Tamping compaction　　(c) Vibrating compaction

Figure 1-42　Compaction methods

(a) Flat roller　　　　　　　　　　(b) Sheep foot roller

(c) Vibration roller　　　　　　　　(d) Tyre roller

Figure 1-43　Rollers

compaction effect is good, which is suitable for compaction of cohesive soil. Vibration roller is a kind of high-efficiency compacting machine which acts simultaneously with vibration and compaction. It is suitable for large-scale filling engineering with blasting stone slag, gravel soil, mixed fill or silt. Tyre roller compacts soil by multiple inflatable tyres. The compaction process of tyre roller has kneading effect, which can make the soil particles embed with each other without destroying, uniform and dense in compaction layer. Tyre roller has good mobility, fast travel speed (up to 25km/h). Tyre roller is suitable for base layer, sub-base layer of different material and asphalt surface but can not be used to crush sharp-edged stones.

When rolling method is used for large area filling compaction, appropriate method "thin filling layer, low speed, more time" should be used.

2. Tamping method

Tamping method uses the impact force of the free falling rammer to compact soil. The commonly used tamping machines are hydraulic rammer, frog rammer, internal combustion rammer and so on, as shown in Figure 1-44. These rammers are widely used in backfilling of foundation pit (trench), pipe ditch, indoor floor and other small area filling work because of their small size, light weight, flexible operation, large tamping energy and high tamping efficiency. Tamping method can be used to compact cohesive or cohesionless soil, which has strong adaptability to soil quality.

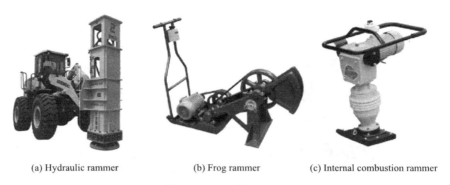

(a) Hydraulic rammer (b) Frog rammer (c) Internal combustion rammer

Figure 1-44 Rammers

3. Vibrating method

Vibrating compaction machines compact soil particles by vibrating force, so that the soil particles have relative movement and reach to a dense state, and it is suitable for cohesionless soil. Commonly used machinery includes plate vibrator and vibration roller. The plate vibrator is small, light and easy to operate, but the vibration depth is limited (Figure 1-45). It is suitable for the vibration of thin layer soil backfilling and sand pebble or gravel cushion.

Figure 1-45 Plate vibrators

1.8.3 The influence factors of filling soil compaction

The factors that affect the compaction of filling soil include internal factors and external factors. Internal factors refer to soil quality and soil water content, external factors refer to compaction work and other external natural and artificial factors during compaction, etc.

1. Water content

The effect of water content on compaction is significant. The relationship between dry density and water content of soil is shown in Figure 1-46. When water content increases gradually, the friction between soil particles decreases due to the lubricating effect of the water, the soil particles are easier to move and the filling soil is easier to compaction. When the water content increases to a certain value, the compaction effect of soil reaches the best, the water content in the soil at this time is called the optimum water content. The soil compacted under the optimum water content has the best water stability and the maximum compactness (dry density). When the water content in the soil is too large, the voids between soil particles are filled with water, and the compaction function is partly offset by water, which reduces the effective pressure and the soil can not be compacted densely. The maximum dry density of soil at the optimum water content can be determined by compaction test. Table 1-6 lists some reference values of different soils.

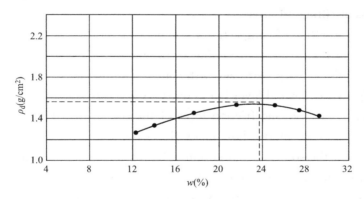

Figure 1-46 Dry density-water content relationship

Reference for optimum water content and maximum dry density of soil   Table 1-6

Item	Types of soil	Range	
		$w_{opt}(\%)$	$\rho_{dmax}(g/cm^3)$
1	Sand	8~12	1.80~1.88
2	Clay	19~23	1.58~1.70
3	Silty clay	12~15	1.85~1.95
4	Silt	16~22	1.61~1.80

In compaction construction, the filling soil should have suitable water content, where the soil can be held together and broken to the ground. If the water content is high, some measures such as turning over in the sun, evenly mixing dry soil or water-absorbing materials into soil can be adopted, and if the water content is low, measures such as sprinkling water or increasing the number of compaction times can be used.

2. Compaction effort

Compaction effort on site refers to the comprehensive action of compaction machine force and compaction times, which is another important factor affecting filling compaction. The relationship between the density and the work applied by the compaction machine to the filling soil is shown in Figure 1-47. It can be seen from the diagram that when the water content of the soil is certain, the density of the soil increases sharply at the beginning of compaction. When the density of the soil is close to the maximum value, the compaction work increases a lot, but the density of the soil does not change obviously. Therefore, in the actual construction, reasonable number of compaction times should be determined for different soil quality based on the compactness requirements and selected compaction machinery, see Table 1-7. In addition, the loose soil should not be compacted directly by heavy roller, otherwise the soil layer will have strong fluctuation phenomenon and the compaction effect is not good. The loose soil should be compacted by light compaction first and then by heavy compaction, a better compaction effect can be obtained.

3. Thickness of fillingsoil layer

The compaction effect of compaction machinery on soil decreases with the increase of depth, as shown in Figure 1-48. According to the measured density of different depths of soil layer on site, the density decreases with depth, and the density of surface layer within 50mm is the highest. If the soil layer is too thick, the compaction force of the lower soil is less than the adhesion force and friction force of the soil itself, the soil particles will not move with each other, no matter how many times the soil is pressed, the filling soil can not be compacted. If the soil layer is too thin, the total compaction times of the machinery would be increased due to more sublayers, and the lower soil may be subjected to shear failure because of too many compaction times. The optimal soil thickness should make the filling soil compacted and the mechanical power consumption minimum. The effective

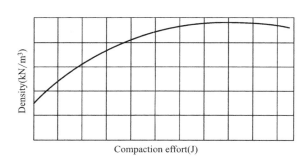

Figure 1-47 Density and work relation curve

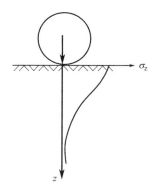

Figure 1-48 Influence depth of compaction effect

compaction depths of different compaction machinery are different. For different soil quality, the thickness of each layer should be determined according to compactness requirement and the types of compaction machinery, see Table 1-7.

Layer thickness and compaction times in filling construction Table 1-7

Compaction machine or method	Clayey soil		Sandy soil	
	Layer thickness (m)	Times per layer	Layer thickness (m)	Times per layer
Heavy flat roller(12t)	25~30	4~6	30~40	4~6
Middle flat roller(8~12t)	20~25	8~10	20~30	4~6
Light flat roller(<8t)	15	8~12	20	6~10
Frog rammer(200kg)	25	3~4	30~40	8~10
Manual compaction(50~60kg)	18~22	4~5		

4. Properties of soil

Under the action of certain compaction work, the maximum dry density of soil with more coarse particles is larger, and the better grading of soil particles, the easier it is to compact. The maximum dry density and optimum water content should be determined according to different soil quality.

1.8.4 Filling construction requirements

(1) Before backfilling construction, the garbage, tree roots and other debris on the base, water and mud in the pit should be removed. The elevation of the base should be checked. If the filling construction on cultivated soil or loose soil, the base shall be compacted first.

(2) The same filling project shall be filled with similar filling soil as far as possible. If different types of soil with different water permeability are used for filling work, the soil must be layered according to the types, and the soil layer with large permeability coefficient should be placed under the soil layer with small permeability coefficient. If the soil with small permeability coefficient has been filled in the lower layer, before filling the soil layer with large permeability coefficient in the upper layer, the combined surface of the two soil layers should be made into a circular arc drainage slope with high center and low sides or set with blind ditch, so as to avoid the formation of water sac at the joint surface. Different types of soil shall not be mixed filled.

(3) Backfilling and compaction should be done simultaneously on both sides and around the foundation to avoid cracking caused by squeezing the foundation.

(4) Filling work shall be carried out layered according to the entire width. When the fill is located on the terrain with big slope, slope should be built into 1∶2 step slope before construction, in order to prevent lateral movement of the filling soil. Each step is preferably 50cm high and 100cm wide. For section filling, the slope of each seam shall be

greater than 1 : 1.5, the rolling or compaction track shall overlap 0.5~1.0m, and the distance between the upper and lower seams shall not be less than 1.0m. The joint part shall not be set at the foundation corner, column base and other important parts.

(5) Drainage measures shall be taken during filling construction to prevent surface stagnant water or rainwater from flowing into the filling area. If the filled earthwork is soaked in water, the mud should be removed before the next process. The filling area should maintain a certain horizontal slope or high center and low sides for drainage.

(6) The filling construction should increase a certain height to compensate sinking height of the soil due to gradually sink and compact under the action of natural factors such as vehicular load, stacking weight or alternating dry and wet conditions. The reserved height shall be determined according to the nature of the project, filling height, filling soil type, compaction coefficient and foundation etc. When compacted by machinery in layers, the reserved height is calculated as the percentage of the filling height, with 1.5% for sand soil and 3%~3.5% for silty clay.

1.8.5 Quality control of compaction

The quality of earth filling and compaction work can be controlled by the degree of compaction which is measured in terms of dry density.

$$\rho_d = \frac{\rho}{1+\omega} \tag{1-41}$$

The compaction characteristics (maximum dry density ρ_{dmax} and optimum water content w_{opt}) of a soil can be assessed by standard compaction test in laboratory. Because the construction conditions on site are not same with lab conditions of standard compaction test, the actual dry density ρ_d on site is usually smaller than the maximum dry density ρ_{dmax} in lab, the ratio of ρ_d to ρ_{dmax} is called compaction coefficient λ_c which can be used to control the quality of actual compaction on site.

$$\lambda_c = \frac{\rho_d}{\rho_{dmax}} \tag{1-42}$$

The compaction coefficient is generally determined by the design according to the nature of the project, filling position and the nature of soil. For example, the compaction coefficient of the general site leveling is about 0.9, and the foundation filling is 0.91~0.97.

The actual dry density ρ_d after compaction can be measured by ring sampling method. The number of sampling group is one group per 20~50m³ filled soil for foundation pit backfilling, per 20~50m lenth of one layer for foundation groove and trench backfill, per 100~500m² area of one layer for indoor backfilling and per 400~900m² area of one layer for site leveling filling. Generally, the sampling position shall be in the lower half part of each layer after compaction.

If the actual dry density of the soil is $\rho_d \geqslant \rho_{d0}$ (construction control dry density $\rho_{d0} =$

$\rho_{dmax} \times \lambda_c$), the compaction is qualified. If $\rho_d < \rho_{d0}$, the compaction is not enough. The actual dry density ρ_d of more than 90% of inspected soil samples should meet the requirements, the difference between the minimum value of the remaining 10% samples and the control dry density ρ_{d0} shall not be more than 0.08g/cm³, and the sampling location shall be scattered and not concentrated. Otherwise, remedial measures should be taken to improve the compactness of the filling soil to ensure the quality of the filling.

Example 1-2

A road subgrade is filled by silty clay, the control λ_{c0} should be 0.95 according to the requirement of relevant code. The maximum dry density of the silty clay $\rho_{dmax}=1.9$g/cm³ in lab standard compaction test. On the job site, after compaction by flat roller, the density of compacted soil $\rho=2.1$g/cm³, and the water content $w=15\%$, so does the density satisfy the requirement or not?

Solution:

Method 1: calculate actual λ_c and compare with control λ_{c0}.

$$\rho_d = \frac{\rho}{1+w} = \frac{2.1}{1+0.15} = 1.83 \text{g/cm}^3$$

$$\lambda_c = \frac{\rho_d}{\rho_{dmax}} = \frac{1.83}{1.9} = 0.96 > \lambda_{c0} = 0.95 \text{ (ok)}$$

Method 2: calculate the actual dry density ρ_d and compare with control dry density ρ_{d0}.

$$\rho_d = \frac{\rho}{1+w} = \frac{2.1}{1+0.15} = 1.83 \text{g/cm}^3 > \rho_{d0} = \lambda_{c0} \times \rho_{dmax} = 0.95 \times 1.9 = 1.805 \text{g/cm}^3 \text{ (ok)}$$

Questions

1. What are the engineering properties of the soil? Try to explain their respective meanings.

2. What preparations should be made before earthwork excavation?

3. What factors should be considered in determining the site design elevation?

4. What are the main types of earth moving machinery? Try to describe their work characteristics and their scope of application respectively.

5. What are the construction key points of deep foundation pit excavation?

6. What are the checking contents of foundation pit and their purposes?

7. Try to analyze the causes of instability of earthwork slope, how to make earthwork slope?

8. What are the foundation pit support methods and applicable range?

9. What are the dewatering methods of foundation pit and what range is each applicable?

10. How does the quicksand happen and how to prevent and control it?

11. What are the filling compaction methods and their applicable range?

12. What are the main factors affecting filling compaction and how to test the quality of filling compaction?

13. The exterior wall of a building is made of rubble foundation, and its section size is shown in Figure 1-49. The foundation soil is clay, and the slope of the earthwork slope is 1 : 0.33. The looseness coefficients of the soil are $K_s = 1.30$ and $K_s' = 1.05$.

(1) Try to calculate the excavation volume per 100m length of foundation.

(2) After the backfill is left, all the remaining soil will be transported away. Try to calculate the volume of reserved backfill and abandoned soil.

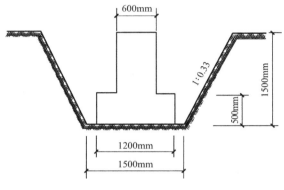

Figure 1-49

14. A foundation pit is filled by clay, the control λ_{c0} should be 0.95. The dry density of clay in lab standard compaction test $\rho_{dmax} = 2.1 \text{g/cm}^3$. On the job site, after compaction by flat roller, the density of compacted soil $\rho = 2.3 \text{g/cm}^3$, water content $w = 18\%$, so whether the density satisfied the requirement or not?

Chapter 2　Foundation Engineering

Mastery of Contents: construction key points of reinforced shallow foundation; the hammer driving method and static pressure driving method for precast pile construction, quality control for precast pile; the construction key points of cast-in-place pile by drilling in slurry method.

Familiarity of Contents: construction of non-reinforced shallow foundation; the requirements of production, hoisting, transporting and stacking of precast concrete pile, the pile driving machines; the influence of pile driving to the surround environment and preventive treatment, the method of driving pile by water jet and vibration; the construction of cast-in-place pile by tube sinking method; the common quality problems and treatment method in the construction of precast pile and cast-in-place pile.

Understanding of Contents: the construction method of dry boring, artificial digging, pedestal pile; post grouting of cast-in-place pile.

Key Points: construction technology of precast pile and cast-in-place pile.

Difficult points: positive circulation and reverse circulation of slurry.

2.1　Introduction

Any building is built on the ground. Therefore, the whole load of the building is borne by the soil under it. The part of the soil affected by the building is called the ground and the substructure that transfers the load from the superstructure to the ground is called the foundation. The foundation is an enlarged portion of a building with a wall or column buried into the ground, which should be strong enough to prevent strength failure, instability and control the settlement within allowable value.

The foundation needs to be buried into the ground to a certain depth to reach a suitable soil layer as bearing stratum and resist horizontal load. So foundation can be classified into shallow foundation and deep foundation just according to the embedded depth. When the embedded depth is less than 5m or the width of foundation, the foundation belongs to shallow foundation, otherwise, it is a deep foundation.

The ground under foundation can be classified into natural ground and manual ground. When the ground can be directly used to support the foundation, it is called natural ground. If natural ground can't satisfy the bearing capacity or the settlement requirements, it needs to be treated to improve the strength, then it is called manual ground. Some foundation treatment methods can refer to the ground improvement books.

Since foundation should be built below ground surface, a foundation pit will be dug first to remove the soil above the embedded depth within the range of foundation area. After the foundation pit is excavated to design depth and checked, the foundation can be directly constructed on the ground or a cushion. The cushion can be made of low grade plain concrete, lime-soil mixture, sand or sand-gravel mixture, lime-earth-broken brick mixture paved on the ground which provides a dry, smooth surface for construction and protection for the foundation. After cushion reaches to enough strength, mark axis lines and outlines according to design map, then can begin the construction of foundation.

2.2 Construction of shallow foundation

According to the material, shallow foundation can be divided into non-reinforced foundation and reinforced concrete foundation. According to the structure and stress performance, it can be divided into isolated foundation, strip foundation, raft foundation and box foundation etc.

2.2.1 Construction of non-reinforced foundation

Non-reinforced spread foundation is also called rigid foundation, it is made of bricks, stones, plain concrete, rubble stone concrete etc. Generally, non-reinforced foundations is built into an enlarged portion of a wall or a column, so it is also called non-reinforced spread foundation. The characteristics of these foundations are high compression strength and poor tensile, bending, shear strength resistance, they are only suitable for low-rise buildings and light industrial plants. The section size of rigid foundation should be decided by controlling material strength and the width-height ratio of steps to reduce shear force and tension stress. The cross section can be uni-thickness, stepped, or sloped shape.

1. Brick and stone foundation

The brick and stone masonry foundation can be constructed according to the methods and requirements introduced later in Chapter 3.

1) Brick foundation

The brick foundation should be stepped section shape in English bond with 1/4 brick length overlap. The step can be set with 2 layers for 1 step or 2 layers and 1 layer for interval steps as shown in Figure 2-1.

2) Stone foundation

The first course of ashlar stone foundation should be the header layer of English bond over mortar. The overlap of upper step to lower step should be 1/3 of the lower stone.

For rubble stone foundation, pave mortar first, the bigger surface of stone should be put downwards. Flat rubble stone should be used at corner and junction position. The overlap of upper step to lower step should be 1/2 of the lower stone. Through stones should be set with 2m space in one course.

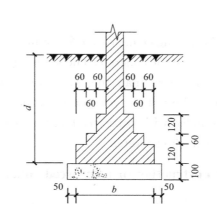

(a) 2 layers for 1 step (1:2)　　(b) 2 layers and 1 layer for interval step (1:1.5)

Figure 2-1　Brick foundation

2. Concrete foundation and rubble stone concrete foundation

The construction processes of plain concrete foundation include formwork erecting, concrete casting, concrete curing and formwork dismantling. The construction methods and requirements can refer to Chapter 4. The main construction key points are as following lists.

(1) If the bottom soil has good upright ability, a small soil pit can be cut same as the design size of the foundation as formwork of the first step, but the size should be right and no leaking of mortar. The formwork of upper step can be wood or steel formwork and should be fixed tightly on the ground.

(2) When casting height of concrete does not exceed 2m, the concrete can be directly filled into the formwork. When the casting height exceeds 2m, the concrete should be filled through funnel, tandem tube or inclined groove.

(3) For stepped foundation, concrete should be cast layered within one step in one time. Corners and sides should be cast first, center later. Two methods can be used to control the integrity of the concrete at the junction of the upper and lower steps. One is casting the concrete of upper step after finishing concrete cast of lower step 0.5~1h. The second method is making concrete slope around the bottom of upper formwork after vibrating the concrete of lower step densely, then casting concrete for upper step, removing the slope and smooth the surface of lower step after that.

(4) For sloped foundation, when the slope is small, concrete can be cast without slope formwork, but the concrete at corner, side and slope position should be vibrated densely, the slope can be modified and smoothed manually. When the slope is large, slope formwork should be erected while casting concrete and kept from floating due to concrete pressure.

(5) After concrete casting, the exposed surface should be properly covered and cured with water for more than 7 days. Backfilling foundation pit in time after dismantling

formwork.

(6) For saving cement, 30% volume of concrete can be filled with rubble stone which is called rubble stone concrete. The stone should be solidunweathered rock. The biggest size should not be more than the width of foundation and 300mm. The surface should be clean. When rubble stones are put into concrete, care should be taken to guarantee that the stone surrounded with concrete. The top stones of the foundation should be covered with lest 100mm thick concrete.

2.2.2 Construction of reinforced concrete foundation

Reinforced concrete shallow foundation has good anti-bending, anti-shearing performance and is suitable for the building with bigger load and low-bearing ground. The construction processes include formwork erecting, rebar laying, concrete casting, concrete curing, and formwork dismantling.

1. Reinforced concrete spread foundations

Reinforced concrete spread foundation includes isolated foundation under column and strip foundation under wall. The plane of isolated foundation can be square one or rectangular one, the section may be stepped, sloped or cup shape, see Figure 2-2. The section of strip foundation can be uni-thickness or sloped shape, with or without inner beam, see Figure 2-3. In addition to the following points, the main construction key points of reinforced concrete and plain concrete construction points are the same.

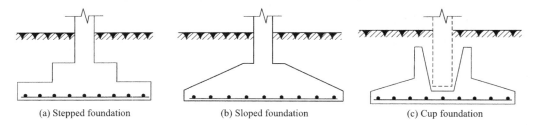

(a) Stepped foundation　　　(b) Sloped foundation　　　(c) Cup foundation

Figure 2-2　Isolated foundation under column

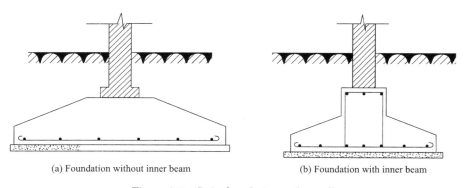

(a) Foundation without inner beam　　　(b) Foundation with inner beam

Figure 2-3　Strip foundation under wall

(1) Reinforcement should be correctly laid and tied firmly. The hook of bottom rebar should be upward. The lower end of column rebars with 90° bending hook should be firmly tied with the foundation rebar. The column rebars can be fixed with formwork by a square or rectangular frame after checking the axis position. Some precast cement mortar blocks with same thickness as protective cover can be placed at the bottom of the steel mesh to ensure the correct position.

(2) Before concrete casting, the garbage, mud, oil and other impurities on the formwork and steel bars should be removed, andthe formwork should be watered and wet.

(3) For concrete casting of foundation under column, 50~100mm thick concrete can be cast and vibrated first to fix the position of steel bars of column and foundation to prevent movement and tilt, then cast concrete symmetrically.

2. Cup-shaped foundation

Cup-shaped foundation is generally used as foundations of prefabricated columns. The prefabricated column can be inserted into a groove reserved in the foundation, and then, after temporary fixation, the surrounding space is filled with fine stone concrete. The forms of cup-shaped foundation are general cup, double cup and high cup foundation. The main construction key points are as follows.

(1) Cup formwork can be made of wood or finalized steel formwork, it can be made into a whole shape or two half form connected by 2 wedge-shaped boards in middle. When dismantling formwork, take out the wedge board first, and then the two half forms can be taken out. The cup mouth formwork should be fixed firmly and exhaust vent should be reserved at the bottom to avoid emptiness of concrete.

(2) Cast the concrete at the bottom of the cup layered and vibrate densely. For high cup-shaped foundation, the concrete should be cast layered in the whole section without any construction joint, the cup mouth formwork can be erected after the concrete reaches the cup bottom.

(3) Because the cup mouth formwork is only fixed at top position, concrete should be cast around it symmetrically and uniformly to avoid movement due to lateral pressure.

(4) Before final setting of the concrete, remove the cup formwork and cut the inside concrete surface, reserve 50mm thickness fine stone concrete leveling layer at the bottom of the cup mouth.

3. Strip foundation under column

Thestrip foundation under the column is composed of one-way beam or cross beam and its transverse protruding wing plate. The cross-section is inverted "T" shape. The construction processes are marking outline and axis on the cushion after it reaches to certain strength, laying the rebar according to the design, erecting formwork and casting concrete.

4. Mat foundation

Under some conditions, spread foundations would have to cover more than half of the building area, and pile foundation or mat foundation might be more economical. The mat foundation, which is sometimes called raft foundation, is combined footing that covers the entire area under a structure supporting several columns or walls. Mat foundation is used for soils that have low load-bearing capacity but have to support high column or wall loads. Generally speaking, mat foundation is a concrete slab of even thickness, sometimes, bigger thickness or beams will be used under columns or walls. Some common types of mat foundation are shown in Figure 2-4. The main construction key points are as follows.

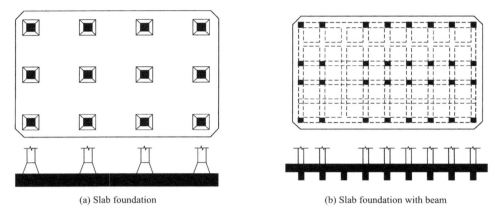

Figure 2-4　Mat foundation

(1) The construction points of rebar and formwork are same with reinforced concrete spread foundation.

(2) The concrete should be cast along length direction or parallel to the secondary beam of mat foundation. For flat plate with uniform thickness, if the concrete can't be cast in one time, vertical construction joints should left at any position but outside the influence range of column foot. When there are beam in the mat foundation, the joints should be left in the middle 1/3 span of secondary beam.

(3) For mat foundation with beams, the concrete of beams above flat slab should be cast layer by layer, the thickness of each layer should not be more than 200mm. When there are columns on flat slab or beams, the concrete can be cast to the top surface of column foot as horizontal construction joint.

(4) After concrete casting, the exposed surface should be properly covered and cured with water for more than 7 days. When the concrete reaches to 25% design strength, side formwork can be dismantled, when reaches 30% design strength, the foundation pit can be backfilled.

5. Box foundation

Box foundation is a closed box mainly composed of reinforced concrete mat, roof, side wall and a certain number of internal columns and walls, as shown in Figure 2-5. Box foundation has good integrity and high stiffness, strong ability to adjust the uneven

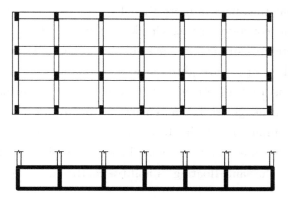

Figure 2-5   Box foundation

settlement and resist seismic which can eliminate the building cracking caused by foundation deformation. At the same time, the box foundation can be also used as basement when setting openings on the side and internal walls, so it is commonly used as the foundations of multistory or high rise buildings nowadays. It is suitable for soft soil, but also often used in non-soft soil areas when considering the basement, seismic or civil air defense requirements. The main construction key points are as follows.

(1) Waterproof layer should be paved on the cushion first and well sealed, and the overlap of junction around the mat and the external wall shall meet the design requirements. During the construction process, attention should be paid to prevent the piercing of waterproof layer and another layer concrete cushion can be cast to protect it before later construction.

(2) The rebar of mat, walls and columns should be laid with right position, types and numbers. The pre-embedded pieces should be fixed tightly at right position. When there are pipes pass through the external wall, water proof steel sheets should be welded to prevent seepage. Waterproof pull bolts can be used for the external wall formwork.

(3) The formwork and concrete casting of slab, wall and roof can be constructed separately. A stretcher brick wall can be used as side formwork of the mat and the bottom of external wall. The construction joint of external wall can be 300~500mm above the mat and 30~50mm below the flat roof. Mortise and tenon or waterstop should be set at the external wall joints.

(4) Concrete casting of mat, wall and roof should be continuously for each part. For big box foundation engineering, when the length of foundation is more than 40m, post-cast gap should be left to prevent concrete cracking due to theheat of concrete hydration. The width of the gap should not be less than 700mm and rebars should pass through the gap. When the roof is finished and the time interval is more than 28 days, fill the gap with micro-expansion fine-stone concrete of one grade more than the design grade and take care of curing. When there are reliable anti-cracking measures, it is not necessary to set the post-cast gap.

(5) When the mat thickness is more than 1m, the concrete work of box foundation belongs to big volume concrete casting. Crack control calculation should be carried out before casting concrete to estimate the possible maximum hydration heat temperature, internal and external temperature difference and temperature contraction force after concrete casting, so as to take effective technical measures during the construction to prevent temperature shrinkage cracks and ensure the quality of concrete engineering.

(6) After the completion of the box foundation construction, the foundation pit should be backfilled in time, and the anti-floating stability of the box foundation is calculated. Only after the anti-floating stability coefficient is greater than 1.1, can dewatering stopped.

2.3 Construction of pile foundation

Pile foundation belongs to deep foundation and composed of pile and cap. Piles are structural members of timber, concrete, or steel used to transmit surface loads to lower levels in the soil mass, their cross-section sizes are much smaller than length. Pile foundation is one type of deep foundation.

Piles may be classified into different types based on the following principles:

(1) The load transmit principle: friction pile and end-bearing pile.

(2) The material of piles: concrete pile, timber pile, steel pile and composite pile.

(3) The nature of placement: displacement pile, partial displacement pile and non-displacement pile.

(4) The construction method of pile: precast pile, cast-in-place concrete pile.

Precast piles are precast in factory or on job-site, and then transported to the design position and driven into the ground by hammer, static pressure, vibration or water jetting etc. The precast piles are commonly made of reinforcement concrete, steel or timber.

Cast-in-place piles are built in the field by boring a hole in the ground at the design position first and then filling it with rebar cage and concrete. The hole can be made by different methods such as bit drilling with or without slurry, impact drilling, tube sinking or manual digging and so on.

The construction methods of different type piles are decidedby the soil properties of the ground, local construction equipment, technical conditions, environment conditions and construction experience etc.

2.3.1 Construction of precast concrete pile

Precast concrete piles can be prepared by ordinary reinforcement or prestressed steel cable and their cross-section may be solid square, hollow square or hollow round shape, see Figure 2-6. The solid square pile are generally made on site, the cross-section side length is 200~500mm, the maximum length of a single pile is limited within the frame height of driving machine. If the length of pile is more than 30m, the pile can be made in several sections and connected during the driving. Most of hollow square and round piles are prestressed concrete piles, they are made in factory by pretensioning method. The cross-section side length of hollow square pile is general 300~700mm, the diameter of hollow part is 160~440mm. For hollow round pile, the outside diameter is general 300~500mm, and the wall thickness is 70~150mm. For convenient transportation, these piles are made in sections, and the length of each section may be 8~15m. The different sections can be connected by welding or flange bolts, the bottom of first section can be sealed, opened, or connected with pile tip. Different driving methods for precast concrete pile will be introduced later in section 2.3.4~2.3.6.

(a) Solid square pile　　　(b) Hollow square pile　　　(c) Hollow round pile

Figure 2-6　Precast concrete pile

1. Precast of solid square piles on site

When there is no factories nearby, solid square piles can be made on site by interval and superimposed casting method. The production process includes: field arrangement→ground leveling→ground concrete casting→rebar cage assembling→formwork erection→concrete casting→curing to 30% strength→formwork removal→formwork erection for upper layer pile→separants brushing→production of upper layer piles→curing→hoisting→transportation→stacking.

The production site of pile should be smooth and solid to prevent uneven settlement. The surface between adjacent piles should be separated by plastic membrane or separant. Only when the adjacent and lower layer pile's strength reach to 30% of the design strength, the adjacent and upper layer piles can be produced. The layers of pile should not be more than 4, and the separants should be brushed between different layers.

The size of formwork should be accurate and the surface is flat and straight, the top formwork should be perpendicular to the pile axis. The shape of the pile tip is a regular four-face pyramid, and the pile tip is located on the axis of the pile. After the formwork is erected, separent should be brushed on the inner surface of the formwork.

The diameter of main rebar should not be less than 14mm, the rebar ratio has relation with driving method: no less than 0.8% for hammering method and 0.4% for static pressure method. The main rebar should be butt-welded or sleeve-connected. The number of main rebar joints in the same cross section should not exceed 50%, the distance between the two joints of the adjacent rebars should be greater than $35d$ (d is diameter of main rebar) and no less than 500mm. There should not be any joints within 1m distance to pile top. Special treatment should be made for the rebar arrangement of pile head and pile tip, see Figure 2-7. The position of longitudinal rebar, steel bar mesh and pile tip should be correct.

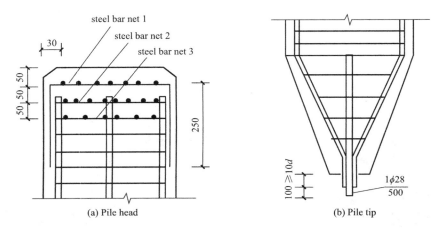

Figure 2-7 Reinforcement arrangement of pile head and tip

The concrete strength should not be lower than C30 for ordinary pile and C40 for prestressed pile respectively. The concrete should be machine-mixed, machine-vibrated and cast continuously from pile head to tip without any construction joint. After casting, the concrete should be cured to prevent cracks due to shrinkage of concrete, the curing time should not be less than 7 days.

2. Precast of hollow square and round piles in factory

Hollow piles are produced by centrifugal method and cured by steam under atmosphere pressure or high pressure in factory. According to the strength of concrete, the piles can be classified into prestressed concrete pile and prestressed high strength concrete pile (Code PS and PHS for hollow square pile, PC and PHC for hollow round pile). The concrete strength used for prestressed concrete pile and prestressed high strength concrete pile are C60 and C80 separately. The main rebars used for the prestressed concrete are special steel bar and steel strand.

3. Hoisting, stacking and transporting

The precast concrete pile can be hoist when the strength of concrete reaches to 70% of design strength. The hoisting points should be decided according to the minimum bending moment principle, see Figure 2-8.

The stacking place should be smooth and solid, the spacing between the stow-woods should be same as that of the hoisting points, the stow-wood position of different layers

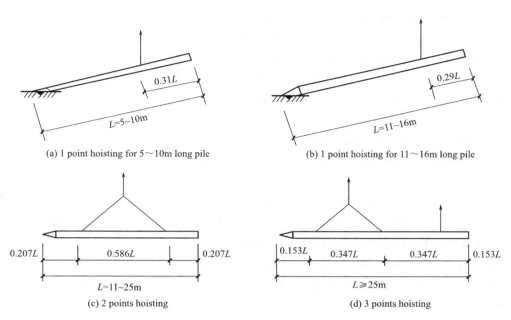

Figure 2-8 Methods of hoisting precast pile

should be in the same vertical line. The stacking layer should not be more than 4 layers for solid pile and 8 layers for hollow piles. Piles of different specifications shall be stacked separately.

Only when the concrete strength reaches 100% of design strength that the precast concrete pile can be transported and driven. During the driving of piles, piles should be transported to the site according to the driving requirement to avoid secondary transportation.

4. Preparation work before pile driving

Before the beginning of pile driving, some following preparation works should be done first.

(1) Ground leveling and obstacles clearing.

(2) Bench mark setting, the measurement of the axis pile and its marking and setting the pile position.

Set 2~4 bench marks around the pile driving ground for the ground leveling and checking of pile driving depth. The bench marks and pile position points should be placed from the influence of pile displacement.

(3) Preparation of pile hat, cushion and driving machine.

5. Hammer driving method

Thehammer driving method uses the impact force of the pile hammer on the top of the pile to overcome the resistance of the soil to the pile, and to make the pile sinking to the predetermined depth or the bearing soil layer.

Hammer driving is a common method for driving precast pile. It has high construction speed, high mechanization and wide application, but it has bump noise and vibration on the ground surface during construction. It has some restrictions in urban and night construction (the noise should not be more than 70dB in day and 55dB at night).

1) Pile driving equipment and selection

A pile driving equipment includes hammer, pile-driving frame and powerplant, the pile hammer and pile-driving frame are mainly considered in the selection.

(1) Hammer

Hammer is the main equipment to give impact force to the pile, there are many types of hammer as drop hammer, air or steam hammer, diesel hammer, hydraulic hammer.

① Drop hammer

Drop hammer is the oldest type of hammer used for pile driving. It is usually made of cast iron, 0.5~2 ton, raised by a winch and drops freely from a certain height. The pile head is rammed by the impact force of the hammer and then the pile is driven into soil. The drop hammer has the advantages of simple structure and convenient use, and the disadvantages of low driving speed, inefficiency and easy destroy of pile head. The drop hammer is suitable for the construction of small diameter concrete pile and steel pile in soft stratum.

② Steam hammer

Steam hammer uses high pressure steam or compressed air as power to raise hammer. According to the working principle, steam hammer can be classified into two types: single acting air or steam hammer and double-acting and differential air or steam hammer.

Figure 2-9 (a) shows the diagram of single-acting air or steam hammer. The hammer is raised by air or steam pressure and then drops freely under gravity. The hammer weight is 3~15 ton, the rate of blows is 40~60 times/min. It has the advantages of small drop distance, high driving speed and efficiency. It is suitable for all kinds of piles.

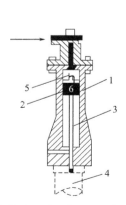 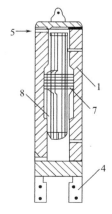

(a) Single-acting air or steam hammer   (b) Double-acting air or steam hammer

Figure 2-9 Working sketch of steam hammer

1—Cylinder; 2—Piston; 3—Piston rod; 4—Pile; 5—Inlet valve; 6—Cavity;
7—Hammer cushion; 8—Impact part

Figure 2-9 (b) shows the diagram of double-acting and differential air or steam hammer. Air or steam is used both to raise the hammer and to push it downward, thereby increasing the impact velocity of the hammer. The hammer weight is 1~7 ton, the rate of blows is 100~135 times/min. It has the advantages of high driving efficiency. It is suitable for all kinds of piles and can also be used for pulling pile, driving inclined pile and pile under water.

Since the steam hammer should be equipped with an extra equipment like boiler or air compressor to provide high pressure steam or compressed air, it is not convenient on site, so it is seldom used nowadays.

③ Diesel hammer

Diesel hammer consists essentially of a hammer, an anvil block and a fuel-injection system, see Figure 2-10. First, the hammer is raised and fuel is injected near the anvil, then the hammer is released. When the hammer drops, it compresses the air-fuel mixture, which ignites. This action, in effect, raises the hammer again and pushes the pile downward at the same time.

The hammer weight is 2~10 ton, the rate of blows is 40~80 times/min, refer to Table 2-1. The diesel hammer has the advantages of simple structure, convenient use and high efficiency. It is suitable for all kinds of piles but not for soft stratum. In soft soils, the downward movement of the pile is rather large, and the upward movement of the hammer is small. The drop height may not be enough to ignite the air-fuel mixture, so the hammer can not be raised again, the working cycle is interrupted.

Figure 2-10 Diesel hammer

Due to public hazards such as vibration, noise, waste gas pollution and so on, the use of diesel hammer in city is increasingly restricted.

④ Hydraulic hammer

Hydraulic hammer uses hydraulic pressure as power to raise hammer, see Figure 2-11. Like steam hammer, the hydraulic hammer can be divided into single acting hydraulic hammer and double acting hydraulic hammer. For the former, the hammer is raised by the hydraulic pressure and released freely, the hammer drops and blows the pile head. For the latter, the hammer is both raised and dropped by the hydraulic pressure, so the hammer has larger acceleration, higher speed and greater impact energy. The hydraulic hammer has the advantages of

Figure 2-11 Hydraulic hammer

good performance, no fume pollution and small noise. But it also has the disadvantages of the complicate structure, heavy maintenance work, high price and lower working rate than diesel hammer.

The selection of the pile hammer is the key to the pile driving construction with hammer driving method. If the weight of hammer is too big, it is easy to destroy the pile. If the weight of hammer is too small, the most part of impact energy is absorbed by the pile, the hammer jumps highly but pile sinks little. First of all, the type of pile hammer should be selected according to the construction conditions, and then the hammer weight should be determined, refer to Table 2-2. When the hammer weight is $1.5 \sim 2$ times the pile weight, the effect is ideal. When the pile weight is greater than 2 tons, it is possible to use a hammer lighter than the pile, but it should not be less than 75% of pile weight. The hammer driving should follow the principle of "heavy hammer and light ramming", which means the heavy hammer and small drop height. The pile is easily driven into the soil without spring back and destroying the pile head.

(2) Pile-driving frame

The functions of pile-driving frame are to suspend pile hammer, fix pile shaft, guide the direction of pile hammer, guarantee the verticality, host the pile and move in a small range.

Pile-driving frame is generally composed of chassis, guide bar, lifting equipment, supporting pole and so on. According to the moving way, there are 4 types of pile-driving frame: fixed type, crawler type, walking type and track type.

① Fixed pile-driving frame

Fixed pile-driving frame's chassis cannot move by itself, it is moved on the rolling wood or steel tube by outer force. This type of frame has advantages of simple structure, easy making and low cost. The disadvantage is that the plane steering is not flexible and the operation is complicated.

② Crawler pile-driving frame

Crawler pile-driving frame uses crawler crane as chassis, and vertical and inclined rods for pile driving. This kind of pile-driving frame has the advantages of flexible verticality adjustment, good stability, convenient disassembly and assembly, fast moving, strong adaptability and high construction efficiency. It is suitable for all kinds of precast pile and cast-in-place pile construction, and it is one of the commonly used pile-driving frames at present.

③ Walking pile-driving frame

The walking type pile-driving frame is supported by two movable chassis which are mutually supported, and can walk forward alternately. The frame can rotate by 360°, and is convenient to move and place without laying any track. It has high driving efficiency.

④ Track pile-driving frame

The track pile-driving frame needs to lay tracks, which is driven by multiple motors and controlled centrally. So the frame can only move along the track, the motility is poor and it is inconvenient for construction.

Diesel hammer selection reference

Table 2-1

Types of hammers				Diesel hammer(t)							
				20	25	35	45	60	72	80
Hammer dynamic performance	Impact weight(t)			2.0	2.5	3.0	4.5	6.0	7.2	8.0
	Total weight(t)			4.5	6.5	7.2	9.6	15.0	18.0	19.5
	Impact force(kN)			2000	2000~2500	2500~4000	4000~5000	5000~7000	7000~10000	8000~11000
	Common stroke(m)			1.5~1.8	1.8~2.2	1.8~3.2	2.0~3.2	2.0~3.5	1.8~2.5	2.5~3.4
Size of pile(mm)				250~350	350~400	400~450	450~500	500~550	550~600	600~800
Soil bearing stratum	Cohesive soil Silty soil	Sinking depth(m)		1.0~2.0	1.5~2.5	2.0~3.0	2.5~3.5	3.0~4.0	3.0~5.0	3.5~6.0
		CPT specific penetration resistance(MPa)		3	4	5	>5	>5	>5	>8
	Sandy soil	Sinking depth(m)		0.5~1.0	0.5~1.5	1.0~2.0	1.5~2.5	2.0~3.0	2.5~3.5	3.0~4.0
		SPT(N)		15~25	20~30	30~40	40~45	45~50	50	>50
	Soft rock	Pile tip in the rock	Intense weathering	—	0.5	0.5~1.0	1.5~2.5	2.0~3.0	2.5~3.5	3.0~4.5
			Moderate weathering	—	—	—	0.5	0.5~1.0	1.0~2.0	1.5~2.5
Control penetration of the hammer				—	2~3	2~5	3~5	3~6	3~7	3~8
Design value of vertical bearing capacity of single pile				—	600~1200	800~1600	1300~2400	1800~3300	2200~3800	2600~4500

Static pressure pile driver selection reference　　Table 2-2

Maximum driving force(kN)	1600~1800	2400~1800	3000~3800	4000~4600	5000~5600
Suitable pile size(mm)	250~400	300~500	400~500	400~550	450~600
Bearing capacity of pile(kN)	1000~2000	1700~3000	2100~3800	2800~4600	3500~5500
End bearing soil layer	Medium dense~dense sand, Hard plastic~hard clay, Residual soil	Dense sand, hard clay, fully weathered rock	Dense sand, hard clay, fully weathered rock	Dense sand, hard clay, fully weathered rock, strong weathered rock	Dense sand, hard clay, fully weathered rock, strong weathered rock
SPT of bearing soil layer(N)	20~25	20~25	30~40	30~50	30~55
Driving depth in medium dense and dense sand(m)	About 2	2~3	3~4	5~6	5~8

The selection of pile-driving frame depends on several factors

a. Material of pile, cross section shape and size of pile, length of pile and connection method.

b. Numbers of pile, pile space and layout.

c. Type, size and weight of pile hammer.

d. Construction condition on site, working space and surrounding environments.

e. Construction period and driving speed requirements.

The height of pile-driving frame should satisfy the requirement of construction.

Frame height=pile length+height of pile hammer+height of pulley group+starting height of hammer.

(3) Powerplant

The powerplant of the pile driving machine depends on the selected pile hammer. When air hammer is used, air compressor should be equipped, when steam hammer is selected, steam boiler and winch should be equipped.

2) Driving order

Precast piles are displacement piles because they move some soil laterally and make densification of soil surrounding them during pile driving, which maybe destroy the former piles and increases the resistance to latter pile driving. So reasonable driving order should be planned before pile construction. Whether the pile driving order is reasonable or not will affect pile driving speed, pile driving quality and surrounding environment.

The selection of pile driving order should be determined in combination with the properties of soil stratum, pile type, pile space, performance of the pile driving machine and requirements of the construction period and so on.

When pile space $D \leqslant 4d$ or $4b$ (d—diameter of round pile, b—width of square pile), the displacement effect is great, the driving order should be from center of the ground to two sides symmetrically, see Figure 2-12 (c), or from the center to four sides of the ground, see Figure 2-12 (b). When there are buildings or underground pipelines near the pile driving ground, the driving order should be from the side close to the buildings or pipelines to another side to protect the buildings or pipelines from cracking. When the pile driving ground is too large, the ground can be separated into several sections, and the driving order can be same with former 2 methods in each section.

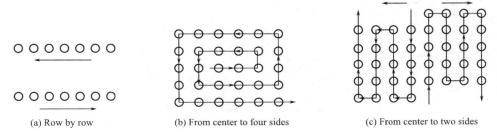

(a) Row by row   (b) From center to four sides   (c) From center to two sides

Figure 2-12 Pile driving order

when $D>4d$ or $4b$, the displacement effect is small, driving order can be selected flexibly. The former 2 methods or unidirectional driving row by row, see Figure 2-12 (a), can all be selected.

According to the design elevation and specifications of piles, the driving order generally should be longer and larger size pile first, shorter and smaller size pile later, which can decrease the influence of later driving piles to the former piles.

3) Pile driving technology

The pile driving construction is important to insure the quality of pile foundation, the main construction technology process is as follows: ground preparation→bench mark of pile position→pile driver in position→hammer lifting→pile hoisting and position alignment→pile verticality adjustment→hammer fixing→pile verticality adjustment→pile driving→pile connecting→pile driving→pile cutting.

(1) The pile-driving frame should be vertical and stable, and the central line of the guide bar is consistent with the driving direction.

(2) A short falling distance of hammer should be selected at the beginning of ramming, after the pile tip enters into ground a certain depth (1~2m), and the pile is stabilized in soil, the ramming can be made at the specified distance.

(3) The principle of the pile driving is heavy hammer with short distance ramming.

4) Pile connection

When the designed pile length is bigger than one section precast pile, different sections can be connected during driving by welding, flange bolts, mechanical quick connection, slurry anchor. The former three methods are suitable for all types of soil stratum, the last one is only suitable for soft soil.

(1) Welding connection

Welding connection is the most widely used at present, as shown in Figure 2-13. Before welding, the veritcality of upper and lower piles should be checked. Then 4 corners of square pile or 4 points of round pile can be spot welded to fix pile sections first, then two people weld the pile simultaneously and symmetrically to prevent uneven welding

(a) Square pile connection

(b) Round pile connection

Figure 2-13　Weld connection

deformation later. The weld should be continuously full, when there is space between upper and lower piles, it can be filled with steel sheets and welded firmly. After connection, the difference of 2 section axes should not be more than 10mm, the bending vector height of the joint shall not be greater than 1‰ of pile length.

(2) Flange connection

Flange connection connects piles by flange and bolts, and is commonly used for prestressed round pile connection, the connection speed is fast. The node structure is shown in Figure 2-14.

(3) Mechanical quick connection

Mechanical quick connection methods include 2 common methods. One method is screwed sleeve connection method, as shown in Figure 2-15 (a). For this method, steel ring with external spiral thread is embedded in pile head, when the upper pile and lower pile head are aligned by a bigger

Figure 2-14　Flange connection

diameter sleeve with inner spiral thread, the joints shall be cleaned and coated with grease, and a special chain wrench shall be used to tightened the sleeve. After the locking, the two plates shall have 1~2mm gap.

Another method is joggle connection method, as shown in Figure 2-15 (b). For this method, several tooth-clamps are embedded in one pile head with same number bolt holes in another pile head, when the upper pile and lower pile head are aligned, clean the upper and lower pile head plates, turn the anchor pins into the bolt holes of the upper pile, fill the tooth-clamp grooves with asphalt paint and coat the asphalt paint of 20mm width and 3mm thickness around pile head sheet, align the anchor pins with every grooves, then put down the upper pile and the anchor pins insert into the grooves, continue add driving pressure to the upper pile, 2 pile head plates contact with each other tightly, pile connection is completed.

(a) Screwed sleeve connection　　　　　　(b) Joggle connection

Figure 2-15　Mechanical quick connection

(4) Slurry anchor connection

The node structure of slurry anchor connection is shown in Figure 2-16. For this method, some holes in lower pile head and some steel bars at the bottom of upper pile should be preformed, the preformed holes and the surface of lower pile head are filled with molten sulfur slurry before connection, then the steel bars and holes are aligned with each other and the upper pile is dropped down, the steel bars are inserted into holes and sealed by set sulfur slurry. This method is not conducive to earthquake resistance, it should be carefully selected for 1st grade building and uplift pile.

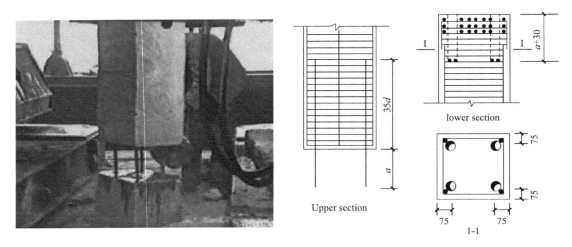

Figure 2-16 Slurry anchor connection

5) Quality control of pile driving

Quality control of pile driving includes the ramming number per meter during driving, ramming number for last 1m driving, the penetration depth of last 30 ramming (average value of every 10 ramming), the elevation of pile tip, verticality and position of pile.

The construction quality of end bearing pile is mainly controlled by the last penetration depth (average value of 3 penetration depth per 10 ramming), and the depth of pile tip or the elevation of pile tip can be used as reference. When the last penetration depth satisfies the design requirement but the pile tip does not reach to the design depth, 30 more ramming should be continued and the penetration depth of each 10 ramming should not be more than the design value.

The friction pile is mainly controlled by the design elevation of pile tip, andthe last penetration depth can be used as reference.

6. Static pressure driving method

Piles can be driven into the soil by static pressure without noise and vibration, it is often used in area with dense buildings and pipelines. This method is suitable for uniform soft soil ground.

Static pressure driving machine uses counterweight and self-weight of the pile-driving frame to balance the resistance of soil to pile, pushes the precast pile downwards by static pressure. According to the difference of static pressure, the pile driver can be mechanical or hydraulic pile driver. The mechanical static pressure pile driver, as shown in Figure 2-17 (a), consists of chassis, pile-driving frame and power unit, it uses pulley block to push the pile into soil. It has low efficiency and is seldom used nowadays. The hydraulic pile driver is shown in Figure 2-17 (b), it consists of pile-driving frame, hydraulic pile clamping device, crane and power unit. During the construction of pile, the hydraulic pile driver can hoist the pile by crane, clamp the pile shaft by clamping device and push the pile into the soil. When clamping device move upward, it can also pull the pile.

(a) Mechanical static pile driver (b) Hydraulic static pile driver

Figure 2-17 Static pressure driving method

7. Vibration driving method

The vibratory pile driver consists essentially of two counter-rotating weights. The horizontal components of the centrifugal force generated as a result rotating masses cancel each other. As a result, a sinusoidal dynamic vertical force is produced on the pile and helps driving the pile downward, as shown in Figure 2-18.

8. Water jetting driving method

Water jetting is a technique sometimes used in pile driving when the pile needs to penetrate hard soil layers such as sand and gravel. For this technique, water is discharged at the pile end by means of a pipe 50~75mm in diameter to wash and loosen the sand and gravel. During the construction, the pile is driven into soil along with water jetting, when there is 1~2 m pile length left, stop water jetting, the pile is driven to the design elevation by hammer to ensure the end bearing capacity. Two different jetting driving methods for solid pile and hollow pile refer to Figure 2-19.

9. Influence of pile driving to surrounding environment and prevention measures

Pile driving will generate some bad influences to thesurrounding environment such as

Figure 2-18　Vibratory pile driver

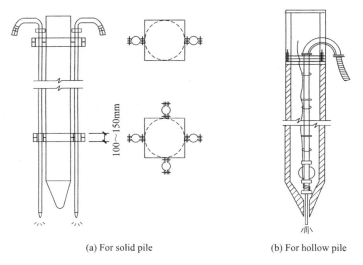

(a) For solid pile　　　　(b) For hollow pile

Figure 2-19　Water jetting driving method

soil displacement effect, noise and vibration during pile driving.

1) Soil displacement effect

Even if a suitable driving order is selected, pile driving still generates excessive pore water pressure and lateral earth pressure, lateral displacement of soil and ground heaving which may make deformation of buildings and pipelines nearby. When it is serious, it even produces a serious accident of cracking or tilting. Some measures can be taken to reduce displacement effect.

(1) Pre-bored method. For this method, the pile is driven into a pre-bored hole in the ground. The hole is 1/3~1/2 of pile length and the diameter is 50~100mm smaller than

pile. Since a big part of soil is removed, this method can decrease 30%~50% of soil displacement, 40%~50% of excessive pore water pressure.

(2) Set up sandbag wells or plastic drainage plates to eliminate partial excessive pore water pressure and reduce displacement effect. The diameter of sandbag well is 70~80mm, the space is 1~1.5m and the depth is 10~12m. The space and depth of plastic drainage plate is same with the sandbag well.

(3) Decrease the ground water level by wells to reduce the water pressure.

(4) Select suitable pile driver and driving order, control the driving speed or drive the pile after excavation of foundation pit.

(5) Combined with foundation pit retaining structure, use steel sheet pile or concrete diaphragm wall to prevent soil displacement and water seepage.

2) Vibration during pile driving.

Vibration can cause dangers like cracking of adjacent buildings, rising of driven piles, and some measures can be taken to reduce these dangers.

(1) Dig trench to prevent vibration and displacement. The width of the trench is general 0.5~0.8m, the depth is decided by the stability of slop, and is better deeper than the foundation of protected building. This method can be used with other measures like sandbag well.

(2) Add special cushion or bumper between hammer and pile head.

(3) Adopt construction technology based on the combination of pre-bored method, water jetting and static pressure method.

(4) Install vibration isolation facilities like vibration isolation wall, the thickness of the wall is 500~600mm, the depth is 4~5m, the distance to the pile driving area is 5~15m. The measure can reduce the vibration of 1/10~1/3.

3) Noise during driving.

The noise hazard is decided by the decibel, the decibel should be controlled under 75dB in residential area and 80dB in commercial area. When the decibel is higher than 80dB, some measures should be taken as follows:

(1) Control the sound source like selecting suitable pile driver, improving pile cushion and clamping device.

(2) Seal the hammer by muffler cover.

(3) Erect shelter protection, the shelter wall is economic and reasonable around 15m high.

(4) Time control. Stop driving in the noon and at night to ensure the normal life and rest.

10. Quality problems of precast pile and treatment methods

Some quality problems such as pile broken, pile lifting, pile bogging, torsion, tilt and displacement of pile shaft, quick sinking of pile etc. may commonly happen during the construction of precast pile. The reasons and treatment methods can refer to Table 2-3.

Reasons and treatment methods of quality problems　　　　Table 2-3

Common problems	Main reasons	Preventive and treatment methods
Broken pile head	The strength of the pile head is low, the reinforcement is improper, the protective layer is too thick, and the pile top is uneven; the hammer is not perpendicular to the pile with eccentricity, the hammer is too light, the drop hammer is too high, the ramming is too long, and the impact force on the pile head is uneven; the deformation of the head plate is too large, uneven	The pile shall be made in strict accordance with the quality standard, pile pad shall be added and pile head shall be leveled; measures such as correction of perpendicularity or low hammer slow ramming shall be adopted; deformation of pile head shall be corrected
Broken pile shaft	The pile quality does not meet the design requirements; excessive hammering in case of hard soil layer	After the steel clamp is tightened with bolts, it shall be welded and reinforced; if it has met the penetration requirements, it can not be treated
Floating pile	Soil squeezing and uplift of adjacent piles in soft soil	Re-drive the pile with a large amount of rise. If the pile fails to satisfy the static load test, re-drive it
Stagnant pile	Stop driving for a long time, improper driving order; in case of underground obstacles, hard soil layer or sand inter-layer	Select the driving order correctly; drill through the hard soil layer or obstacles with a drilling machine, or drive with jetting water
Torsion or displacement of pile shaft	Pile tip is asymmetry, pile shaft is not vertical	It can be corrected by crowbar and slow hammer with low blow. If the deviation is not large, it may not be handled
Tilt or displacement of pile shaft	The pile tip is not straight, the pile head is uneven, the head plate is not in the same line with the pile shaft, the pile distance is too close, and the soil mass is squeezed when the adjacent pile is driven; in case of lateral obstacles, the soil layer has a steep inclination angle	If depth into the soil is not big and the deviation is not large, it can be fixed with a wooden frame and then driven by a slow hammer for correction; if the deviation is too large, it should be pulled out to fill sand for re-driving or adding pile; if the obstacle is not deep, it can be dug out to fill sand for re-driving or adding pile
Sharp sinking of pile	The connection or the pile tip is broken, the pile body is bent or there is serious transverse crack; the drop hammer is too high, the pile connection is not vertical; in case of soft soil layer and soil hole	Improve the inspection before pile sinking; pull out the pile for inspection, correct and re-drive or add a new pile near the original pile position
Pile shaft and hammer jump	The pile shaft is too curved, the connected pile is too long, and the drop hammer is too high; the pile tip meets the tree root or hard soil layer	Take measures to pass through or avoid obstacles, or change pile and re-drive. If depth into soil is not big, the pile shall be pulled out and re-driven
Loosening and cracking at pile connection	The surface of the pile joint is not clean, with impurities and oil stains; the iron parts or flanges of the pile joint are uneven, with large gaps; the welding is not firm or the bolts are not tightened, the proportion of sulfur mastic is improper, and the operation is not in accordance with the regulations	Clean up the connection plane; correct the plane of iron parts; after welding or bolt tightening, check whether it is qualified by hammering, and test and check the sulfur cement ratio

2.3.2 Construction of steel piles

Steel piles generally are either pipe piles or rolled steel piles, as shown in Figure 2-20. They can also be driven into ground by static pressure or vibration force like precast concrete piles. Pipe piles can be driven into the ground with their ends opened or closed. In many cases, the pipe piles are filled with concrete after they have been driven. Rolled steel piles include H-section, I-section, wide-flange and steel sheet piles. However, H-section piles are usually preferred because their web and flange thickness are equal.

(a) Steel pipe pile   (b) H-section steel pile

Figure 2-20   Construction of steel piles

Steel piles are usually used as temporary structures for retaining soil or water. When they are used as permanent structures, they may be subject to corrosion under the ground. To offset the effect of corrosion, an additional thickness of steel is generally recommended. In many circumstances factory-applied epoxy coating on piles work satisfactorily against corrosion. Concrete encasement of steel piles in most corrosive zones also protects against corrosion.

2.4 Construction of cast-in-place pile

For cast-in-place pile, a hole needs to be directly bored at the design position by machinery or manpower first, then a rebar cage is put into the hole and concrete is cast later. Compare with precast pile, cast-in-place pile has the characteristics such as no limit of soil layers, no squeezing of soil, no connection of pile, smaller vibration and noise, and can be built in downtown area with dense population and buildings. But it also has the shortcomings as strict operation requirements, no easy control of quality, producing much slurry or debris of soil during boring, and long time concrete curing before next process.

The general construction process can refer to Figure 2-21. According to the boring technology, the boring methods include dry drilling, drilling in slurry, tube sinking, artificial digging and rotary drilling.

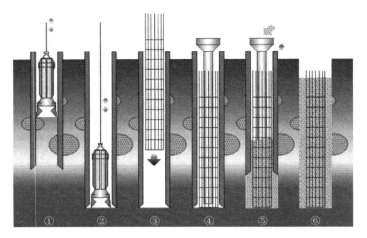

Figure 2-21   Construction diagram of cast-in-place pile

1. Dry drilling method

At present, continuous flight auger is often used for dry drilling construction of cast-in-place pile, which cuts soil into debris by drilling bit and discharge the soil debris out of the hole along the continuous spiral blade, as shown in Figure 2-22. The diameter of the dry drilling pile is 300~900 mm, and the drilling depth is 8~30 m. It is suitable for the general clay layer, sandy soil and artificial fill without groundwater in the depth of boring, but not suitable for muddy soil and soil with underground water.

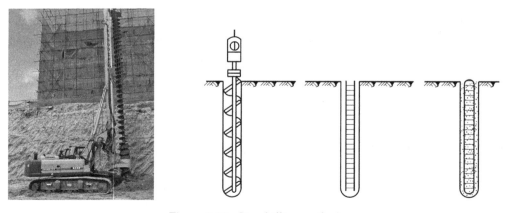

Figure 2-22   Dry drilling method

Dry drilling operation requires verticality, stability and correct position of the drill pipe. During the drilling, the soil debris discharged out of the hole should be cleaned at any time, and the abnormal conditions such as hole collapse and shrinkage should be analyzed and solved in time. When the drilling bit drills to the design elevation, continue turning the drill pipe without sinking until there is no soil debris discharged out of the hole. After the hole is cleaned, rebar cage should be carefully put into the hole to prevent hole wall collapse and cast concrete continuously.

2. Drilling in slurry method

Slurry boring is a mechanical boring method which uses slurry to protect the stability of hole wall. The soil is cut into debris and suspends in slurry, and then discharged out of the hole by circulation of slurry. It is suitable for soil layers with or without groundwater. The construction process of slurry drilling method for cast-in-place pile is shown in Figure 2-23.

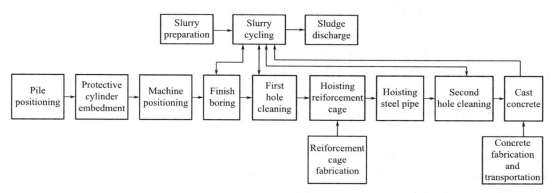

Figure 2-23　Construction process of slurry drilling method

1) Protective cylinder embedment

Before drilling a hole, a steel cylinder should be embedded at the pile positionr to protect the hole opening. The protective cylinder is made of 4～8mm thick steel sheet, the inner diameter should be 100mm bigger than the drill bit. On the top of the cylinder, there are 1～2 openings for the flowing of slurry. Generally, the cylinder is about 1.5～2m high, the embedded depth is better not less than 1m in clayey soil, 1.5m in sandy soil. When the upper soil layer is soft soil or loose sand, the cylinder can reach to the bottom of the soil layer to prevent the collapse of hole wall. The space between cylinder and hole wall should be sealed with clay tightly. The cylinder should be 0.3～0.4m higher than ground surface, and the slurry surface should be kept more than 1m above the groundwater table inside the protective cylinder.

The functions of protective cylinder are fixing the position of pile hole, protecting the hole opening, preventing inflow of surface water, keeping slurry head in the hole, preventing hole collapse, and guiding the direction of drill bit in drilling process.

2) Slurry preparation and circulation

During the drilling process, slurry can be injected into the hole to keep the stability of hole wall. Since the slurry density and head are all bigger than the groundwater, slurry will infiltrate into the hole wall and form a thin mud layer on the wall which can avoid the leakage of slurry in the hole, and to keep the stable slurry head in the protective cylinder. In addition, the functions of slurry are to carry soil debris, cool and lubricate drilling bit.

(1) Preparation of slurry

When drilling in clayey soil layer, the original soil can be used to make slurry, that is, during drilling, clean water is injected into the holeand mixed with clay into slurry during the drilling process.

When drilling in sandy soil layer, the slurry shall be prepared specially in slurry tank or pool on site, see Figure 2-24. The slurry is a mixture of high plasticity clay or bentonite and water. Other admixtures, such as weighting agent, dispersant, tackifier and plugging agent, can also be added into the slurry to improve its properties. The properties of slurry includes relative density, viscosity, sand content, PH, etc., in which the relative density is the key control parameter. The relative density of injected slurry should be controlled at about 1.1, and the relative density of discharged slurry should be 1.15～1.25.

Slurry pools should be constructed for mixing, sedimentation and circulation of slurry on site. There are usually two slurry pools, one is mixing pool, one is sedimentation pool, they areconnected for slurry circulation. The slurry pool can be directly excavated on the ground with natural slop or with retaining structures such as masonry wall and steel sheet. The volume of the slurry pools must ensure the stable slurry head in the borehole and discharged volume during the circulation.

Protective rails and safety wire mesh should be installed on the ground around the slurry pool and safety warning signs must be hanged.

(a) Slurry pool   (b) Bentonite powder

Figure 2-24   Slurry preparation

(2) Circulation of slurry

During drilling, soil will be cut into soil debris and discharged out by slurry circulation. According to the circulation direction, the circulation routes can be divided intopositive circulation and reverse circulation.

For positive circulation, see Figure 2-25(a), slurry or high-pressure water is pumped into the hollow drill pipe and ejected from drilling bit. The slurry flows upward along the hole wall, and the soil debris is taken out of the hole with slurry and flows into the sedimentation pool. After sedimentation, the slurry with required density can be injected into the hole again for circulation. This method relies on the upward flow of slurry to discharge soil debris, with less lifting force and more sediment at the bottom of the hole.

For reverse circulation, see Figure 2-25 (b), slurry flows into hole, soil debris with slurry is sucked out by slurry pump through the drill pipe and ejected into sedimentation pool. Due to the pumping effect, the slurry of the reverse circulation process has higher up-flow speed and can carry larger soil debris, but for holes with poor soil quality or easy to collapse, this method should be used carefully.

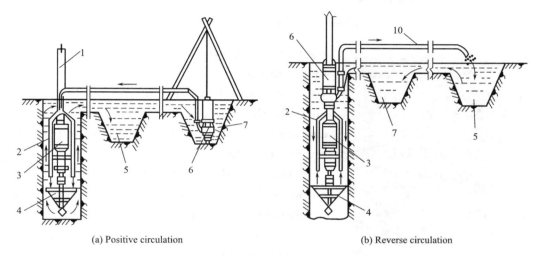

(a) Positive circulation (b) Reverse circulation

Figure 2-25 Slurry circulation
1—Drill pipe; 2—Slurry pipe; 3—Diving motor; 4—Drill bit; 5—Sedimentation pool;
6—Slurry pump; 7—Slurry pool

3) Drilling

According to the types of drill bit, drilling machines can be divided into rotary drilling machine, diving drilling machine, punching machine, grabbing cone machine etc..

(1) Drilling by rotary bit

Rotary drilling machine is driven by powerplant above ground, which rotates drill pipe with bit forcibly, see Figure 2-26. The soil is cut into debris by rotary bit and discharged out of the hole by positive or reverse slurry circulation.

The performance of the rotary drilling machine is reliable, the noise and vibration are small, the drilling efficiency is high, and the drilling quality is good. It is widely used in many geological conditions, such as loose soil layer, clay layer, sand gravel layer, soft rock layer, etc..

(2) Drilling bydiving bit

Diving bit is a kind of rotary drilling machine, its water-proof motor and bit are sealed

(a) Rotary drilling machine   (b) Rotary drilling bit

Figure 2-26　Drilling by rotary bit

Figure 2-27　Diving bit

together, as shown in Figure 2-27. After positioning by pile-driving frame and drill pipe, it can dive into water and drill holes in slurry. After the slurry is injected, the soil or stone is cut into debris and discharged out of the hole with slurry by positive or reverse circulation.

Diving bit drilling machine is suitable for silt, mucky soil, cohesive soil, sandy soil and strongly weathered rock stratum, not for gravel soil.

(3) Boring by punching bit

The punching machine raises the heavy bit with blade (percussion hammer) to a certain height through the frame and hoist, cuts and breaks rock or soil layer to form holes by the free-falling impact force, see Figure 2-28. The soil debris can be discharged by the slag grabbing cone or slurry circulation. It can be used in clayey soil, silty clay, especially for hard soil layer, sand gravel, pebble boulder and rock stratum.

Punching bits are cross-shaped, I-shaped, herringbone and so on. Generally, cross-shaped bit is commonly used.

After the boring machine is in place, the center of the bit should be aligned with the center of the protective cylinder. In the range of stroke 0.4～0.8m, the density of the slurry should be lowered. Stone and slurry should be added in time to protect the wall. The stroke can be increased to 1.5～2.0m after the protective cylinder sinks 3～4m. The relative density of the slurry can be measured and controlled at any time.

(4) Boring by punching claw cone

There is a heavy iron block and movable claw pieces on the punch-claw cone head, as shown in Figure 2-29. The punch-claw cone is lifted to a certain height through the frame

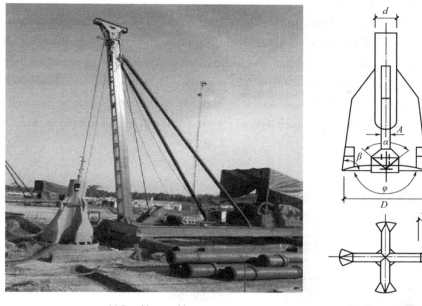

(a) Punching machine   (b) Cross-shaped bit

Figure 2-28 Boring by punching bit

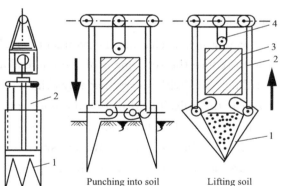

Figure 2-29 Boring by punching claw cone
1—Claw blades; 2—Connection rod; 3—Additional weight; 4—Pulley block

and hoist. When it falls, the drum brake is loosened and the claw blades are opened. The cone head falls freely into the soil, and then is lifted by the hoist, the claw blades close and grab the soil. The punching and grabbing cone is lifted to the ground as a whole to unload the soil slag, and the hole is formed in turn.

The construction process, the installation requirements of protective cylinder, the circulation of slurry, wall-protecting of punching and grabbing cone boring are the same as that of impact boring.

It is suitable for punching in soft soil (sand and clay), but it is better to use punching bit when encountering hard soil.

(5) Rotary boring machine

Rotary boring machine has a hollow bit, as shown in Figure 2-30, the bit is driven by powerplant above ground, which rotates drilling pipe with the bit forcibly. The soil is cut into debris by rotary bit and stored in the hollow tube. After the bit is full of soil debris, the bit will be lifted out of the hole and discharges the soil on the ground. The rotary boring machine can be used in different soil layers by flexible changing of cutting bits and can drill a hole with or without slurry, which are widely used in pile construction now.

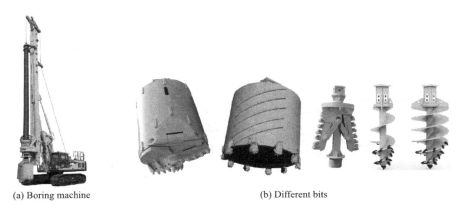

(a) Boring machine   (b) Different bits

Figure 2-30   Rotary boring machine

4) Hole cleaning

Hole cleaning means clearing the sediment and slurry floating soil at the bottom of the hole, so as to reduce the settlement of the pile end and improve the bearing capacity.

For drilling holes with slurry made by original clay, the method of water ejection can be used to clear the hole. The drill bit continues rotating without sinking at the bottom of hole, and the relative density of slurry can be reduced to about 1.1 by water ejection. When clear the hole in the slurry, the air suction machine can be used to clean the pile holes which are not easy to collapse. The air pressure is 0.5MPa, which makes a strong high-pressure airflow upward in the pipe. At the same time, fresh water is continuously replenished. The agitated slurry is discharged from the nozzle along with the airflow until the clear water is ejected.

For the hole wall with poor stability, the slurry circulation method should be used to clear the hole by replacing thick slurry with dilute slurry, and the relative density of the slurry after the hole cleaning should be controlled between 1.15 and 1.25.

Pile holes can be cleaned once before hoisting rebar cage or again before casting concrete. After hole cleaning, the thickness of residual sediment at the bottom of the hole should be measured and controlled no more than 50mm for end bearing pile, 100mm for friction pile, and 200mm for anti-uplift and horizontal force resistant piles.

5) Fabrication and hoisting of rebar cage

Rebar cages can be made in factory or on site. The stirrup can be connected with the main steel bar by welding. Except stirrup, Φ12～16mm reinforcement rings should be connected with the main steel bars at an interval of 1.5～2.0m to increase the stability and rigidity of the cage and used as hoisting points during the hoisting. The rebar cage can be made in sections when the length is too big. The length of each section will be decided by the hoisting height and capacity of hoisting machine. After the first hole cleaning, the rebar cage can be hoisted with 2 or 3 hoisting points, and different cage sections can be connected by welding or screw sleeve. After the rebar cage was lowered to the design elevation, a steel beam can be passed through the cage horizontally to suspend the cage on the opening of the hole. The fabrication and hoisting process can refer to Figure 2-31.

(a) Welding   (b) Hoisting   (c) Suspending

Figure 2-31   Fabrication and hoisting of rebar cage

6) Concrete casting underwater

When slurry is used to retain the soil wall during drilling the hole, concrete is cast in the water or slurry, so it is called underwater concrete cast (Figure 2-32).

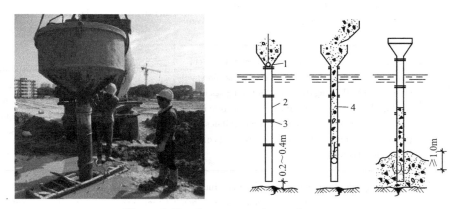

Figure 2-32   Concrete casting underwater
1—Water-stop ball; 2—Tremie section; 3—Joint; 4—Concrete

Underwater concrete should be increased by one strength grade compared with the design strength, and it must have good workability. The mix proportion should be determined by test.

Underwater concrete can be cast by tremie method. After rebar cage is put into the hole and the hole is cleared secondly, steel pipe sections are connected and put into the hole as tremie. A water-stop ball can be suspended in the pipe by steel wire, then the funnel at top and the pipe will be filled with enough concrete to ensure that the end of the pipe can be buried below the concrete surface more than 0.8m at one time when concrete rushes out of pipe. Then the steel wire is cut off, the water-stop ball will be rushed out by the concrete pressure. The pipe end will be buried into the concrete so that fresh concrete will be discharged inside the old concrete to avoid pollution by slurry during the following casting. The pipe will be lifted and dismantled section by section with the rise of concrete surface. The volume of concrete cast and pipe lifting speed should be controlled properly to ensure the pipe embedment depth in concrete ≥1m throughout casting process.

Finally, the concrete should be cast 0.8~1m higher than the design elevation so that the strength of pile head can be guaranteed after the top concrete-slurry mixture is chiseled.

7) Analysis and treatment of common problems

Because cast-in-place piles are constructed below the ground surface, the quality can not be seen directly and some construction defects often happen during the construction process. The analysis and treatment methods can refer to Table 2-4.

Analysis and treatment of common problems of pile construction in slurry   Table 2-4

Common problem	Main reason	Preventive and treatment method
Hole collapse	The protective cylinder is not buried tightly and water leaks or is buried too shallow; the slurry surface in the hole is lower than the water level outside the hole or the slurry density is not enough; drilling in the quicksand, soft mud and loose sand layer, the drilling and rotation speed is too fast, etc.	The surrounding of the protective cylinder shall be tightly sealed with clay; slurry shall be added in time during drilling to make it higher than the ground water level; in case of quicksand and loose soil layer, the slurry density shall be increased appropriately and the drilling speed shall not be too fast
Suspended pile	The debris is not cleaned clearly and the debris sediment is too thick; the slurry density is too small after the hole cleaning, the hole wall collapses or the sand gushes into hole bottom, or the concrete is not cast immediately; rebar cage or tremie collides with the hole wall during lifting, causing the soil collapse to the hole bottom	Pay attention to the slurry density and timely clean up the debris; do well in hole cleaning, and immediately cast concrete when meeting the requirements; pay attention to protect the hole wall from heavy objects collision during construction
Broken pile	The first batch of concrete is cast unsuccessfully for many times, and a mud inter-layer appeared when the upper concrete is cast, which resulted in pile breaking; the collapse of hole wall stuck the tremie, and the slurry mixed into the concrete when the tremie is pulled out forcefully; the pipe joint is poor, and the slurry entered the tremie	Try to cast concrete successfully for one time; select slurry with high density, viscosity and good colloid rate to protect the wall; control the drilling speed and keep the hole wall stable; connect the tremie joint with square screw and set rubber ring to seal tightly

(1) Necked pile

Necked pile means the cross section area of pile shaft in a certain range is smaller than the design value due to collapse of hole wall.

(2) Broken pile

Broken pile refers to the partial separation or fracture of the pile shaft. More serious condition is that there is no concrete in a section of the pile. The pile breaks easily on the interface between soft and hard soil layer 1~3m below ground surface.

(3) Suspended pile

Suspended pile means there is no concrete at the bottom of the pile or the concrete is mixed with debris at the bottom of the hole and the concrete becomes soft, so the pile has no solid end contacts to the soil layer at the bottom of the hole.

3. Pipe-sinking method

Pipe-sinking method uses hammer-driving equipment or vibration equipment to sink a steel pipe into the ground with its end closed by pile tip. Then a rebar cage is put into the pipe, the pipe is withdrawn along with the casting of concrete. The steel pipe can be withdrawn by vibration equipment to tamp concrete at the same time. After the pipe is pulled out, the required cast-in-place pile is formed. The construction process can refer to Figure 2-33.

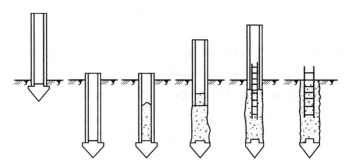

Figure 2-33　Construction process of pipe sinking method

1) Pile tip

The role of pile tip is preventing soil and water from entering the pipe. Pile tips can be reinforced concrete pile tip, steel plate pile shoes or valve-type pile shoes and have enough strength to bear the resistance of the soil, see Figure 2-34.

2) Pipe sinking

During the construction of pipe sinking, there are displacement and vibration effects on the soil. In the construction, the sinking order of holes should be considered according to the site construction conditions. When the center distance of the pile is less than 5 times the outer diameter of the pipe or less than 2m, it should be constructed with jumped order, the inter-space can be at intervals of one or two piles, sink the pipe after final setting of adjacent pile concrete. If there are more than 5 piles under one pile cap, the inner

(a) Concrete tip (b) Steel tip

Figure 2-34 Pile tips

piles should be built first and the outer piles later.

3) Concrete casting and pipe pulling

After the pipe is sunk to the design depth, check the bottom of pipe without water or slurry, then put rebar cage into the pipe and cast concrete. Ensure that concrete has flowed out of tremie before pipe pulling. The concrete in the pipe should be filled as far as possible, and the pipe should be evenly pulled out. The pulling speed is better no more than 1.2~1.5m/min in general soil layer and 0.6~0.8m/min in soft soil layer.

To compact the concrete and increase bearing capacity, the pipes can be sunk and pulled out one or more times.

(1) Single sinking method (also known as one-time pipe pulling method)

During pulling the pipe, the pipe should be vibrated 5~10s after pulling out every 0.5~1.0m, repeat the process until the pipe is totally pulled out.

(2) Double sinking method

Double sinking means sinking the pipe in one hole two times in the whole or partial depth. First, sink pipe and cast concrete the same as single sinking method without rebar cage, and then sink the pipe again before the initial setting of concrete with same axis of the first sinking. Put rebar cage, cast concrete and pull out the pipe like first one.

(3) Repeated sinking method

During pulling the pipe, sink 0.3~0.5m after pulling out every 0.5~1.0m, repeat the process until the pipe is totally pulled out. The concrete in tremie should be always above the ground surface and the pulling speed should not be more than 0.5m/min.

The cross section area of pile with single sinking, double sinking and repeated sinking methods can be 30%, 80%, 50% bigger than that of pipe respectively, therefore the bearing capacity of pipe sinking cast-in-place pile can be increased by 50%~80% higher than that of bored pile.

4) Quality problems and treatment of pipe sinking method

The analysis and treatment methods of quality problems and treatment of cast-in-place pile with pipe sinking method can be referred to Table 2-5.

Analysis and treatment of pipe sinking method   Table 2-5

Common problem	Main reason	Preventive and treatment method
Necked pile	When sink the pipe in the mud and soft soil layer, the soil displacement will generate excessive pore water pressure. After the pipe is pulled out, the pore water pressure will squeeze the newly cast concrete, so the cross section of pile will be decreased at different degrees like necking. When the pipe is withdrawn too fast, the volume of concrete is too small, or workability of the concrete is poor, the concrete spreads slowly after flowing out the tremie, the surrounding mud will fill part of the hole and form pile necking	During withdrawing of the tremie, the concrete surface in the tremie should be kept higher than the ground so that it has sufficient diffusion pressure, and the slump of the concrete should be controlled at 50~70mm. Upside-down method can be adopted during pulling of the pipe, and the pulling speed should be strictly controlled
Broken pile	The space between piles is small, during the construction of new pile, the displacement of soil generate horizontal force and lifting force. The horizontal force transmits unevenly in soft and hard soil layer, which will generate shear force to pile on the surface between the two soil layers. The concrete of adjacent piles does not have sufficient strength, and is crushed and broken by outer force	Reasonable construction order and driving route are determined to reduce the influence to the new cast-in-place concrete pile. After the concrete of former pile reaches to 60% of the design strength or using jump-driving method, the latter pile can be constructed
Suspended pile	When the groundwater pressure is high, or the prefabricated pile tip is damaged, or when the gap of the valve of the pile boot is large, water and slurry enter the steel pipe and mix with concrete. When the valve of the pile boot is not opened under the earth pressure at the beginning of pulling, the pipe is pulled out to a certain height before it opens, then concrete falls down, so the pile end is not solid	Try to prevent the valve from unopened, the pipe should be withdrawn slowly with dense stroke at beginning. The pipe can be sunk and pulled several times locally at the bottom of pile, then it can be pulled out normally. Cushion material with better performance is used at the interface between pile boot and pipe opening to prevent the infiltration of groundwater and slurry

4. Artificial digging method

When the diameter is bigger than 0.8m, general 1~5m, the piles are called large diameter pile. The large diameter pile is used singly under one column and has big strength, rigidity and bearing capacity, so it is commonly used in the project. In some areas, when it is difficult to transport drilling machine to the job-site or the soil stratum is too hard to drill a hole by machine, the hole can be dug manually.

There are many advantages of artificial digging method. Since the pile diameter is large, the pile has high bearing capacity. During digging, the quality of the hole and the change of geological soil quality can be directly checked in the hole, the depth of the hole is decided by the actual situation of the soil layer, the hole can be cleaned thoroughly at the bottom, and the quality of cast concrete is easy to guarantee.

Workers need to dig soil in the hole, so ventilation and lighting system should be provided. With the increase of depth, some bracing structures should be adopted to prevent collapse of upper soil, as shown in Figure 2-35.

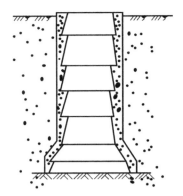

Figure 2-35　Artificial digging method

1) Cast-in-place concrete wall protection method

Dig the hole and cast concrete retaining wall segmentally. The wall can not only prevent the collapse of the hole wall, but also proof water.

The excavation height of each section depends on the erection ability of the soil wall. Generally, 0.5~1.0m is a suitable section height, and the diameter of excavation hole is designed with pile diameter plus the thickness of concrete wall.

Construction of retaining wall includes erecting inner formwork of retaining wall (movable steel formwork) and casting concrete. Generally, the concrete strength is not less than C15, and the concrete should be vibrated and compacted. When the strength of concrete reaches 1MPa, the formwork can be removed and begin the construction of next section. Do the cycle works until dig to the design depth.

2) Precast concrete wall protection method (Precast concrete tube sinking method)

When pile diameter is large, excavation depth isbig, geology is complex, soil quality is poor (soft soil layer), and groundwater level is high, caisson sinking method can be used for excavation.

For this method, a reinforcement concrete tube with edge foot can be precast just at the pile position, and then dig soil in the tube. Relying on the self-weight of tube or additional pressure, the tube overcomes the friction resistance between the outer wall and the soil, and sinks along with excavation. Finally, it sinks vertically to the required depth of design.

3) Notice in construction

(1) The deviation of the plane position of the center line of the pile hole should not exceed 50mm, the deviation of the vertical degree of the pile should not exceed 0.5%, and the diameter of the pile should not be less than the design diameter of the pile.

(2) Excavated soil and building materials shall not be piled up around the excavation area, and local concentrated loads and mechanical vibration shall be prevented.

(3) Pile end must be located on the supporting layer required by the design. The excavation depth of pile hole should be decided by the designer according to the actual

situation of soil layer on site.

(4) Manual excavation should be carried out continuously and concrete casting should be carried out immediately after checking.

(5) Carefully remove the residual soil from the bottom of the hole, drain the accumulated water, and prevent groundwater from flowing into the hole during the concrete casting.

(6) In the process of manual excavation, the construction should strictly follow the operating rules. Safety fence should be set up around excavated hole. When the net distance between pile holes is less than 2 times the diameter of pile and less than 2.5m, the hole should be excavated at intervals.

5. Construction of pedestal pile

When the bearing soil layer is good, the end of the pile can be enlarged to increase the bearing capacity, the pile with larger diameter end is called pedestal pile.

There are three methods for enlarging pile end.

(1) Enlarge pile end by artificial digging.

Artificial digging method is suitable for hard plastic to hard clayey soil, medium dense to dense sand or gravel, weathered rock without water or with small quantities of groundwater. The center position at the bottom of the hole should be measured before digging, then dig the soil around the hole evenly to the design shape and size. The enlarged part should have regular surface, good shape and correct size. After the enlarged part is finished, the soil debris should be cleaned, and then the rebar cage can be hoist into the hole and cast concrete. During the concrete casting of enlarged part, some measures should be taken to prevent disintegration and the concrete should be compacted layer by layer.

(2) Enlarge pile end by rotary drilling bit.

(3) Enlarge pile end by explosion.

2.5 Post grouting of cast-in-place pile

Post grouting of cast-in-place pile means that after the pile is completed for a certain period of time, cement slurry is injected through the grouting pipes embedded in the pile shaft and the grouting valves at the pile end and pile side connected with the pipes, so that the sediment soil at the pile end and mud skin around the pile shaft can be strengthened to improve the bearing capacity of single pile and reduce the settlement, see Figure 2-36.

The post grouting pipe shall be steel pipe,

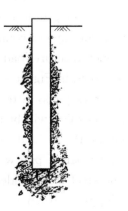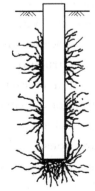

Figure 2-36 Post grouting

and bound or welded with the rebar cage. When the pile diameter is ⩽1.2m, 2 grouting pipes can be set symmetrically along the circumference of the rebar cage, for piles with a diameter of more than 1.2m, 3 or more grouting pipes shall be set uniformly along perimeter. The bottom of the grouting pipe should extend more than 15cm beyond the rebar cage.

If the pile length is more than 15m and the required increment of bearing capacity is big, the pile end and pile side compound grouting should be used. The number of post grouting pipe valves on the pile side shall be determined according to the soil stratum conditions, pile length, required increment of bearing capacity and other factors. Generally, one grouting valve on the pile side shall be set every 6~12m. The grouting valve shall be able to withstand more than 1MPa water pressure. The protective layer of the grouting valve shall be able to resist the impact of sand and other materials from damaging the grouting valve. The grouting valve shall be one-way valve (non-return valve).

The grouting can be started 3 days after the pile is completed. The grouting sequence of the same pile is as follows: upper side pipe→lower side pipe→end pipe. The grouting sequence of pile group under one pile cap is side piles first, center piles later.

The post grouting technology is mainly controlled by grouting volume, supplemented by pressure control. The grouting parameters shall be reasonably selected according to the geological conditions. For example, if the pile end is a dense gravel and pebble layer, large grouting volume and large grouting pressure may be considered, with the grouting volume as the main control index. If the pile side is a dense sand layer, the grouting pressure may be taken as the main index, and the grouting volume as the reference index. After total grouting volume meets the requirement or stabilizedgrouting pressure is greater than 3.0MPa for 1min, grouting can be stopped.

2.6 Construction of pile cap

In most cases, piles are used in groups to transmit the structural load to the soil. A pile cap is constructed over group piles, the cap can be an isolated one under columns or connected into a raft. Pile cap can connect each pile together as a whole, convert and adjust the allocation the superstructure load to each pile. It is also a connection of pile head and superstructure, as shown in Figure 2-37.

For precast concrete piles, after the piles are driven into the ground to the design elevation, a concrete cushion will be cast 50~100mm lower than the pile head. Generally, the cushion is 100mm thick, and 50~100mm wider than the size of pile cap in all directions. Anchorage rebar of pile with cap can be a rebar cage inserted into the hollow part of piles or some steel bars welded with pile head, the length of the anchorage rebar should satisfy the design requirements.

For cast-in-place concrete pile, since the concrete is often cast higher than the design

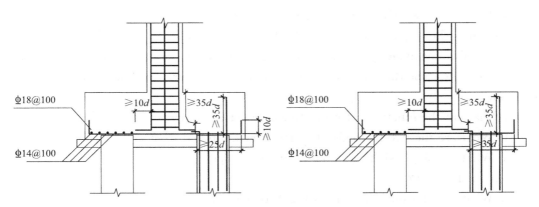

Figure 2-37 Construction of pile cap

elevation to guarantee the quality of pile head. After the concrete reaches to the design strength, the concrete of pile head should be chiseled and cut to the design elevation, the main rebar of the pile will be straighten up. The length of rebar should also satisfied the requirements of anchorage.

Different anchorage rebars can refer to Figure 2-38.

(a) Precast pile   (b) Cast-in-place pile

Figure 2-38 Anchorage steel bar

After the pile heads are treated, the rebar of pile cap and superstructure are laid out on the cushion, then concrete is cast into design size by formwork. When concrete reaches to the design strength, the piles and superstructure are connected by the pile cap.

Questions

1. What are the construction key points of reinforced concrete spread foundation?

2. What are the construction key points of reinforced concrete box foundation?

3. Briefly describe the main technical requirements on manufacturing, lifting, transport, and stacking of precast reinforced concrete piles.

4. How to choose hammer weight and pile-driving frame?

5. How to choose driving orders of precast pile?

6. What is the displacement effect of pile driving and how to reduce it?

7. Describe advantages of static pressure driving method ant applicable conditions.

8. Briefly describe the construction technology of the slurry drilling method for cast-in-place pile.

9. What are the functions of protecting cylinder and slurry?

10. What are the two methods of slurry circulation and what are their effects?

11. What are single sinking, double sinking, repeated sinking of pipe sinking method and their final effects?

12. What are the purposes of post grouting of cast-in-place pile and quality control methods?

Chapter 3 Masonry and Construction Facilities Engineering

Mastery of Contents: master the brick masonry construction technology; bricklaying quality standards.

Familiarity of Contents: raw materials of mortar, preparation and use of building mortar; types and technical requirements of mortar, bricks and other big blocks of masonry work; the technical requirements and application of landed scaffold, adhesive self-lifting scaffold.

Understanding of Contents: the characteristics and application of other types of scaffold; the characteristics and application of head frame, gantry frame and construction elevator.

Key Points: brick masonry construction technology.

Difficult points: quality standards of brick and big block masonry work.

3.1 Introduction

Masonry structures are the structures of wall and column made of blocks and mortar as the main bearing components or partition wall of a building. The masonry structures mainly include brick masonry, concrete block masonry and stone masonry structure. Masonry structure engineering includes the construction process of preparing masonry materials, laying masonry blocks and erecting construction facilities. The construction quality of masonry engineering is an important guarantee for the strength of masonry structure, scaffold and vertical transport equipment are auxiliary facilities for masonry construction. This chapter mainly introduces the masonry materials, construction of brick masonry, concrete block masonry, stone masonry, construction facilities like scaffold and vertical transport equipment.

3.2 Masonry materials

The main materials of masonry work are blocks and mortar, concrete and rebar are needed when necessary. Concrete strength grade is generally C20, HPB300, HRB335 and HRB400 strength grade steel bar or cold drawn low carbon steel wire is generally adopted as reinforcement. The materials used in masonry work should meet the design requirements, have the product certificate and performance test report, and the quality shall meet the requirements of the current relevant national standards. Blocks, cement, rebar, concrete and admixture shall also have the re-inspection report of the main performance of the materials on site. It is strictly prohibited to use materials that are explicitly eliminated by the government.

3.2.1 Masonry mortar

Masonry mortar is made of cementing materials (cement, lime, etc.) mixed with fine aggregate (sand) and water. When there is only cement used with sand and water in mortar, the mortar is called cement mortar, when other cementing materials like lime or flyash used to replace part of cement, it is called cement mixture mortar. In order to save cement and improve performance, appropriate lime paste or flyash and other admixtures are often mixed into the mortar, and corresponding additives can also be added. Mortar is divided into field mixed mortar and ready mixed mortar. Ready mixed mortar can be divided into wet mixed mortar and dry mixed mortar further. Wet mixed mortar is a mixture of cement, fine aggregate, additives and water, mixed in the mixing station, transported to jobsite by mixing transport vehicle, stored in a special container, and used in a specified time. Dry mixed mortar is a dry mixture of cement, fine aggregate, additives, transported to jobsite in packing or bulk and mixed with water before using on site.

1. Requirements of raw materials

1) Cement

The cement for masonry mortar should be ordinary Portland cement or slag Portland cement. The strength grade of cement shall be appropriate and meet the design requirements. Generally, it shall not be less than grade 32.5, and grade 42.5 is suitable. Too high strength of cement will affect the workability and water retention of mortar, which is not only conducive to construction, but also affects the growth of mortar strength.

When transported to the site, the types, grades, packing or bulk warehouse number and manufacture date of the cement shall be inspected, and its strength and stability shall be retested. For continuous incoming cement of the same manufacturer, same variety, same grade and same batch number, each batch of bagged cement shall not exceed 200t, and bulk cement shall not exceed 500t, and each batch of sampling shall be no less than one time. Its quality must comply with the relevant standards of the current national codes "General Portland Cement" (GB175, Chinese code). When there is doubt about the quality of cement in using or the cement is more than 3 months out of factory (1 month for quick hardening Portland cement), the cement should be reinspected and used according to the retest results. But if the initial setting time or stability is not qualified, it is prohibited to use. Different types of cement shall not be mix-used.

2) Sand

The sand for masonry mortar should be sieved medium sand and river sand is preferred. At the same time, the sand shall not contain harmful substances, including impurities, soil content, organic matter, sulfide, sulfate and chloride content, etc., which shall comply with the relevant standards of the current industry code "Standard for Quality and Inspection Methods of Sand and Stone for Ordinary Concrete" (JGJ 52, Chinese code). Artificial sand, mountain sand and extra fine sand shall meet the technical requirements of masonry mortar after trial mixing.

3) Slaked limepaste

Lime slaking is the process of producing calcium hydroxide ($Ca(OH)_2$) by hydration of quicklime (CaO) and water. Slaked lime paste can be made of quicklime blocks or quicklime powder by slaking, and the slaking time should not be less than 7 days and 2 days respectively. For slaking of quicklime blocks, the blocks should be mixed with water uniformly to be lime slurry, then the lime slurry flows through 3mm×3mm sieve and deposits in the sedimentation tank for 7 days before use to prevent undesirable expansion. The lime paste stored in the sedimentation tank should be prevented from drying, freezing and pollution. It is strictly prohibited to use the dehydrated and hardened lime paste. Quicklime powder and slaked lime powder can not be directly used for preparing cement lime mortar, they should be mixed with water into paste and stored 2 days in a sealed container before use.

4) Flyash

The quality index of flyash used in masonry mortar should conform to the relevant standards of the current industry code "Technical Code for the Application of Flyash in Concrete and Mortar" (JGJ 28, Chinese code). Generally, class II or class III flyash can be used. The replacement ratio of flyash in mortar should not exceed the standard value (replace ratio for cement $\leqslant 40\%$, lime paste $\leqslant 50\%$).

5) Water

The water for mortar mixing should comply with the relevant standards of the current industry code "Concrete Water Standard" (JGJ 63, Chinese code). Generally, potable water can be directly used, and treated industrial waste water can also be used to mix mortar after being qualified by laboratory analysis or trial mixing.

6) Additives

According to the needs of the project and in order to improve the performance of the mortar, plasticizer, early strength agent, retarder, antifreeze, waterproof agent and other additives can be added into the mortar to facilitate the construction. However, the type and quantity of the additives should be determined by the inspection and trial mixing by a qualified testing party. The technical performance of the additives should meet quality requirements of the current industry code "Masonry Mortar Plasticizer" (JG/T 164, Chinese code) and other relevant national standards.

2. Technical performance of mortar

1) Fluidity

Mortar fluidity refers to the performance that mortar mixture is easy to flow under gravity or external force. It is expressed by consistency value and measured by mortar consistency meter. The higher the consistency value is, the better the fluidity of mortar is, and the easier it is to be paved into a dense and uniform thin layer on the brick or block. Different masonry work has different requirements for mortar fluidity which can refer to Table 3-1.

Consistency value for different masonry work　　　　　　Table 3-1

Masonry type	Mortar consistency(mm)
Sintered ordinary brick masonry Autoclaved flyash brick masonry	70~90
Ordinary concrete small hollow block masonry Autoclaved lime sand brick masonry	50~70
Sintered perforated brick Hollow brick masonry Light aggregate concrete small hollow block masonry Autoclaved aerated concrete block masonry	60~80
Stone masonry	30~50

2) Water retention

Water retention of mortar refers to the ability that the mortar mixture can hold water and is not easy to bleed, it reflects the property that each component material in mortar is not easy to separate in layers. It is expressed by the stratification value and measured by the degree of stratification tester. The stratification value should be between 10~20mm and not more than 30mm. The large stratification value indicates that the water retention of mortar is poor, and the mortar is prone to bleed, which reduces its fluidity and makes it difficult to pave. Inorganic plasticizers such as lime paste, flyash or micro-foaming agent can be added into mortar to improve the water retention in the project.

3) Strength grade

According to the compression strength, cement mortar is divided into 7 grades: M5, M7.5, M10, M15, M20, M25, M30, cement mixture mortar is divided into 4 grades: M5, M7.5, M10, M15. The selection of strength grade shall meet the design requirements. The determination of mortar strength is based on the average value of compression strength measured after 28 days of curing under the standard curing conditions (The temperature is 20 ± 3℃, the relative humidity for cement mortar test block is required more than 90%, and 60%~80% for cement mixture mortar test block). Mortar test blocks shall be randomly sampled and manufactured at the outlet of mixer or storage container for wet mixed mortar (For mortar mixed on site, only one set of 6 test blocks can be made for the same batch).

4) Bonding strength

The bonding strength of masonry mortar can directly affect the shear strength, durability, stability and seismic performance of masonry structure. Masonry mortar must have good cohesive force to cement the blocks into an integral structure, so as to improve the bearing capacity of the masonry structure and achieve the required design strength. During the construction, some measures such as cleaning the masonry base, wetting the surface of the blocks and improving the construction and curing conditions can be taken to improve the bonding strength of mortar and ensure the quality of the masonry structure.

5) Durability

Mortar durability is the premise and guarantee of masonry durability. The mortar used for hydraulic masonry structure must meet the requirements of impermeability and anti-erosion, and the mortar used in severe cold area must meet the requirements of anti-freezing. In order to get the mortar with good durability, the amount of cement in the cement mortar should not be less than $200 kg/m^3$, and the total amount of cement and additive in the cement mortar should be $300 \sim 350 kg/m^3$.

3. Preparation and use of mortar

Cement mortar and cement lime mixed mortar are mostly used in masonry work. Cement mortar has higher strength but slightly worse consistency, which is suitable for masonry with higher strength requirements and wet environment, such as foundation, basement and other underground masonry structures, reinforced brick beams etc. Cement lime mixed mortar is suitable for masonry in dry environment, such as load-bearing or non-load-bearing masonry structures above the ground.

The mixing ratio of masonry mortar should be calculated and tested by qualified laboratory according to the actual situation of the job site, and meet the requirements of compression strength, consistency and stratification. When the composition of masonry mortar is changed, its mixing ratio should be re-determined. Cement mortar with strength grade less than M5 should not be used instead of cement mixture mortar with the same strength grade. If necessary, the cement mortar should be raised by one strength grade.

3.2.2 Masonry blocks

Masonry blocks occupy the most volume of masonry structures and are the main parts of masonry structures to bear the load, so each block should have certain volume for convenient masonry work and enough strength to support load. The most common blocks used in masonry work are bricks, concrete blocks, and stones.

1. Bricks

There are many kinds of bricks used in masonry work, which are divided into solid bricks, perforated bricks and hollow bricks according to the different void ratio in the bricks. According to their hardening method, these bricks can be classified into sintered bricks, steam curing bricks and autoclaved bricks further.

1) Solid bricks

Solid bricks include sintered ordinary bricks (clay bricks, shale bricks, coal gangue bricks, building slag bricks, silt bricks, etc.) and autoclaved or steam cured bricks (lime sand bricks, flyash bricks, cinder bricks, etc.), as shown in Figure 3-1. The nominal dimension of the modular brick is 240mm×120mm×60mm, while the actual size of the brick is 240mm × 115mm × 53mm. According to mechanical performance, solid bricks are divided into five strength grades MU10, MU15, MU20, MU25 MU30 (autoclaved or steam cured bricks without MU30), which can all be used as load-bearing masonry structure. The strength grades of the solid bricks refers to Table 3-2.

(a) Clay bricks

(b) Shale bricks(shale/coal gangue)

(c) Autoclaved sand-lime bricks
(lime and quartz sand, sand or fine sandstone)

(d) Flyash bricks

Figure 3-1　Solid bricks

Strength grades of the solid bricks　　　　Table 3-2

Strength grade	Compression strength(MPa)		Anti-break strength(MPa)	
	Average value of five bricks	Minimum of single brick	Average value of five bricks	Minimum of single brick
M30	30	22	13.5	9.0
M25	25	18	11.5	7.5
M20	20	14	9.5	6.0
M15	15	10	7.5	4.5
M10	10	6	5.5	3.0

2) Perforated bricks

Perforated bricks refer to the bricks with some small holes perpendicular to the bearing surface, the hole ratio is $\geqslant 25\%$, see Figure 3-2. There are mainly sintered perforated bricks and concrete perforated bricks in the market. Sintered perforated bricks mainly include sintered clay, shale, coal gangue, building slag perforated bricks and other varieties. Sintered perforated bricks have many specifications, such as 290mm, 240mm, 190mm in length and width, and 90mm in height. The hole size of sintered perforated brick is small, and the distribution is reasonable, the solid part of brick is as dense as solid bricks. It can be used as load-bearing masonry structure with the same strength grades as sintered ordinary bricks.

3) Hollow bricks

Hollow bricks have the same variety as perforated bricks, withlarger hole size and hole ratio ($\geqslant 40\%$), see Figure 3-3. Hollow bricks have fewer but larger holes. The holes are along the horizontal direction, that is, parallel to the bearing surface of the bricks. Sintered hollow bricks have length of 390mm, 290mm, 240mm, width of 240mm, 190mm, height of 115mm, 90mm and other specifications. Its strength is low and can only be used for non-load-bearing masonry structure. Brick grades are MU2.5, MU3.5, MU5.

Figure 3-2 Sintered perforated brick   Figure 3-3 Sintered hollow brick

2. Big blocks

As shown in Figure 3-4, the types of big block mainly include ordinary concrete hollow block, thermal insulation block, light aggregate concrete hollow block, flyash concrete hollow block, autoclaved aerated concrete block, gypsum block, etc.

Ordinary concrete hollow blocks are mainly used as load-bearing masonry structures. The block has vertical square holes, the main specification size is 390mm × 190mm × 190mm, the length of auxiliary blocks are 290mm, 190mm, 90mm, and the width and height are both 190mm, so as to cooperate with the main specification block.

There are a variety of lightweight blocks. The specifications and dimensions of light aggregate concrete hollow block and flyash concrete hollow block are exactly the same as

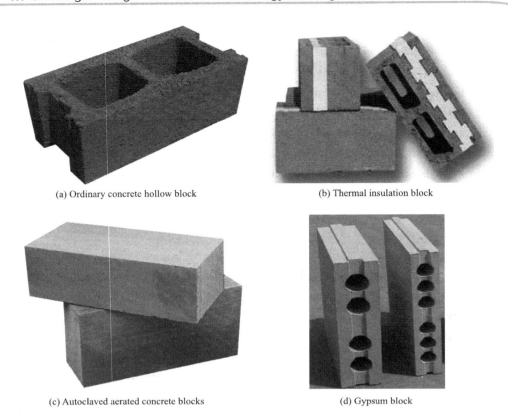

(a) Ordinary concrete hollow block (b) Thermal insulation block

(c) Autoclaved aerated concrete blocks (d) Gypsum block

Figure 3-4 Big blocks

ordinary concrete hollow block. For the autoclaved aerated concrete block, the length is 600mm, the width is 100~300mm, the height is 200~300mm and other kinds. Gypsum block is a new light material block, the length is 666mm or 600mm, the height is 500mm, the thickness is 60~200mm, the best block size is three blocks to form $1m^2$ wall. Grooves and tenons are made alternatively around the thickness sides for convenient connection of adjacent blocks. Lightweight blocks are mainly used for non-load-bearing masonry structures such as infilling walls and enclosure walls.

According to mechanical performance, concrete hollow blocks can be divided into 5 strength grades as shown in Table 3-3.

Strength grades of concrete hollow blocks   Table 3-3

Strength grade	Compression strength(MPa)	
	Average value of five blocks should be no less than	Minimum value of single block is no less than
M15	15	12
M10	10	8
M7.5	7.5	6
M5.0	5	4
M3.5	3.5	2.8

3. Stone blocks

According to the shape of the stone block, it can be divided into rubble and ashlar (dressed stone), see Figure 3-5. While rubble can again be divided into random rubble and plane rubble. Random rubble is irregular shape stone. Plane rubble is irregular shape stone but has two almost parallel planes, and the middle thickness is no less than 150mm.

(a) Rubble　　　　　　　　　　　　(b) Ashlar

Figure 3-5　Stone block

According to the degree of dressing, ashlar can be divided into fine ashlar, semi-fine ashlar, coarse ashlar and square rubble, as listed in Table 3-4. The width and length of ashlar are better no less than 200mm, and the length is better not bigger than 4 times of thickness.

Classification of ashlar　　　　　　　　　　　Table 3-4

Kinds of ashlar	The concave depth of exterior surface	The concave depth of joint surface
Fine ashlar	2 mm	10mm
Semi-fine ashlar	10mm	15mm
Coarse ashlar	20mm	20mm
Square rubble	Ratherish finishing	25mm

3.3　Brick masonry work

Brick masonry work uses sintered ordinary bricks, sintered perforated bricks, concrete perforated bricks, concrete solid bricks, autoclaved lime sand bricks, autoclaved flyash bricks, etc. as the main block of the masonry structure. It is widely used in civil engineering because of its low cost, simple construction and its applicability to buildings, structures and sporadic small masonry of various shapes and sizes.

3.3.1 Bonds in brick masonry work

Bond is the interlacement of bricks, which is formed by the bricks and those immediately below or above them. It is the method of arranging the bricks in layer so that all bricks are bonded together and the vertical joints of 2 successive layers are broken at same vertical line. Good bonds help in distributing the concentrated loads over a large area. An unbonded wall, with its continuous vertical joints has little strength and stability. Bonds of various types are distinguished by their elevation or face appearance. Bricks used in masonry are all of uniform size. If they are not arranged properly, continuous vertical joints will result. But since bricks are small blocks with uniform dimensions, the process of bonding is easily performed in construction.

1. Technical terms

Following are some of the technical terms used in masonry work. Since these terms are frequently used in the description and procedures, it is essential to understand the meaning of these terms first.

1) Course or layer

A course is a horizontal layer of masonry unit. Thus, in brick masonry, the thickness of a course will be equal to the thickness of modular brick plus the thickness of one mortar joint. Similarly, in stone masonry, the thickness of a course will be equal to the height of the stones plus the thickness of one mortar joint.

2) Header

A header is a full brick unit or stone which is so laid that its length is perpendicular to the face of the wall. Thus, the surface of brick with width and height lies parallel to the wall surface. In the case of stone masonry, header is sometimes known as through stone.

3) Stretcher

A stretcher is a full brick unit or stone which is so laid that its length is along or parallel to the face of the wall. Thus, the surface of brick with length and height lies parallel to the wall surface.

4) Bond

Bond is a term in masonry applied to the overlapping of bricks or stones in alternate courses, so that no continuous vertical joints are formed and the individual units are tied together.

5) Joint

The junction of adjacent units of bricks or stones is known as a joint. Joints parallel to the bed of bricks or stones is known as bed joint. Bed joints are thus horizontal mortar joints upon which masonry courses are laid. Joints perpendicular to the bed joint are known as cross-joint or vertical joints. All joints are formed in mortar. A joint which is parallel to the face of the wall is known as wall joint.

6) Closer

It is the portion of brick cut in such a manner that its one long face remains uncut. Thus, a closer is a header of small width. There are queen closer, king closer, beveled closer and mitred closer.

7) Bat

It is the portion of the brick whose one end is cut across the width. Thus a bat is smaller in length than the full brick. If the length of the bat is equal to the half length of the original brick, it is known as half bat. A three quarter bat is the one having its length equal to 3/4 of the length of a full brick. If a bat has its width beveled, it is known as beveled bat.

8) Toothing (racking bond)

These are the bricks left projecting in alternate courses for the purpose of bonding future masonry.

9) Offset

These are the narrow horizontal surface which are formed by reducing the thickness of the wall. It is often used in footing masonry.

10) Quoins

The exterior angle or corner of a wall is known as quoin. The stones or bricks forming the quoins are known as quoin stones or quoin bricks. If the quoin is laid in such a manner that its width is parallel to the face of the wall, it is known as quoin header. However, if the length of the quoin is laid parallel to the face of the wall, it is known as quoin stretcher. Quoin stone are selected sound and large and their beds are proper dressed.

11) Bed

This is the lower surface of a brick or stone in each course. This is the surface of stone or brick perpendicular to the line of pressure.

12) Through stone

A through stone is a stone header. Through stones are placed across the wall at regular interval. If the thickness of the wall is small, through stone may be the same length as the full width of the wall. However, if the wall is considerably thick, two through stones with an overlap are provided. Through stones should be strong, non-porous, and should be of sufficient thickness.

2. Rules for bonding

For getting good bond, the following rules should be observed:

(1) The bricks should be uniform size. The length of the brick should be twice of its width plus one joint, so that uniform overlap is obtained. Good bond is not possible if overlap is non-uniform.

(2) The amount of overlap should be minimum 1/4 length of a brick along the length of the wall and 1/2 length of a brick across the thickness of the wall.

(3) Use of brick closer should be discouraged because it is difficult to cut on site,

except in special locations.

(4) The center line of header should coincide with the center line of the stretcher in the course below or above it.

(5) The vertical joints in the alternate courses should be along the same vertical line but not continuous.

(6) The stretchers should be used only in the facing, they should not be used in the hearting. Hearting should be done in headers only.

3. Types of bonds

Following are the types of bonds provided in brick work:

1) Stretcher bond

Stretcher bond or stretching is the one in which all the bricks are laid as stretchers on the faces of walls, as shown in Figure 3-6. The length of the bricks are thus along the direction of the wall, with each vertical joint lying between the centers of stretchers above and below. This patterns is used only for those walls which have thickness of half brick length, such as those used as partition walls, sleeper walls and division walls. The bond is not possible if the thickness of the wall is more.

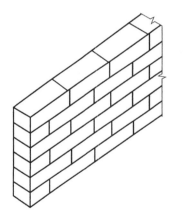 

Figure 3-6　Stretcher bond

2) Header bond

Header bond or heading bond is the one in which all the bricks are laid as headers on the face of walls, as shown in Figure 3-7. The width of the brick is thus along the direction of the wall. The pattern is used only when the thickness of the wall is equal to one brick length. The overlap is usually kept equal to half the width of brick. This is achieved by using three-quarter bats in each alternate courses as quoins. This bond does not have strength to transmit pressure in the direction of the length of the wall. As such, it is unsuitable for load bearing walls. However, the bond is specially useful for curved brick work like brick chimney and brick arch lintel.

3) English bond

This is the most commonly used bond for all wall thickness. This bond is considered to

Figure 3-7 Header bond

be the strongest. The bond consists of alternate courses of headers and stretchers, as shown in Figure 3-8. In this bond, the vertical joints of the header courses come over each other, similarly, the vertical joints of the stretcher courses also come over each other, as shown in Figure 3-9. In order to break the vertical joints in the successive courses, it is essential to place three-quarter bat in each stretcher course. Also, only headers are used for the heart of thicker walls. Sometimes this bond can also consist of several courses of stretchers with one course of header, as shown in Figure 3-10.

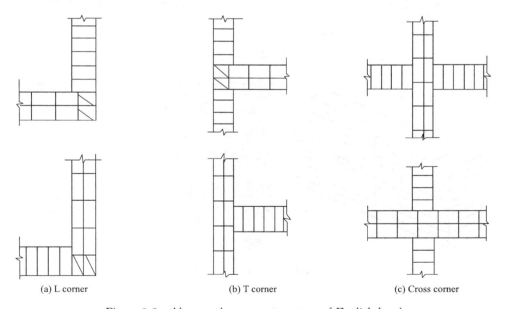

(a) L corner   (b) T corner   (c) Cross corner

Figure 3-8 Alternated courses at corners of English bond

4) Flemish bond

In this type of bond, each course is comprised of alternate headers and stretchers, as shown in Figure 3-11. 3/4 bats and headers can be placed at the corner of alternate courses to develop the face overlap. Every header is centrally supported over the stretcher below it.

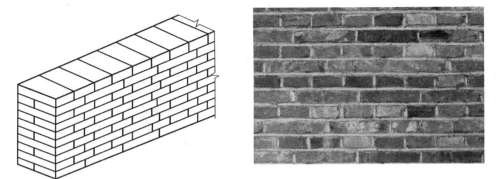

Figure 3-9　English bond with alternate courses of headers and stretchers

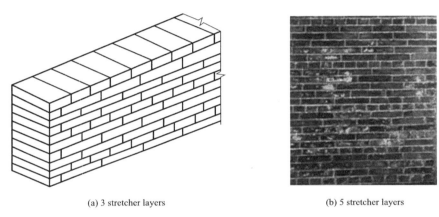

(a) 3 stretcher layers　　　　　　　　　(b) 5 stretcher layers

Figure 3-10　English bond with multiply stretcher layers

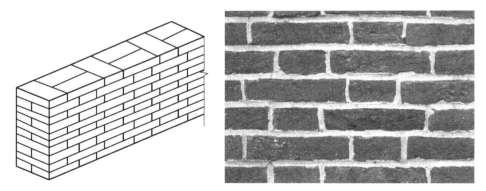

Figure 3-11　Flemish bond

3.3.2　Preparation for masonry work

1. Material preparation

1) Bricks

(1) The variety, size and strength grade of bricks must meet the design requirements.

and the specifications should be consistent. Bricks used on the surface of fair-faced walls and columns should also have neat corners and uniform color.

(2) When masonry work is at normal temperatures, the bricks should be moderately wetted 1~2 days in advance, so as to prevent absorbing excessive water from the mortar and affect its bonding strength. It is strictly prohibited to use dry bricks or full saturated bricks. If a brick absorbs too much water, it will form a layer of water film on the surface of the brick, which will make the cement slurry running out and polluting the wall surface, and the brick moving or sliding. The water content of sintered ordinary bricks and perforated bricks should better be 10%~15%. Concrete perforated bricks and solid concrete bricks do not need to be watered and wetted, but in the case of dry and hot climate, they should be sprayed and wetted before masonry construction. The water content of non-sintered blocks such as autoclaved lime-sand brick and flyash brick should better be 8%~12%. In construction, after cutting off the bricks, the water absorption depth around the section should better reach 15~20mm.

(3) When concrete perforated bricks, concrete solid bricks, autoclaved lime-sand bricks, autoclaved flyash bricks etc. are used in masonry work, their product age should not be less than 28 days.

2) Mortar

(1) The masonry mortar shall be mechanically mixed. When the mortar is mixed on site, the materials should be measured by mass. The mixing time of cement mortar and cement mixture mortar shall not be less than 2min. For cement flyash mortar and mortar mixed with additives, the time should not be less than 3min. For mortar mixed with organic plasticizer, the mixing method and time should compliance with current industry standards and the mixing time is generally in 3~5min. For dry-mixed mortar, the mixing time can be same with additive mortar or follow the product instructions.

(2) The mortar prepared on job site should be used just after mixing, and the mixed mortar should be used up within 3h. When the maximum temperature exceeds 30℃ during construction, the mortar shall be used out within 2h. The time for ready-mixed mortar can be determined according to the instructions provided by the manufacturer.

(3) The wet-mixed mortar used in masonry work must be stored in special containers that do not absorb water except for direct use, and measures such as sun shading, heat preservation, rain and snow protection should be taken according to climatic conditions. It is strictly forbidden to add water at will during the storage process.

2. Technical preparation

Brick masonry construction is a great art since blocks laying must be systematically done with respect to bonding, jointing and finishing. Technical preparation includes following steps:

1) Leveling

Before laying brick foundation or walls of each floor, the bottom elevation of masonry work shall be determined on the top of foundation or the floor. The masonry base should be leveled by M7.5 cement mortar or C10 fine stone concrete. If the thickness of the base joint is more than 20mm, C10 fine stone concrete should be used to make the elevation of brick foundation and the bottom of brick wall meet the design requirements.

2) Setting outline

Before masonry construction, the masonry area should be cleaned. The positioning axis of the foundation or the first floor can be measured according to the axis control datum around building. The positioning axis of upper floors can be measured according to the axis control points on the external wall surface which is measured in advance with the help of theodolite or line plumb. Then the position of the axis, sidelines, doors, windows and other holes of each wall can be set out according to the dimensions on drawing.

3) Making layer-pole or height board

In order to control the vertical height of each course and the elevation of brick foundations or walls, a layer-pole can be made of straight square wood in advance. According to the design dimension, the thickness of every course of brick and joint, the elevation of the doors, windows, lintels, ring beams, slabs and embedded parts etc. should be marked on the board. It is the sign of the vertical dimension and can ensure the verticality of the wall.

3.3.3 Brick masonry construction technology

The general construction process of brick masonry engineering is as follows: laying the brick sample→erecting the layer-pole→setting the corner and hanging the line→laying bricks→floor elevation control, etc.

1. Laying brick sample

Laying brick sample is to place one layer of dry bricks with width of vertical joints without mortar on the surface with set lines according to the type of bonds and dimensions on drawing. The purpose of brick sample laying is to make the longitudinal and horizontal wall be built accurately according to the position of setting out, and try to make the door and window openings, attached wall buttresses and other structures conform to the brick modulus, which can reduce the cutting of bricks as much as possible. At the same time, laying brick sample can make the vertical mortar joints uniform and satisfy the thickness requirements.

2. Erecting layer-pole

The layer-pole shall be erected at the corners, junctions and positions with different height for the foundation, and the corners and junctions for the brick wall, as shown in Figure 3-12. The vertical dimension of the masonry shall be controlled according to the layer-pole, and the thickness of the mortar joints shall be uniform to ensure that the bricks of each layer are at the same horizontal height. When erecting the pole, the datum elevation line marked on the pole should be consistent with the design elevation determined by leveling, and the distance between the poles shall not exceed 15m.

3. Setting corners and hanging alignment line

The corner of the foundation and brick wall is the main basis to ensure the horizontal and vertical of the masonry. Therefore, during the brick laying, bricks should always be laid a few layers higher at the corners and the junctions first, which is called setting corner. The corners and junctions should be vertical and neat, with uniform joints and shall not exceed five layers. Then alignment line shall be drawn between them, and the bricks of the middle part shall be laid one by one according to the line, as shown in Figure 3-13. When laying walls with thickness of one and a half brick length or above, the alignment line shall be hung on both sides, and for smaller thickness wall, the line can be hung on one side.

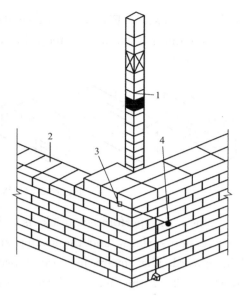

Figure 3-12   Erecting layer pole
1—Layer pole; 2—alignment line;
3—wood sheet; 4—steel nail

Figure 3-13   Brick laying according to the line

4. Laying bricks

When laying bricks, the first thing is to determine the bonding method. Generally, the brick foundation is built in English bond. Solid brick wall can select bonding method according to its thickness. The stretcher bond is suitable for half a brick thickness (115mm) wall. "Two flat with one side bond" is composed 2 course stretchers and one brick laid on one side alternately with its width parallel to the height of wall. The upper and lower vertical mortar joints are staggered with each other for more than 1/4 of the brick length (60mm), it is suitable for laying 3/4 of the brick thickness (178mm) wall. Header bond, English bond, Flemish bond are all suitable for 240mm or more thickness walls, as shown in Figure 3-14. In addition, the top layer of 240mm thick load-bearing wall, the outer brick on the stepped brick masonry and the overhanging layer shall be laid header bond.

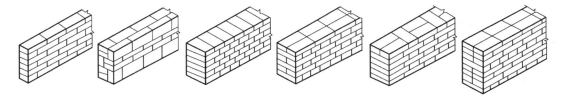

Figure 3-14 Walls with different thickness

When the perforated brick is laid, the holes should be perpendicular to the compression surface, and the sealing surface of the semi-blind perforated brick shall be laid upward. The square perforated brick is generally laid in stretcher bond, with an overlap 1/2 of the length of the brick, as shown in Figure 3-15 (a). The cuboid perforated brick can be laid in English bond or Flemish bond, with an overlap of 1/4 of the length of the brick. For hollow bricks, the holes should be parallel to the compression surface, and the bricks are laid in stretcher bond with an overlap 1/2 of the length of the brick, as shown in Figure 3-15 (b).

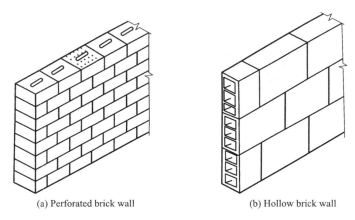

(a) Perforated brick wall   (b) Hollow brick wall

Figure 3-15 Walls with perforated brick

"Three ones" masonry method or spreading mortar method can be adopted for brick laying. "Three ones" masonry method means spreading one trowel of mortar, putting one brick, squeezing and kneading mortar one time, and scraping the extruded mortar subsequently. This bricklaying method can easily get full mortar joints, strong bonding force, clean wall surface. Therefore, it is suitable for bricklaying, especially for the projects with anti-seismic requirements. Spreading mortar method means spreading mortar to a certain length on the brick bed first, then laying bricks one by one later. The spreading length of mortar shall not exceed 750mm, but 500mm when the temperature exceeds 30℃.

The openings, slots, pipes required by the design should be properly reserved or embedded during the masonry work. Without the permission of designer, chiseling on walls or horizontal slotting shall not be allowed. Reinforced concrete lintels shall be provided at the upper part of the opening when its width is more than 300mm. Pipelines shall not be buried in bearing walls or independent columns with the long side of the section less than 500mm.

Under normal construction conditions, the daily brick masonry height should be controlled within 1.5m or one-step scaffold height to ensure the stability of masonry structure. The height difference of masonry wall in adjacent working sections should not exceed the height of one floor, and no more than 4m.

5. Cleaning and pointing of wall

Clean the mortar on the wall and floor after finishing the laying of the bricks. For the fair-faced wall, it needs joint pointing by 1∶5 or 1∶2 cement mortar.

3.3.4 Quality standard of brick masonry

1. Horizontal along length direction and vertical along height direction

The brick masonry should be perfectly in level and be truly in vertical. The horizontal joint of each course is required to be straight, and each brick must be leveled. In addition to leveling the top surface of floor or foundation, the alignment line should be hung strictly according to the layer-pole layer by layer and the line should be tightened, and each brick should be laid according to the line. The vertical faces should be checked by means of plumb bob and guiding rule. The vertical joints in alternate courses should be in the same vertical line.

2. Joints full of mortar

The mortar should cover completely the bed and the sides on the bricks. The bricks should be lightly pressed into the bed mortar so that uniform joint thickness is obtained. The brick, while laying, should be pushed sideways to make uniform vertical joint thickness and filled with mortar. All joints should be filled with mortar of proper consistency so that no cavity is left.

The full degree of horizontal joint for brick wall should not be less than 80%, while vertical mortar joint should not pass light. For brick column, both full degrees are 90%. Generally, the thickness of joint is 10mm, but should not be less than 8mm or more than 12mm.

3. Proper bonding

The brick work should be done in proper bond suggested by the designer. The overlap of the bricks of different courses should not be less than 60mm and there should be no continuous vertical joint. For brick column, there should be an overlap in the center as shown in Figure 3-16 (a).

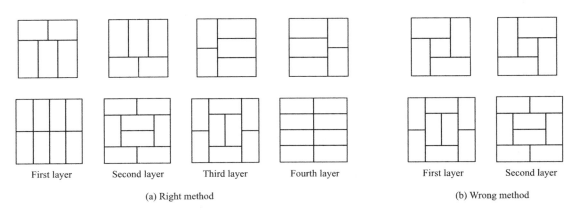

| First layer | Second layer | Third layer | Fourth layer | First layer | Second layer |

(a) Right method   (b) Wrong method

Figure 3-16 Column masonry work

4. Reliable junction toothing

The walls at junction and corner should be constructed at the same time, if it is impossible, steps or toothing or recesses should be provided during the construction as a temporary break to get a reliable junction bonding between the former and the later masonry work. It is strictly forbidden to construct the inner and outer walls at junctions separately without reliable measures. In regions with seismic fortification intensity of 8 degrees and above, the junctions should be built into inclined toothing. The horizontal length of ordinary brick toothing should not be less than 2/3 of the height, and the ratio of length to height of perforated brick toothing should not be less than 1/2, as shown in Figure 3-17 (a), (b). The height of inclined toothing shall not exceed the height of the one-step scaffold.

In the region of non-seismic and seismic fortification intensity of 6 and 7 degrees, when inclined toothing can not be left, vertical toothing can be left at other junctions except corners. For the vertical toothing, outward protruding toothing and tie bar should be adopted, as shown in Figure 3-17 (c). Some regulations about the tie bar are as follows:

(1) The number of tie bar is one 6mm diameter rebar per 120mm thickness of wall and no less than 2 at least.

(2) The interval along the height of the wall should not exceed 500mm.

(3) The tie bar should be bent into L shape with same side length. The embedded length counted from the toothing should not be less than 500mm. In 6 and 7 degrees earthquake fortification intensity region, the length should not be less than 1000mm and there should be a right angle hook at each end.

In order to ensure the integrity of brick masonry, the surface of the toothing must be cleaned and wetted before the construction of later wall. The mortar should be full filled to keep straight joints. The allowable deviation and inspection of brick masonry size and position shall conform to Table 3-5.

Chapter 3 Masonry and Construction Facilities Engineering

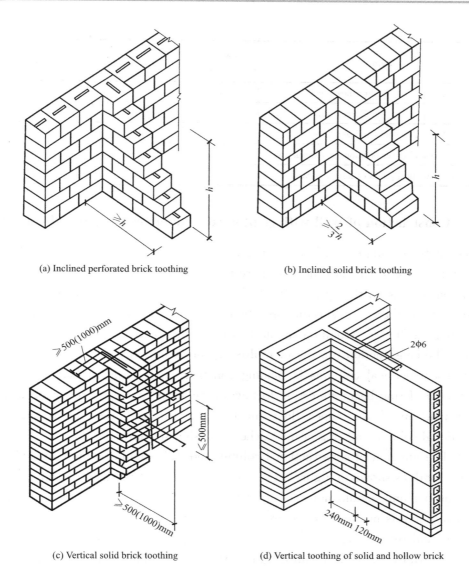

(a) Inclined perforated brick toothing
(b) Inclined solid brick toothing
(c) Vertical solid brick toothing
(d) Vertical toothing of solid and hollow brick

Figure 3-17 Toothing and tie bars used at T junction

The allowable deviation and inspection of brick masonry size and position Table 3-5

Item			Allowable difference(mm)	Checking method
Axis offset			10	Theodolite and ruler or other apparatus
Verticality	Every floor		5	2m line board
	Total height	≤10m	10	Theodolite, plumb and ruler or other apparatus
		>10m	20	
Elevation of foundation and floor			±15	Level and ruler
Surface evenness	Fair-faced wall and column		5	2m guiding rule and wedge-shaped feeler
	Furred wall and column		8	

111

Continued

Item		Allowable difference(mm)	Checking method
Evenness of horizontal joint	Fair-faced wall	7	Stretch 10m line and ruler
	Furred wall	10	
Opening of doors and windows		±5	ruler
Offset of upper and lower openings of external wall		20	Theodolite, plumb, based on bottom opening
Offset of vertical joints of fair-faced wall		20	Plumb and ruler, based on first course

3.3.5 Construction of wall with reinforced concrete constructional column

Reasonable setting of reinforced concrete constructional columns and ring beams is a necessary measure to enhance the seismic resistance of brick masonry structures in seismic fortification regions. In the current "Code for Seismic Design of Buildings" (GB 50011, Chinese code), there are specific provisions on the arrangement and construction of constructional columns. Attention should be paid to the following construction points.

(1) The construction process of multistory brick masonry structure with reinforced concrete constructional columns is binding constructional column rebar → bricklaying → binding rebar of ring beam → erecting formwork → casting concrete → removing formwork. The construction of the upper floor must be carried out after the concrete of the constructional column and ring beam reaches to design strength.

(2) The junction of constructional column and wall should be built by laying bricks like horse tooth as shown in Figure 3-18, so it is called horse toothing in China. The height of one toothing unit should not exceed 300mm along the height direction, the extended width of toothing is not less than 60mm. The toothing unit should start from the foot of the column and retreat first. 2Φ6 tie bars of L shape should be set every 500mm along height of the wall, the length embedded into the wall should not be less than 1000mm per side as shown in Figure 3-18. The bars must not be arbitrary bent during the construction.

(3) The section size of the constructional column, types, specification and quantity of steel bars shall conform to the design requirements. Constructional column section size should not be less than 240mm × 240mm, its thickness should not be less than the thickness of wall. The section width of side column and corner column can be increased appropriately. The vertical rebars of column can be HRB300 and should not be less than 4Φ12, the diameter should not be greater than 16mm. The diameter of stirrups should be Φ6@200mm along height. In the ranges 700mm above and 500mm below floor, the space of stirrups should be 100mm. Concrete strength grade should not be lower than C20 and should meet the design requirements.

(4) The anchoring length and the lapping length of the vertical rebar of the constructional column should not be less than 35 times the diameter of the steel bars.

Stirrup spacing in joint sections should not be greater than 200mm. A protective layer of rebars is generally 20mm thick.

(5) Before casting concrete, the toothing and formwork must be watered wet, the falling mortar, brick slag and other debris at the bottom inside the formwork should be cleaned, and a thin layer of cement mortar with the same ratio of the concrete can be cast first. Vibration should avoid touching the wall, it is strictly prohibited to conduct vibration through the wall.

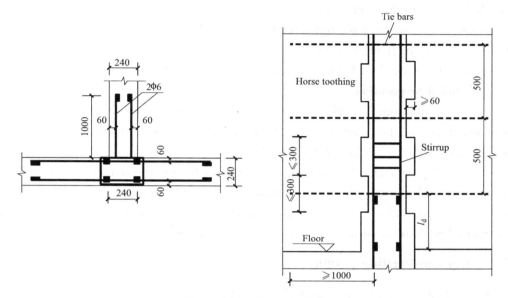

Figure 3-18   Horse toothing

3.4 Concrete block masonry construction

The main size of small concrete hollow block is 390mm×190mm×190mm and bigger than brick. So the construction is convenient and with less joints, which can save mortar and improve the efficiency of masonry work. It is commonly used in multistory buildings as load bearing wall.

3.4.1 Preparation work

1. Material preparation

(1) The strength grade of the small blocks shall meet the design requirements. Production age shall be checked before use, the products age of the small blocks used for construction should not be less than 28 days.

(2) Try to adopt the main specifications of the block, small block shall be complete, no damage, no cracks.

(3) For the masonry below the indoor ground floor or below the moisture-proof

layer, the holes of blocks shall be filled by concrete no less C20 (or Cb20) in advance. Small blocks built at the clamp installation place of heat radiator and facility of kitchen, toilet, etc. should also be filled with concrete no less than C20 (or Cb20) before construction.

(4) Ordinary concrete small hollow block does not need to be water wetted. In case of dry and hot weather, it is appropriate to spray water to wet it before the masonry construction. For light aggregate concrete block, since it has more voids, it should be wetted before 2 days. For aerated concrete block, the bed can be properly wetted before laying upper blocks without too much water. In rainy days and when there is floating water on the surface of small blocks, no construction shall be allowed. For this reason, small blocks should be stacked with rain prevention and drainage treatment.

2. Technical preparation

The size of the small blocks is large and should be used as the whole block, and cannot be cut. Before the construction, the facade arrangement drawing of different block sizes should be drawn according to the building design drawing, and construct according to it. The technical preparation for leveling and line stretching is the same as that of the brick masonry works. Moreover, the course number of the blocks shall be determined according to the height of the wall, the specification of the size of the small blocks and the thickness of the joint. The distance of layer pole for block laying should not be more than 15m.

3.4.2 Key construction points

The method of constructing the wall with concrete blocks is the same as that used for brick masonry. First, the corners or ends of the wall are constructed with few courses of blocks. Mortar is applied to the bottom of the concrete block at the horizontal face member only. For vertical joints, the mortar is applied to the projection at the sides of the block. For building the portion between the corners, the string is spread between the two horizontal end blocks of a course, and the blocks are laid in between. The final closing block is fitted carefully.

The following points should be kept in mind while supervising the construction work:

(1) The small block can be laid in stretcher bond. Block of successive course should be so laid that vertical joints are staggered. The overlap of upper and lower block should be hole to hole and rib to rib. For block with single-row holes, the overlap is half length of the block. For blocks with multi-row holes, the overlap should not be less than 1/3 length of the block, and 90mm and 120mm for ordinary concrete blocks and light aggregate concrete blocks respectively. When it can not accord with this requirement, tie bar or rebar net should be adopt in the horizontal joint. The 6mm tie bar or the rebar net welded by 4mm steel bar should be embedded at the position. The length should not be less than 700mm and continuous vertical joint should not be more than 2 courses.

(2) The block should be laid bottom up. The paving length of mortar should not be

more than 2 main specification block length. The vertical joint mortar can be paved on one side that is erected upwards and squeezed tightly after laying the block. If blocks are disturbed after laying, re-lay the block.

(3) Generally, the thickness of joint is 10mm, but should not be less than 8mm or more than 12mm. The full degree of horizontal joint and vertical joint should not be less than 90% and 80% respectively.

(4) The walls at corner and junction should be laid at the same time, when temporary break needed to be left. Inclined toothing can be built with length more than height, see Figure 3-19 (a). When it is difficult to build inclined toothing, vertical toothing can be left except for corner of exterior wall and earthquake fortification area. The extended block with 200mm length every 2 courses can be built as vertical toothing and 2 tie bars or steel bar net per 600mm high should be set and the length is no less than 600mm from toothing, as shown in Figure 3-19 (b).

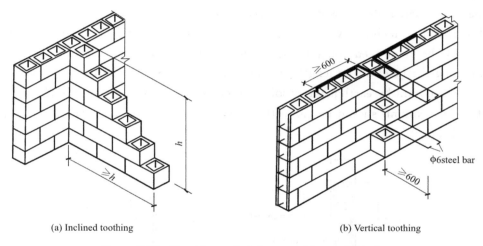

(a) Inclined toothing   (b) Vertical toothing

Figure 3-19 Toothing and tie bars used in small blocks

(5) At junction of block wall and later partition wall, a steel bar net of 2ϕ4 straight bars with transverse bars per 400mm can be embedded in horizontal joint, the length into later wall is no less than 600mm.

(6) The openings, pipes, trenches and embedded parts required by the design shall be correctly reserved or embedded when laying the walls, and the already built walls shall not be excavated at will. Scaffold holes should not be set in small block walls. If necessary, single hole small block with auxiliary specifications (190mm×190mm×190mm) can be laid with hole perpendicular to the wall surface for scaffold. Finally, the holes can be filled with concrete with strength grade no lower than C20.

(7) The wall under the beam should be 3 layers of standard brick lower than the bottom of beam. After 14 days, fill the space with inclined bricks.

(8) Construction height should be within 1.8m per day, for lightweight aggregate concrete block is less than 2.4m.

3.4.3 Construction of reinforced concrete core column

Concrete blocks have high thermal expansion, due to which walls crack at the corners. Long walls may have cracks even at its mid-length. Hence at the junction of walls, solid concrete blocks or hollow blocks filled with concrete should be used.

(1) It is better to set concrete core column at the corner of outer wall, four corners of stair room, transverse and longitudinal junction.

(2) For 5 and more story buildings, reinforced concrete core column should be set in the above parts.

(3) The core column section should not be less than 120mm×120mm, the fine-stone concrete strength is not lower than C20.

(4) The vertical rebar of the core column should not be less than 1Φ10 or 1Φ12 for building more than 5 stories when seismic fortification intensity is 6 and 7 degrees or building more than 4 stories when 8 degrees or 9 degrees, it should be stretched 500mm into the ground or anchored with foundation ring beam at the bottom and top ring beam.

(5) In the reinforced concrete core column, Φ4 steel bar net should be set every 600mm along the height. The length extend into the wall should be not less than 600mm per side or 1000mm by earthquake-proof requirement as shown in Figure 3-20.

(6) Core column should be along the story overall height, and cast in-situ together with ring beam.

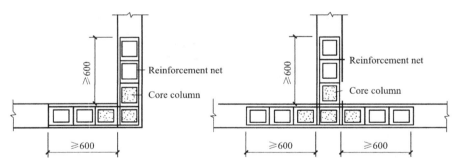

Figure 3-20 Concrete core column

3.5 Stone masonry

Depending upon the arrangement of stone in the construction, degree of refined used in shaping the stone and finishing adopted, stone masonry can be classified as rubble masonry and ashlar masonry.

3.5.1 Rubble masonry

In the rubble masonry, the blocks of stone that are used are either undressed or comparatively roughly dressed. The masonry has wide joints, since stones of irregular are used. Rubble masonry may be out of the following types:

1. Random rubble

1) Uncoursed

This is the roughest and cheapest form of stone walling. In this type of masonry, the stones used are wildly different sizes.

Since stones are not of uniform size and shapes, greater care and ingenuity have to be exercised in arranging them in such a way that they adequately distribute the pressure over the maximum area and at the same time long continuous vertical joint are avoided.

Sound bond should be available both transversely as long as longitudinally. Transverse bond is obtained by the liberal use of headers. Larger stones are selected for quoins and jambs to give increased strength and better appearance. This type of masonry is also known as uncoursed rubble masonry.

2) Built to course

The method of construction is the same as above except that the work is roughly leveled up to form course varying from 30cm to 45cm thick. All the courses are not of the same height. For the construction of this type of masonry, quoins are built first and line (string) is stretched between the tops of quoins. The intervening walling is brought up to this level by using different size of stones. This form of masonry is better than uncoursed random rubble masonry.

2. Square rubble

1) Uncoursed

Square rubble masonry uses stones having straight bed and sides. The stone usually squared and brought to hammer dressed orstraight cut finish.

In the uncoursed square rubble also sometimes known as square-snecked rubble, the stones with straight edges and sides are available in different sizes (height). They are arranged on face in several irregular pattern. Good appearance can be achieved by using risers (a large stone, generally a through stone), leveler (thinner stones) and sneck or check (small stones) in a pattern, having their depths in the ratio of 3:2:1 respectively. Snecks are the characteristics of this type of construction, and hence the name. This prevents the occurrence of long continuous vertical joints.

2) Built to courses

This type of masonry uses the same stones as used for uncoursed square rubble. But the work is leveled up to courses of varying depth. The courses are of different heights. Each course may consist of quoins, jamb stones, bonders and thoughts of the same height, with small stones built in between them up to the height of the larger stones, to complete the course.

3) Regular coursed

In this type of masonry, the wall consists of various courses of varying heights, but the height of stones in one particular course is the same.

3. Polygonal rubble masonry

In this type, the stones are hammer finished on face to an irregular polygonal shape. These stones are bedded in position to show face joints running irregularly in all directions. There are two types of polygonal walling. In the first type, the stones are only roughly shaped, resulting in only rough fitting. Such a work is known as rough picked. In the second type, the faces of stones are more carefully formed so that they fit more closely. Such a work is known as close-picked work.

4. Flint rubble masonry

The stones used in this masonry are flints or cobbles, which vary from 7.5cm to 15cm in width and thickness and from 15cm to 30cm in length. These are irregularly shaped nodules of silica. The stones are extremely hard. But they are brittle and therefore may break easily. The face arrangement of the cobbles may be either coursed or uncoursed. Strength of flint wall may be increased by introducing lacing courses of either thin long stones or bricks at vertical interval of 1m to 2m.

5. Dry rubble masonry

Dry rubble masonry is that rubble masonry, made to courses, in which mortar is not used in the joints. This type of construction is the cheapest, and requires more skill in construction. This may be used for non-load bearing walls, such as compound wall etc.

For rubble masonry with mortar, the joint of rubble masonry is better to be 20~30mm and should be full filled with mortar. The height of the rubble masonry for everyday should not be more than 1.2m.

There must be a through stone at an interval of 2m in one course or in $0.7m^2$ area. When the wall is equal or less than 400mm, the length of the through stone is equal to the thickness of the wall. If the wall is more than 400mm thickness, two through stones with an overlap are provided. The overlap length should be no less than 150 mm and the length of through stones must be no less than the 2/3 of the thickness of the wall.

3.5.2 Ashlar masonry

Ashlar masonry consists of blocks of accurately dressed stone with extremely fine bed and end joints. The blocks may be either square and rectangular shape. The height of stone varies from 25cm to 30cm. The height of blocks in each course is kept equal but it is not necessary to keep all the course of the same height. Ashlar masonry may be subdivided into the following categories.

1. Fine-tooled ashlar

This is the finest type of stone masonry work. Each stone is cut to regular and required size and shape so as to have all sides rectangular, so that the stone gives perfectly horizontal and vertical joints with adjoining stone. The beds, joints and faces are chisel dressed, such that all waviness and unevenness is completely removed and a fairly smooth surface is obtained. The face which remains exposed in the final work is so dressed that no point on the dressed face is more than 1mm from a 600mm long straight edge placed on the surface in any direction. The top and bed is also so dressed that no point on it varies by more than 3mm when checked with the straight edge. The side surfaces which are to form the vertical joints are also so dressed that no point on the surface is more than 6mm from the straight edge.

The surfaces forming internal joints which are not visible are also so dressed that no point on the surface is more than 10mm from the straight edge. All angles and edges that remain exposed in the final position are kept as true square and free from chippings. The thickness of course is generally not less than 15mm. The width of stone is not kept less than its height. Headers and stretchers are laid alternately in each course or course of headers and course of stretchers may be laid alternatively or they may be laid as otherwise directed. The thickness of masonry joint is kept uniform throughout and it should not be more than 5mm. The exposed joints are finely pointed.

2. Rough-tooled ashlar

In this type of masonry, the beds and sides of each stone block are finely chisel dressed just in the same manner as for ashlar fine, but the exposed face is dressed by rough tooling. A strip, about 25mm wide and made by means of a chisel is provided around the perimeter of the rough dressed face of each stone. The rough tooled face when tested with a straight edge 600mm in length, should not show any point on the surface to vary by more than 3mm in any direction. This type of masonry is also known as bastard ashlar. The size, angle, edges etc. are maintained in order, similar to that for fine dressed ashlar. The thickness of mortar joint should not be more than 6mm.

3. Rustic or quarry-faced ashlar

In this type of masonry, the exposed face of the stone is not dressed but is kept as such so as to give rock facing. However, a strip of about 25mm wide, made by means of a chisel, is provided around the perimeter of the exposed face of every stone. The projections on the exposed face exceeding 80mm in height are removed by light hammering. Each stone block, however is maintained true to its size, with perfectly straight side faces and beds, and truly rectangular in shape. This type of construction gives massive appearance. The height of each block may vary from 15cm to 30cm. The thickness of mortar joint may be up to 10mm.

4. Chamfered ashlar

This is special form of rock-faced ashlar masonry in which the strip provided around the perimeter of the exposed face is chamfered or beveled at an angle of 45° by means of chisel to a depth of 25mm. Due to this, a groove is formed in between adjacent blocks of stone. Around this beveled strip, another strip of 15cm is dressed with the help of chisel. The space inside this strip is kept rock faced except that large bushings in excess of 80mm projections are removed by a hammer.

5. Ashlar block in course

This type of masonry is intermediate between rubble masonry and ashlar masonry. The faces of each stone are hammer dressed, and the height of blocks is kept the same in any course, though it is not necessary to keep uniform height for all the course. The vertical joints are not as straight as in ashlar masonry. The depth of course may vary from 15cm to 30cm. This type of masonry is adopted in heavy works such as retaining walls, bridges etc.

6. Ashlar facing

Ashlar facing masonry is provided along with brick or concrete block masonry, to give better appearance. The sides and beds of each block are properly dressed so as to make them true to shape. The exposed faces of the stone are rough tooled and chamfered. The backing of the wall may be made in brick masonry.

The joint of fine ashlar masonry, semi-fine-ashlar, coarse ashlar should better be not more than 5mm, 10mm, 20mm respectively. When pave the mortar, it should be thicker than the designed thickness. For the fine ashlar and semi-fine-ashlar, it should be 3~5mm thicker, for the coarse ashlar or square rubble, it should be 6~8mm thicker. The height of the ashlar masonry work should not be more than 1.2m for everyday.

3.6 Scaffold

When the height of wall or column or other structural member of a building exceeds about 1.2m, temporary structures are needed to support the platform over which the workmen can carry on the construction continuously. These temporary structures constructed very close to the wall, in the form of timber or steel framework, are commonly called scaffold. Such scaffold is also needed for the repairing, demolishing or ornamenting work of buildings. The scaffold should be stable and strong enough to support workmen, and other construction materials and tools placed on the platform without tilting, shaking and big deformation. The scaffold should be simple in construction, easy to disassemble, easy to use, and its materials should be utilized repeatedly.

According to the erection location, the scaffold can be divided into outdoor scaffold and indoor scaffold. The scaffold set up on the outside of the building is called the outdoor scaffold, which can be used not only for the masonry work of the external wall, but also for the concrete and decoration construction. As the length and height of the outdoor scaffold are big, the length is built along the perimeter of the building and the service time is long, so the safety requirement is high, the safety checking calculation should be carried out before setting up. The scaffold set up inside the building is called indoor scaffold, which is used for masonry, decoration and other operations on the ground or floor. The length and height of the indoor scaffold are small, and the construction is usually simple.

Chapter 3 Masonry and Construction Facilities Engineering

According to the materials, the scaffold can be divided into wood, bamboo, steel scaffolds. Wood and bamboo scaffold are seldom used nowadays, their components or members can be connected by means of ropes, nails; bolts, steel wires (4mm diameter) etc. Steel tube scaffold is the most commonly used scaffold because of their high bearing capacity and repeating utilization. Unless specified, the scaffold described in this section is steel tube scaffold.

3.6.1 The outdoor scaffold

1. Component parts of outdoor scaffold

As shown in Figure 3-21, the components of outdoor scaffold are as follows:

(1) Standards: these are vertical members of the framework, supported on the ground or drums or embedded into the ground.

(2) Braces: these are diagonal members fixed on standards.

(3) Ledgers: these are horizontal members, running parallel to the wall.

(4) Putlogs: these are transverse members, place at right angles to the wall with one end supported on ledgers and the other end on the wall.

(5) Transoms: these are those putlogs whose both ends are supported on ledgers.

(6) Bridle: this is a member used to bridge a wall opening supports on the end of putlog at the opening.

(7) Boarding: these are horizontal platform to support workmen and materials. They are supported on the putlogs or transoms. The thickness of the board should not be less than 50mm, while the width should be 200~300mm.

(8) Guard rail: this is a rail provided like a ledger, 1.2m above working level.

(9) Toe boards: these are boards placed parallel to ledgers and supported on putlogs to give protection at the level of working platform.

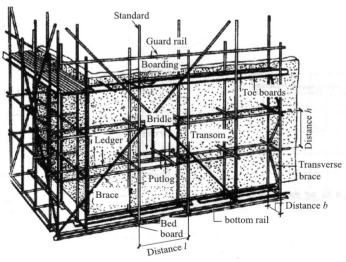

Figure 3-21 Components of outdoor scaffold

2. Types of outdoor scaffold

According to the structural form, outdoor scaffold can be divided into landed scaffold, cantilever scaffold, suspended scaffold and self-lifting scaffold, etc., as shown in Figure 3-22.

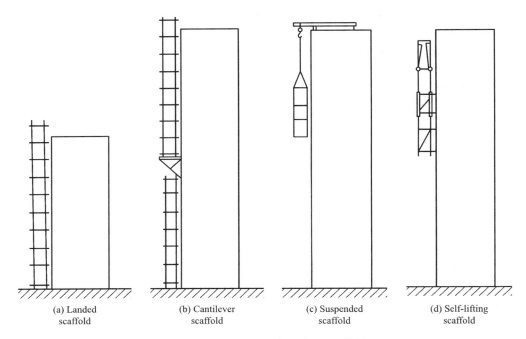

Figure 3-22 Types of outdoor scaffold

1) Landed scaffold

According to the rows of standards, landing scaffold can be divided into single row scaffold and double rows scaffold as shown in Figure 3-23.

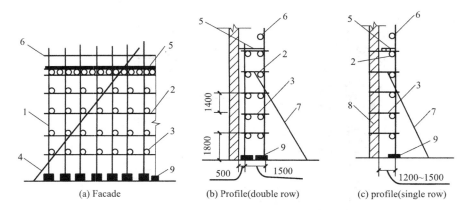

Figure 3-23 Multiple standards scaffold

1—Standards; 2—Ledger; 3—Transome; 4—Bracing; 5—Board;
6—Guard rail; 7—Inclined bracing; 8—Wall; 9—Base

(1) Single-row scaffold

Single-row scaffold consists of a single row of standards, ledgers and putlogs etc. As shown in Table 3-6, the standards are erected parallel to the wall at a distance of about 1.2m. The interval along the length of the wall is less than 1.5m or 1.8m. Ledgers are connected inside the standards at a vertical interval of 1.2m to 1.5m. Putlogs are placed with one end on the ledgers and the other end in the holes left in the wall at an interval of 0.75m to 1m. The length enter into the hole should be $\geqslant$ 180mm. Such scaffold is commonly used for bricklaying, and is also called putlog scaffold. However, due to its poor overall stiffness and low bearing capacity, it is not applicable to the following projects: the wall thickness is less than 180mm, building height more than 24m, mortar strength less than M10, hollow brick, aerated brick wall and other light block wall.

Space of single-row scaffold Table 3-6

Purpose	Interval of standards(m)		Interval of putlogs (m)	Vertical interval of ledgers (m)
	Longitudinal	Transverse		
Masonry	$\leqslant$1.2	$\leqslant$1.5	$\leqslant$0.75	1.2~1.5
Ornamentation	$\leqslant$1.2	$\leqslant$1.8	$\leqslant$1.0	$\leqslant$1.8

(2) Double-row scaffold

When it is very difficult to provide holes in the wall to support putlogs, the load is big or the height of a building is more than 24m, stronger scaffold is used consisting of two rows of standards. Each row forms a separate vertical framework. The first row is placed 0.5m away from the wall, while another framework is placed at a distance no more than 1.5m from the first one. The space of different components can refer Table 3-7.

Space of double-row scaffold Table 3-7

Purpose	Interval of standard(m)		Interval of transomes(m)	Vertical interval of ledgers(m)	The length of transome extend into the wall(m)
	Longitudinal	Transverse			
Masonry	$\leqslant$1.5	$\leqslant$1.5	$\leqslant$0.75	1.2~1.5	0.35~0.45
Ornamentation	$\leqslant$1.5	$\leqslant$1.8	$\leqslant$1.0	$\leqslant$1.8	0.35~0.45

Diagonal bracing can be used to increase the transverse stability of single-row and double-row scaffold. When the height of scaffold is $\leqslant$24m, diagonal bracing can be set at two ends of a framework and at an interval of 15m between them. When the height is >24m, diagonal bracing should be set continuously along the total length. Its width is no less than 4 spans of standards or 6m, the angle is 45°~60°.

The scaffold must be connected with the building to maintain the stability perpendicular to the wall. The putlogs or transomes can connect with exterior wall by bridles such as steel tube, embedded reinforcement ring. Different connection methods are shown in Figure 3-24. The maximum space of connection with wall can refer to Table 3-8.

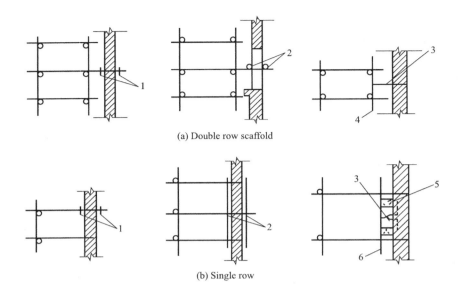

Figure 3-24 Different connection methods with wall
1—Coupler; 2—Short steel tube; 3—Preembedded piece;
4—Ledger; 5—Wood wedge; 6—Short steel tube

Space of connections with wall　　　　　　　　　　Table 3-8

Types	Height(m)	Vertical space h	Transverse space l_a
Double-row	≤50	3	3
	>50	2	3
Single-row	≤24	3	3

(3) Coupler for connection

Different components of steel tube scaffold can be connected at junction points by different types of couplers such as bolt-coupler, bowl-coupler, disc-coupler, wheel-coupler.

① Bolt-coupler

There are 3 types of bolt-coupler: coupler for end to end joint, coupler for right angle connection and coupler for diagonal connection, as shown in Figure 3-25.

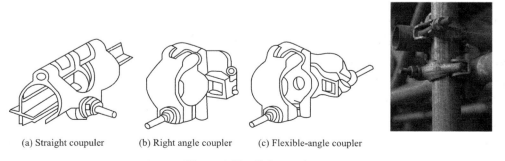

(a) Straight coupuler　　(b) Right angle coupler　　(c) Flexible-angle coupler

Figure 3-25　Bolt-coupler

Bolt-coupler scaffold has some advantages like convenient installation and disassembly, flexible erection, adapting to the changes of the structural plane and height, strong versatility, large bearing capacity, high setting height, strong and durable, and many turnover times, simple material processing, low investment cost, relatively economic. But it also has some shortcomings such as the fasten nuts are easily lost and damaged, bolts are easy to rust to reduce the construction speed, and need regular maintenance, the degree of bolt tightness is quite different, there is an eccentricity or intersection distance between the external forces of the nodes which will generate bending moment to standards and influence stability.

② Bowl-coupler

Bowl-coupler scaffold is a multifunctional scaffold. The joint of standard and ledger are connected by 2 bowl-shape rings on standards and cross-bar joints fixed on both ends of ledger as shown in Figure 3-26. When erecting scaffold, the cross-bar joint can be inserted into the bottom bowl welded on standard, and then the top bowl is covered on and fastened by a limit pin, so workers can finish all the work just by a hammer. The connection is convenient and fast which can be 2~3 times faster and 0.5 times less manpower than bolt-coupler scaffold connection. Since standards and ledgers are axial connections in same frame plane, bowl-coupler scaffold is stable and reliable, the bearing capacity can be 15% larger than bolt-coupler scaffold. Compare with bolt-coupler scaffold, bowl-coupler scaffold has good mechanical properties and reliable connection, good overall stability, strong versatility. However, the length of the steel pipe and the position of the bowl-coupler are fixed, so the distance between the horizontal and vertical poles can only be selected in several specifications, and the size of the frame is limited to some extent. Generally, the length of standard are 1.8m, 3m and vertical space of bowl-coupler on it is 0.6m, the length of cross-bar has 7 specifications: 2.4m, 1.8m, 1.5m, 1.2m, 0.9m, 0.6m, 0.3m, which can be used flexibly according to the space of standards.

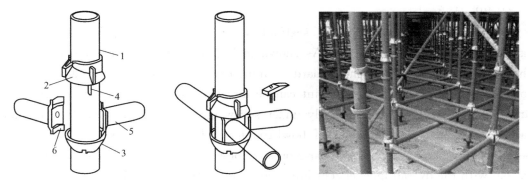

Figure 3-26 Bowl-coupler
1—Standard; 2—Top bowl; 3—Bottom bowl; 4—Limit pin; 5—Ledger; 6—Cross-bar joint

③ Disc-coupler

Disc-coupler scaffold is also called Layher frame because it is invented by this German company. It is often used as lighting frame and background frame for large concert. The disc-coupler scaffold is an upgrade after the bowl-coupler scaffold. The socket of this kind of scaffold is 133mm diameter, 10mm thick disk with 8 holes, which is welded on standard at an interval of 0.60m as shown in Figure 3-27. The cross-bar is made of steel pipe with 2 plugs welded on both ends. The disk and plug can be connected or removed by inserting and pulling out a separate wedge, so the erecting and dismantling speed of disc-coupler scaffold is 4~8 times and 2 times faster than bolt-coupler scaffold and bowl-scaffold respectively.

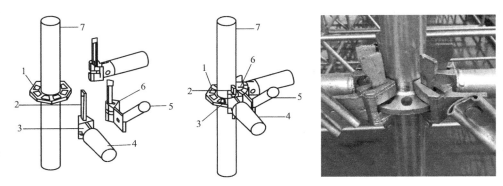

Figure 3-27 Disc-coupler
1—Disc plate; 2—Wedge; 3—Cross-bar joints; 4—Ledger;
5—Diagonal member; 6—Joint; 7—Standard

The separate wedge has self-locking mechanism. Due to interlocking and gravity, the cross-bar plug can not be removed even if the wedge is not tightened. The contact surface between the disk and the plug is large, which improves bearing capacity of vertical shear force. Like bowl-coupler scaffold, all kinds of bars are axis intersection at nodes, but the numbers of connecting cross-bar is 1 times more than the bowl-coupler joint, and the overall stability strength is 20% higher than the bowl-coupler scaffold.

④ Wheel-coupler

Wheel-coupler (straight-insert) scaffold is a new type of straight-insert steel pipe scaffold with self-locking function. As shown in Figure 3-28, like disc-coupler scaffold, 4-hole wheel plate is welded on standard at an interval of 0.6m, but the wedge is fixed on cross-bar joints, so the cross-bar joint can be directly inserted into the wheel hole by a hammer and it will be locked tightly under self-weight and vertical load. Wheel-coupler scaffold combines the advantages of bowl-coupler and disc-coupler scaffold and has no special locking parts and no active parts on steel pipe scaffold. It is the first new steel pipe scaffold with the overall independent intellectual property rights in China and widely used in different projects.

(4) Frame scaffold

A basic unit frame scaffold is composed of gantry framework, diagonal bracing, and horizontal beam or board which can be connected into a unit frame on job site as shown in

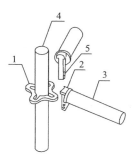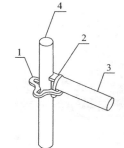

Figure 3-28　Wheel-coupler
1—Wheel plate; 2—Cross-bar joints; 3—Ledger; 4—Standard; 5—Fixed wedge

Figure 3-29. Several unit frames can be connected along height direction to compose a multistory frame, and diagonal bracing and horizontal beam can connect 2 adjacent unit into a whole frame. Some inclined ladder, guard rail and cross-bar can be added to connect traffic of the upper and lower units.

The total height of frame scaffold is not more than 45m, set horizontal beam every 5 units along height and connect with wall every 4~6m in horizontal and vertical direction.

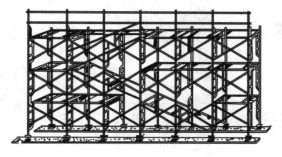

Figure 3-29　Frame scaffold

2) Cantilever scaffold

Cantilever scaffold uses an overhanging structure sticking out of the building structure to support the outdoor scaffold and to transfer the load to the building structure. The structure of the scaffold is same with landing scaffold and the height is decided by the requirement of construction bearing capacity of building structure and hoisting capacity of the tower crane. The height is about 20m at most and can be used for the construction of 2~3 floors. After the construction, the scaffold can be hoist to a higher position and fixed with overhanging structure again. The overhanging structure forms of cantilever scaffold are as shown in Figure 3-30.

3) Suspended scaffold

Suspended scaffold uses the lever principle, the self-weight and load of the suspended platform and guard rail is balanced by 2 cantilever beams and counterweight fixed on the roof. The platform can be lifted and dropped through the motor, wire rope and pulley group fixed on both sides of the platform as shown in Figure 3-31. the suspended scaffold

(a) Supported by cantilevered beam (b) Supported by triangle truss

Figure 3-30 Cantilever scaffold

has low bearing capacity because of the length limitation of cantilever beam and weight limitation of counterweight, it is often used for some simple work like external wall decoration and cleaning work.

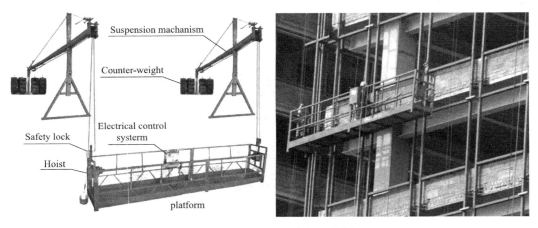

Figure 3-31 Suspended scaffold

4) Self-lifting scaffold

(1) Attached self-lifting scaffold

Attached self-lifting scaffold is a kind of multi-storey scaffold which can be lifted and lowered by hydraulic system or chain hoist along the guide rails attached to the walls or columns around a building. The structure of the scaffold is same with landing scaffold but it doesn't need to erect along full height of building, set hard ground for standards and occupy construction ground. The height of self-lifting scaffold is decided by the height of floor, generally, it is height of 4 standard floors and 1 step safe guard rail as shown in Figure 3-32, so it will save much materials and manpower for erecting and dismantling. After the scaffold moves to the right position along the guide rails, it is fixed with wall or column by some connection joints. The self-weight and load acting on scaffold will transform to the building structures by connection joints, so the connection joints and building structures should have enough strength. The self-lifting scaffold can be lifted up with the raise of height during construction of main structure and lowered down floor by floor when doing external wall decoration work.

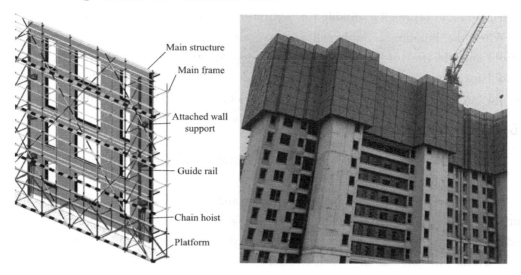

Figure 3-32 Attached self-lifting scaffold

(2) Electric bridge scaffold

Electric bridge scaffold is a kind of working platform which can be lifted and lowered by gear rack transmission along the triangular truss column attached to the building. Electric bridge scaffold consists of three parts: driving system, attached column system and working platform system, see Figure 3-33. The driving system consists of motor, anti-drop device, gear drive group, guide wheel group, intelligent controller and so on. Attached column system consists of standard section with racks, limit section and wall attachment. The column sections can be connected up to 260m. The distance of the columns along the building outline is 6m. Working platform is composed of triangular truss beam, platform, guard rail, widening beam etc. The length of platform can be 30.1m and 9.8m for double columns and single column respectively. The platform width is 1.35m and can be broadened to 2.25m.

Figure 3-33 Electric bridge scaffold

Electric bridge scaffold can replace ordinary scaffold and suspended scaffold and save a lot of materials. The platform runs smoothly, it is safe and reliable. It is used in different construction and especially suitable for decoration work.

When the building is high and construction schedule is tight, attached self-lifting scaffold and electric bridge scaffold can be used together. After the attached self-lifting scaffold for main structure construction raised to a certain height, the electric bridge scaffold can be erected from the ground for outdoor decoration work at the same time.

3.6.2 The indoor scaffold

According to different structure, indoor scaffold can be divided into 3 types.

1. Folding trestle

Folding trestle can be made of steel tube or angle steel as 2 piece of frame, which are connected on the top with hinges as shown in Figure 3-34. It can be folded when it is not used, so it can save store space. Some boards can be put on 2 trestles for the construction. The interval of trestles should be 1.8m for masonry work and 2.2m for ornamentation work. Generally, 1m and 1.65m construction height can be reached when the boards are put on the middle and top position of the folding trestle.

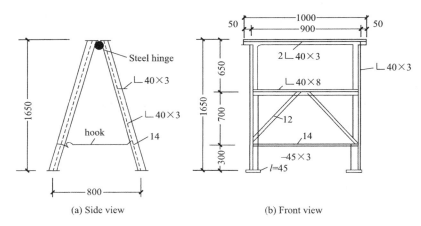

Figure 3-34 Folding trestle

2. Pillar trestle

Pillar trestle is composed of columns and beams, and the platform is laid on the beams, as shown in Figure 3-35. The space of pillar trestles for masonry work is not more than 2m, and 2.5m for decoration work. For casing pipe trestle, a spiling column can be inserted into the casing column and the height is adjusted by the spacing of the pin holes. A square wood or steel beam can be placed in the U shape head of spiling column to support the platform. For socket pipe trestle, some socket pipes are welded along the height of the pillar, and the two ends of the beam are inserted into the socket pipe by latch. Generally, the adjustable heights are 1.2m, 1.6m and 1.9m.

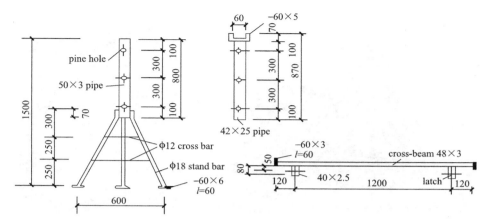

Figure 3-35 Stanchion trestle

3. Split head trestle

The split head trestle can be made of wood, bamboo bars connected by nails, bolts and steel wires or welded angle steel, so the shape is fixed and the height can not be adjusted.

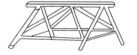 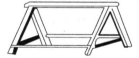 

Figure 3-36 Split head trestle

3.7 Vertical transportation facilities

Vertical transportation facilities refer to mechanical equipment and facilities for vertical conveying of material and personnel. In the process of building construction, various materials (brick, mortar, steel bar etc.), tools (scaffold tubes, couplers, boards etc.) need to be transported to the floors by vertical traffic, the quantity is big. Commonly used vertical transportation facilities are headframe, gantry frame, construction elevator and tower crane, etc.

1. Headframe

Headframe can be made of angle steel or steel pipe into a finalized product, or erected with scaffold components (such as steel tubes, couplers or frame scaffold etc.), as shown in Figure 3-37. Generally, headframe can have a single hole, double holes or three holes sometimes. There is a hoisting platform in the headframe, which is driven by the winch, pulley group and wire rope. The lifting capacity of headframe is up to 1~1.5t, the height is up to 40m. The characteristics of headframe are simple structure, convenient installation, good stability, large transportation capacity and low price.

Figure 3-37 Headframe

2. Gantry frame

The gantry frame is aportal frame composed of two vertical columns and a pulley beam. The gantry frame is equipped with guide rail, suspension platform, safety device and so on, and the winch drives the lifting platform to rise and fall, which constitutes a complete vertical transportation system as shown in Figure 3-38. The column of the gantry frame can be combined with angle steel or steel pipe to improve the strength and stiffness of the frame. The lifting weight is generally 0.6~1.2t, the lifting height is generally 20~30m. The characteristics of the gantry frame are simple structure, easy manufacture and convenient installation and disassembly, but because of its poor stiffness and stability, the erection height should not be large, so it is often used in multi-storey building construction.

Figure 3-38 Gantry frame

3. The construction elevator

The construction elevator is a vertical transportation equipment for transporting materials and persons up and down. Its cage is mounted on the outside of the steel tube truss column, as shown in Figure 3-39. The elevator can be divided into three types according to lifting method: gear rack type, wire rope type and mixed type. The construction elevator can carry 1.0~3.2t or 12~15 people. Because it is attached to the exterior wall of the building or other structural parts, the stability is very good. It can be gradually connected up with the construction of the main structure, and the erection height can reach 150m. At present, the construction elevator has been widely used in the construction of high-rise buildings.

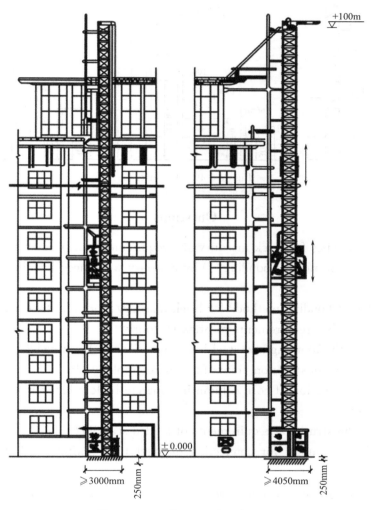

Figure 3-39　Construction elevator

4. Tower crane

Tower crane is an upright tower made of steel tube truss, a lifting arm is installed on the top of the tower and can be used as 360° rotation lifting machinery as shown in Figure 3-40. In addition to widely used in vertical transportation of multi-storey and high-rise buildings, it is also used for structural installation works. Detailed instruction can be found in Chapter 6, Section 4.

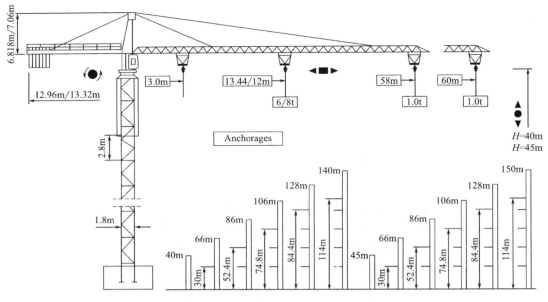

Figure 3-40 Tower crane

Questions

1. List mortar types for brick masonry and requirements of production and use.
2. What is the meaning of bonding? Why masonry work need bonding and how to get good bonding?
3. List different bonding methods for brick masonry work and their characteristics.
4. What are quality requirements of brick masonry?
5. How to set the toothing and steel bar at junctions of brick wall?
6. List some key construction points for concrete hollow blocks.
7. What is the function of core column in concrete hollow block masonry and its key construction points?
8. What are the structural requirements of landed scaffold?
9. List different coupler for connection of steel tube scaffold and their merits and demerits.
10. List some vertical transportation facilities and their suitable conditions.

Chapter 4   Concrete Structure Engineering

Mastery of contents: Structure of formwork, installation and dismantling, rebar blanking, construction technology and methods of concrete engineering, retaining and processing of construction joint.

Familiarity of contents: Technique requirement of formwork, rebar fabrication, connection, binding and installation and its quality acceptance, the meaning and method of concrete construction in winter.

Understanding of contents: Varieties of formwork and rebar, quality acceptance of concrete, technical process to defects of concrete.

Key points: Structure of formwork, sequence and key points while installation and dismantling of formwork, rebar blanking and its connection methods, construction technology of concrete engineering, key points and guarantee measures among it's process, meaning of the construction joint, position and processing of construction joint, meaning of concrete construction in winter.

Difficult points: Structure of formwork, rebar blanking, key points and guarantee measures in concrete construction, thermal calculation of thermal storage method in winter concrete construction.

4.1   Formwork engineering

The formwork board is model plate which ensuring position, shape and size of the concrete structure and its component as required by design.

Formwork system includes: formwork board, supporting and connection joints.

Formwork engineering includes design, installation, removal and other technical works. It is one of the important contents of concrete structure engineering.

Formwork is widely used in the construction of cast-in-situ concrete structures. The amount of formwork used for each $1m^3$ concrete project is as high as $4 \sim 5m^2$. The engineering cost accounts around $30\% \sim 35\%$ of the construction costs, and $40\% \sim 50\%$ of total amount of labor required.

Therefore, the selecting correctly of formwork materials, types and reasonable construction organization, will be of great important for ensuring the quality of the project, improving labor productivity, speed up construction, decreasing the cost of the project and civilized construction.

4.1.1 Overview of formwork

1. Technical requirement

(1) Forms and their supports shall be designed in according with structure type, loads, ground soil type, construction equipment, and materials supplying. Forms and their supports shall have adequate load-carrying capacity, rigidity and stability to bear the weight, side pressure and construction loads reliably during the construction.

(2) Forms and their supports shall ensure the accuracy of the shape, dimension and position of the structure and members, which shall also be convenient for the reinforcement installation and concrete placing and curing.

(3) Formwork shall be simple in structure, convenient in assembly, disassembly, as well as being convenient in binding and installation of reinforcement, conforming to requirements of concrete placing and curing.

(4) Joints among formwork boards shall be tighten firmly to prevent of concrete grout leakages; as for wood formwork adopted in construction, it should be watered and keep moisture but there shall not be accumulated water inside.

(5) The surface of forms exposed to concrete shall be cleaned and painted with release agent, release agent that affecting structure performance or decorating finish is forbidden; no contaminations at the joint of reinforcement and concrete is allowed while painting release agent.

(6) For exposed concrete and decorative concrete, formwork board that satisfying the design requirement effects should be adopted.

2. Categories of formwork

1) Categories according to materials used

It is divided into wood formwork, steel formwork, plywood formwork, steel-wood (bamboo) combination formwork, plastic formwork, glass reinforced formwork, plastic formwork, aluminum alloy formwork, profiled steel sheeting formwork, decorative concrete formwork, prestressed concrete sheet formwork, etc.

2) Categories according to construction methods

It is divided into assembly and disassembly formwork (prefabricated then install formworks), mobilization formwork, permanent formwork, etc.

Assembly and disassembly formwork is composed of prefabricated parts, which are assembled in site, then cleaned and repaired and reused after being disassembled. Wood and combined steel formwork are commonly used, as well as the large-scale tool style customized formwork such as large panel formwork, bench formwork and tunnel formwork.

Mobilization formwork refers to the tool-style formwork made according to the shape of the structure, it will move vertically or horizontally along with the progress of the project after assembling removed at end of the project, such as sliding lifting formwork, lifting formwork and movable formwork.

Permanent formwork is permanently attached to the structural members and becomes parts of them, such as profiled steel sheeting formwork, prestressed concrete sheet formwork, etc.

3) Categories according to structure type

Foundation formwork, column formwork, beam formwork, floor formwork, wall formwork, stair formwork, shell formwork, chimney formwork, bridge pier formwork, etc.

4) Categories according to type

(1) Integrated formwork. It is mostly used in frame building which needs overall shuttering.

(2) Standard size series formwork. Formwork made of standard size series can be reused (including large steel panel formwork).

(3) Lifting formwork. it is mainly used in special structure such ascylindrical silo and chimney, might be used in frame and frame shear-wall structure.

(4) Tools formwork, which is often used at cylindrical shell structure and tunnel structure.

(5) Bench formwork. It is a type of larger size tools formwork adopted in frame and frame shear-wall structure inplacing of concrete slabs.

In the cast-in-place concrete structure, adopting high strength, durable, standard size and tools style formwork, has more advantageous for many times recycling, convenient in assembly and disassembly, as well as the important construction measure for enhancing project quality, decreasing the cost, speeds up the progress and obtaining economic benefit.

4.1.2 Detailing of formwork

1. Composite steel formwork (Standard size series steel formwork)

Composite steel-formwork is designed and fabricated according to several predetermined specifications and sizes. It can meet the requirements of most geometric dimensioning of structure components by its better universality and assembly flexibly, just simply choose the size of the corresponding specification of steel-formwork and combine them according to the size of the structure components.

Composite steel-formwork is composed of steel-formwork with certain modules, tie members and supporting members.

1) Steelformwork

Main type of steel formwork includes: flat formwork, convex formwork, concave formwork and joint formwork, as shown in Figure 4-1.

The flat formwork consists of panels and ribs, made of Q235 steel plate. The thickness of the panel is 2.3mm or 2.5mm, ribs are made of flat steel of 55mm×2.8mm with connecting holes.

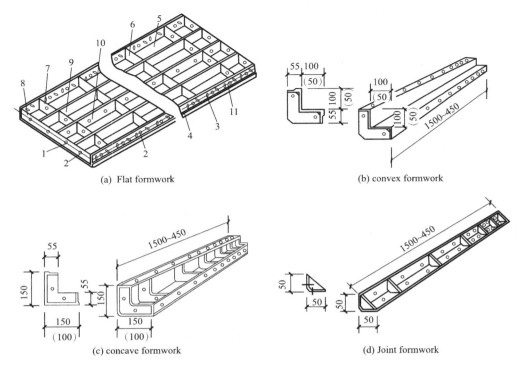

Figure 4-1　Flat formwork (mm)

1—Pin hole; 2—U shape clip hole; 3—Convex point; 4—Convex edge; 5—Pannel;
6—Transverse rib without hole; 7—Transverse rib with hole; 8—Transverse rib at ends;
9—Longitudinal rib without hole; 10—Longitudinal rib with hole; 11—Longitudinal rib at sides

The flat formwork can be used for flatness surface of foundation, column, beam, board and wall; the convex and concave formwork has the same length as that of flat formwork, in which convex formwork can be adopted at convex parts of wall and various other construal components and the concave formwork can be adopted at concave parts of column, beam and wall; joint formwork also can be selected at convex edge part of beam, column and wall instead of convex formwork. The specifications of steel formwork are as shown in Table 4-1.

Varieties of standard steel formwork　　　　Table 4-1

Specs(mm)	Flat formwork	Convex formwork	Concave formwork	Joint formwork
Width	300,250,200,150,100	100(50)×100(50)	150(100)×150	50×50
Length	1500,1200,900,750,600,450			
Height of rib	55			

2) Tie members

The joint elements for steel formwork are U shape clip, L shape pin, hook bolt, fastening bolt, split bolts and couplers, as shown in Figure 4-2.

3) Supporting members

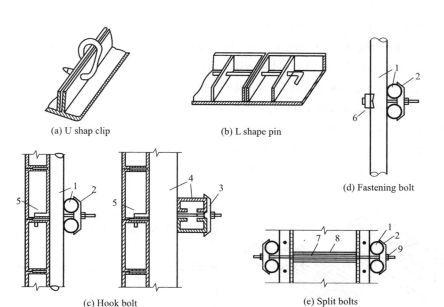

Figure 4-2　Tie members

1—Steel tube; 2—Curve type couple; 3—Straight type coupler; 4—Channel steel tube with inner frill; 5—Hook bolt; 6—Fastening bolt; 7—Bolt rod; 8—Plastic sleeve; 9—Nut

Steel formwork supporting includes steel tube, bracing, oblique bracing, column hoop, plane composite truss, etc.

2. Steel frame formwork

Steel frame formwork includes two types: wooden plywood board and bamboo plywood board with exactly the same structural form, they are new types of formwork after steel formwork as shown in Figure 4-3.

For wooden board steel frame formwork, its popularizing is restricted cause of its high cost. Bamboo board steel frame formwork is made of multi-layers of bamboo sheets by making use of abundant bamboo resources, which has lower cost and better technical characteristics, thus, it is advantages of updating, popularizing and applications in construction site.

Panel part in bamboo plywood formwork is made of 3～5 layers of bamboo sheets or multi-layers bamboo curtain and the steel frame part is made of structural steel with connection holes on both of its sides. Panel is embedded in the steel frame and fixed with bolts or rivets, and the panel can be turned over for reuse or replaced with a new one whenever the panel is damaged. Water proofing shall be done on the surface of panel, and surface of the panel shall align to the side frame.

Bamboo plywood formwork has itself size series of 55, 63 and 75mm in which the series 55mm can be mixed used with the steel formwork because of they are match well for each other's side frame and holes spacing.

Characteristics: light weight (about 1/3 lighter than steel formwork); less steel consumption (about 1/2 less than steel formwork); the area of the single board is 40% larger than that of the steel board of the same weight, thus, the assembly work is small and the joints is less; the heat conductivity of the panel materials is only 1/400 of that of the steel formwork, as a well insulative performance, it is beneficial to the construction in winter; easy maintenance. But its rigidity, strength is less than steel formwork.

At present, with its good construction performances, bamboo plywood formwork has been used in the construction of cast-in-situ concrete foundation, column, wall, beam, slab and tube structure, as well as bridge and municipal engineering.

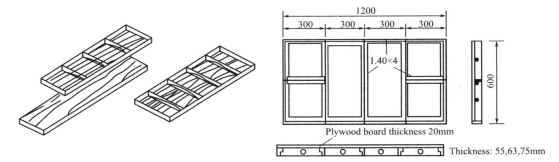

Figure 4-3 Steel frame formwork (mm)

3. Wooden formwork and plywood formwork

Wooden formwork are still used in civil engineering constructions currently. Generally, it forms in a way of assembly and disassembly, also, it can be assembled in construction with prefabricated parts (jointed wooden board).

Wooden formwork can be reused after removal but the turnover times are less. Joint board are made of wooden strip and jointing wooden strip as shown in Figure 4-4. Thickness and width of wooden strip is 25~50mm with width not more than 200mm so as to ensuring seam uniformly while shrinking and making up seam easily after watering, but the width of soffit formwork has no this limitations so as to reduce the slurry in concrete placing process.

The spacing of the ribs in ribbed panels depends on the pressure of the newly placing concrete and the thickness of the wooden strip using as combining, usually is 400~500mm.

Plywood formwork is made of plywood board with wooden girder nailed on. Plywood board has larger area with thickness of 12~21mm; wooden girder arranged in a smaller spacing way which has spacing of 200~300mm with section size 50mm×100mm or 100mm×100mm.

Plywood formwork can be further classified into wooden plywood board and bamboo plywood board, and the bamboo plywood board formwork has more advantages comparing with that of wooden plywood formwork in strength, rigidity and turnover times.

Plywood formwork has many of advantages when used as concrete structure members formwork:

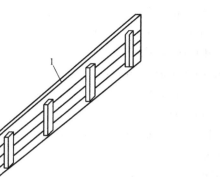 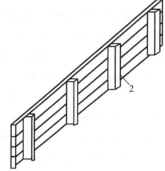

(a) Normal jointed wooden board   (b) Jointed wooden board at beam sides

Figure 4-4 Structure of joint board
1—Wooden strip; 2—Jointing wooden strip

(1) Larger area, lighter weight and flatness board, can not only reduce the installation works, but also be convenient for transportation, stacking, usage and management, ensuring the concrete surface smooth, especially an ideal selection of fairfaced concrete members.

(2) Convenient in sawing cut, easy to be processed into various shapes and can be used at curved surfaces.

(3) Good insulation performance, can prevent the temperature from changing too faster, conducive to heat preservation in winter concrete construction.

4. Aluminum alloy formwork

Aluminum alloy formwork is designed by a certain serial modulus and made of aluminum alloy ribbed panel, end plate and main and secondary ribs welded. Composite aluminum formwork includes flat formwork, flat adjustable formwork, concave angle formwork, concave angle adjustable formwork, convex angel formwork, convex angle adjustable formwork, aluminum alloy beam, support head and special formwork. Composite aluminum formworks system is as shown in Figure 4-5.

Figure 4-5 Composite aluminum formwork system

Standard formwork should be selected priority, the wooden formwork, plywood formwork and plastic formwork could be selected as complement for the rest parts, so as to decrease the proportion of non-standard formworks.

Common specifications of aluminum alloy formworks:

Flat aluminum formwork: width: 100~600mm, length: 600~3000mm, thickness: 65mm.

Concave angle formwork: section: 100mm×100mm, 100mm×125mm, 100mm×150mm, 110mm×150mm, 120mm×150mm, 130mm×150mm, 140mm×150mm, 150mm×150mm; length: 600~3000mm.

Convex angle formwork: 65mm×65mm.

Support adjustable length: 1900~3500mm.

Aluminum formwork has advantages of light weight, high strength, high machining accuracy, larger single block area, less seams, convenient construction, lower cost, higher quality of concrete surface, less construction waste. It meets the requirements of building industrialization, environmental protection and energy conservation.

Aluminum formwork mainly used in concrete structure construction of wall, column, beam and slabs; attached formwork of outer wall in vertical structure; simultaneous formwork construction of beam and slab.

5. Large panel formwork

Large panel formwork is consisted of faceplate, reinforced rib, vertical support, support truss, stabilizing mechanism, operation platform, through wall bolts, etc. It is a large tool-based formwork for cast-in-site reinforced concrete wall, as shown in Figure 4-6.

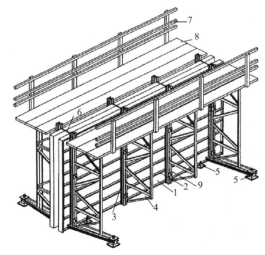

Figure 4-6 Detail structure of large panel forwork (wooden form)
1—Surface board; 2—Horizontal reinforce rib; 3—Vertical support; 4—Support truss;
5—Leveling adjusting screw jack; 6—Fixing clamp; 7—Hand rail; 8—Scaffold board; 9—Through wall bolts

Faceplate, as the part connecting with the concrete directly, might be made of plywood board, wooden board, steel board, etc; reinforce rib is used to fix the faceplate and to transfer the lateral pressure generated by the newly placing concrete to the vertical support, reinforce rib can be horizontal or vertical direction, fixed with metal faceplate by spot welding or with plywood and wooden plywood by bolts.

Vertical support is to bear the lateral pressure of the newly placing concrete by supporting the horizontal reinforce ribs, as well as to strengthen the overall stiffness of the large panel formwork. Vertical support are usually placed in pairs with channel steel (No. 65 or 80) with gaps left between the two channels to pass through the through wall bolts. Support trusses are connected by bolt or welding with vertical supports to bear horizontal force such as wind load and prevent large panel formwork from capsizing.

The stabilizing mechanism is an adjustable screw jack installed at bottom of the extending legs at both ends of the large panel. In serving stage, it is used to adjust the verticality of the formwork and transfer the force to the ground or floor in service stage. In stacking stage, it is used to adjust the inclination of the formwork to ensure the stability in stacking stage.

Operation platform is the place that construction personnel operate, it formed in two kinds of methods:

(1) Scaffold board directly lay on the horizontal chords of the support truss, with a railing setting up on its outside. Characteristics of this method is small operation space but lower cost and convenient in assembly and disassembly.

(2) Scaffold board lay fully on the horizontal grid formed by angle iron members which connected the top part of support trusses at the two transverse walls. It has characteristics of safety in construction but larger steel consumptions.

6. Gliding formwork

Gliding formwork is a tool-based formwork. It is often used for structures with high-rise and vertical structure of buildings such as: chimney, silo, high bridge pier, TV tower, shaft, caisson, hyperbolic cooling tower and high-rise buildings.

Construction characteristic of gliding form:

At the bottom of the structure or the building, assembling gliding board with height around 1.2m along the periphery of the structure, as the concrete is placing into the form continuously layer by layer, a hydraulic lifting device is used to lift the gliding board by the support rod buried in the concrete until it reaches the required placing height.

The construction with sliding formwork can greatly decrease the amount of formwork and supporting materials, reduce the assembly and disassembly work forces, accelerate the construction speed and ensure the integral of the structure; however, it has higher one-time investment and consumption of steel, also there are some restrictions for elevation changes and structural section changes; construction should be carried out continuously and the construction organization should be fulfilled strictly.

The gliding formwork is mainly composed of three parts of formwork system, operation platform system and hydraulic lifting system, as shown in Figure 4-7.

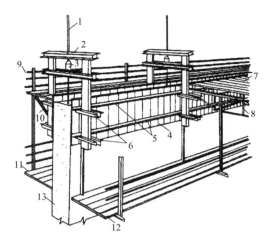

Figure 4-7 Composition sketch of gliding forwork
1—Support rod; 2—Hoisting frame; 3—Hydraulic jack; 4—Form board; 5—Waler;
6—Support of waler; 7—Inside operation platform; 8—Truss of inner operation plat form; 9—Rail;
10—Cantilever triangle truss; 11—Outside hanging scaffold; 12—Inside hanging scaffold; 13—Concrete wall

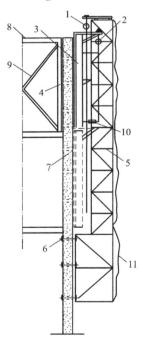

Figure 4-8 Diagram of attached formwork
1—Electric block for lifting external formwork;
2—Electric block for lifting external climbing frame;
3—External formwork; 4—Climbing fram reserved holes;
5—External climbing frame; 6—Bolt; 7—Outer wall;
8—Floor formwork; 9—Floor formwork supporting;
10—Formwork corrector; 11—Safety net

Formwork system is composed of form board, waler and hoisting frame.

Operation system is composed of operation platform and its laying boards, hanging scaffold.

Hydraulic system is composed support rod, hydraulic jack, hydraulic control console and oil system.

7. Attached formwork

The attached formwork lift itself by lifting device to the upper floor after the concrete placing of the lower wall is completed, it is composed of form board, hoisting frame and lifting equipment. Electric block can be used as lifting equipment for outer wall external surface formwork, as shown in Figure 4-8.

The attached formwork adopts one piece large panel formwork which is composed of surface board and rib, supporting system is not needed; lifting equipment can be electric screw hoist, hydraulic jack or chain guide.

The attached formwork is a combination of large panel formwork and gliding formwork technologies, which not only maintains the advantages of smooth wall surface of large panel formwork, but also maintains the advantages of gliding formwork using its own equipment to lift itself upward without tower crane.

The attached formwork is suitable for the construction of concrete walls in high-rise buildings, elevator shaft walls and pipe spaces.

8. Bench formwork

Bench formwork is a large tool-based form for placing reinforced concrete floor, it can be demoulded and transferred as a whole. A crane can be adopted to lift the mold out from lower floor and transfer to the upper floor without fall to the ground during construction, so it is also called "flying mold".

According the support methods of bench form, it can be classified into two types of legs supporting and legless supporting.

Legless supporting type is suspended itself from a wall or column; leg supporting type is composed of panel, waler and support frame, as shown in Figure 4-9.

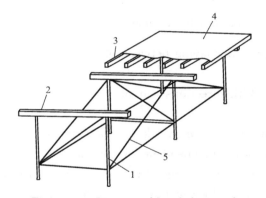

Figure 4-9 Diagram of bench formwork
1—Support legs; 2—Adjustable transverse beam; 3—Waler; 4—Surface board; 5—Brace

Surface board is the component that contacts with concrete directly, and plywood, steel board and plastic board can be adopted as surface board which should be smooth, high strength and stiffness.

The legs of the supporting frame are retractable or collapsible and equipped with wheels at the bottom for movement conveniently.

The covering area of single bench formwork ranges from $2 \sim 6m^2$ to more than $60m^2$.

Bench formwork itself is of good integrity, smoother surface and faster construction progress, suitable for all kinds of cast-in-site concrete structures floors with smaller compartment and smaller deep.

9. Tunnel formwork

A tool-based formwork, a combination of bench formwork and big panel formwork, supporting floor and wall simultaneously.

There are two type of tunnel formwork: integral type and double assembly type. The integral type has heavy weight, is difficult to move, thus, less used current time; double combined type has widely used in high-rise and multi-floors buildings such as "hanging at exterior and placing at interior" and "masonry at exterior and placing at interior".

Double combined type tunnel formwork is composed of twosemi tunnel formworks and one independent formwork, as well as independent braces, as shown in Figure 4-10.

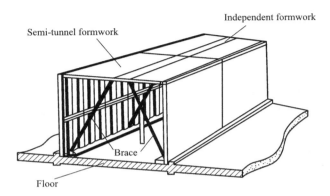

Figure 4-10 Diagram of tunnel formwork

The independent formwork laying between of the two semi tunnel formworks has the functions as following:

(1) Size in room width directions can be changed easily to suit different width of buildings.

(2) The two semi tunnels formwork can be removed in advance to speed up its turnover without removing the independent formwork.

Diagonal bracing is used to connect the vertical wall formwork of the semi-tunnel formwork with the horizontal floor formwork; in the length direction of the formwork, walking wheels and jacks are set along the bottom of the wall formwork; after the formwork is in place, the jack will lift the formwork up and walking wheel will leave from the floor ground, the construction loads shall be beard by only jacks; in the disassembling process, after loosening the jack, the semi-tunnel formwork will be lowered and demoulded under the action of its self-weight, the walking wheel will then fall down to floor ground, the formwork can be moved away from the floor ground and hoisted up to the higher floor to continue construction.

10. Early removing formwork

The early removing form system is a kind of mold supporting device and methods adopted to realize the early removal of floor formwork, and its technologic principle is essentially "removing the board without removing the column".

Vertical supporting system which stays at between upper layer and lower layer, is composed of early removing column cap, upright tube and adjustable pedestal.

In the process of mold removal, the original designed floor slab is under the stress state of short span (the spacing between columns is less than 2m), that is, the span of floor slab formwork shall not exceed the span requirement of mold removal specified in the national code. In this way, when the concrete strength reaches 50% of the designed strength (3~4 day at normal temperature), the floor formwork and parts of its support can be removed, while the inter-column braces, vertical column and adjustable support remain in the supporting state.

When the strength of concrete increases enough to support self-weight and construction load in the condition of full span, all vertical supports are removed, as shown in Figure 4-11.

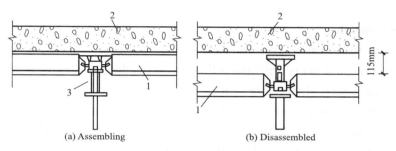

(a) Assembling　　　　　　(b) Disassembled

Figure 4-11　Diagram of early removing form
1—Supporting beam; 2—Cast-in-site slab; 3—Early removing column cap

This method has advantages of less support amounts, less investment, short duration and comprehensive benefit. It is a developing and improving new construction technology at present.

The key part in early removing form system is its early removing column cap as shown in Figure 4-12. The top board size of the cap is 50~150mm, might contact to the concrete directly. Both sides corbel which hold the end part of the supporting beam is attached to the square tube which can be moved up and down in 115mm, when square tube is at upper position, it can lock the supporting beam by its corbels, strike the pin away from the locking hole on square tube will let the square tube fall down so as to dismantle the supporting beam and the form board it supported.

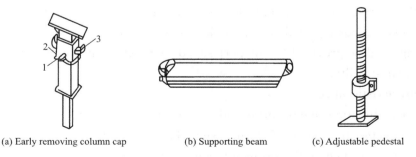

(a) Early removing column cap　　(b) Supporting beam　　(c) Adjustable pedestal

Figure 4-12　Diagram of components of early removing form
1—Plug; 2—Sliding block; 3—Corbel

Adjustable pedestal is inserted into the lower end of the column and with connection to the floor ground in order to adjust the height of the column, adjustable range is 0~50mm, as shown in Figure 4-12.

4.1.3 Design of formwork

Design of formwork system includes type selection, materials selection, load calculation and structure calculation, preparation of installation and removal scheme, drawing formwork diagrams. Design of forms and their supports shall be determined in according with such conditions as structure type, loads, ground soil category, construction equipment and materials supply, etc.

1. Design principle

In addition to meeting the technical requirements mentioned above, the steel formwork design shall comply with the following principles:

(1) The universal, large size formwork should be adopted priority in order to minimum its variety, quantities, as well as the number of wooden splicing strip.

When pull bolts are adopted, in order to reduce the wastage of steel formwork cause of bolt boring on it, a square wooden with size of 55mm×100mm can be used instead of steel formwork board at the bolt boring position, or use the steel formwork board with the bolt boring repeatedly.

(2) The staggered arrangement along the length direction shall be adopted to increase the overall rigidity of the formwork.

(3) The inner steel tube shall be arranged perpendicular to the length direction to bear the loads from the formwork board; outer steel tube shall be perpendicular to the inner steel tube, bear the load from the inner steel tube, strength the overall rigidity of the formwork structure, adjust flatness; specification of outer steel tube shall not be less than that of inner steel tube.

(4) When the steel formwork align at seam, each steel formwork board shall have at least two support tube; when it is staggered, the spacing will not limited by the end position of the steel formwork board.

(5) Supporting column should have enough strength and stability, the slenderness ratio of normal supporting or that of parts in between should be less than 110; horizontal or X bracing shall be provided for support supporting column in continuous or truss form to ensure their stability.

2. Characteristic value

(1) Characteristic value of forms and their supports self-weight G_{1k}:

The characteristic value of forms and their supports self-weight G_{1k} shall be determined in according with formwork construction drawing. As for formworks with beamed floor-slab and beamless floor-slab, the characteristic value of them may be selected according to those specified in Table 4-2.

Chapter 4 Concrete Structure Engineering

Characteristic value of self weight of forms and their supports (kN/m²) G_{1k} Table 4-2

Item name	Timber formwork	Typified combined steel formwork
Formwork and minor ridge of beamless floor slab	0.30	0.50
Beamed floor-slab formwork(formwork containing beams)	0.50	0.75
Floor-slab formwork and bracket(floor height below 4m)	0.75	1.10

(2) Characteristic value of freshly placed concrete self-weight G_{2k}:

As for freshly placed concrete, the characteristic value of its self-weight G_{2k} should be determined in accordance with the actual gravimetric density γ_c of the concrete; as for ordinary concrete, γ_c may be valued at 24kN/m³.

(3) Characteristic value of reinforcement self-weight G_{3k}:

The characteristic value of reinforcement self-weight G_{3k} shall be determined in according with the construction drawing. As for ordinary beam and slab structure, the self weight of floor-slab reinforcement may be valued at 1.1kN/m³; as for the self-weight of beam reinforcement, it may be valued at 1.5kN/m³.

(4) Characteristic value of the side pressure G_{4k}:

If immersion vibrator is adopted, the placing speed is not larger than 10m/h, and the concrete slump is not larger than 180mm, the characteristic value of the side pressure G_{4k} applied on the formwork by freshly placed concrete may be calculated respectively according to the following formulae as shown in Formula (4-1) and the smaller value shall be taken.

When the placing speed is greater than 10m/h or the concrete slump is greater than 180mm, the characteristic value of side pressure may be calculated according to second formula in Formula (4-2).

$$F = 0.22 \gamma_c t_0 \beta v^{1/2} \tag{4-1}$$

$$F = \gamma_c H \tag{4-2}$$

There in:

F——Characteristic value of the maximum side pressure applied to formwork by fresh concrete placed (kN/m²).

γ_c——Gravimetric density concrete (kN/m³).

t_0——Initial setting time of freshly placed (h), which may be determined by measurement; if test date are lack, it may be calculated by using $t_0 = 200/(T+15)$, where T is concrete temperature (℃).

β——Slump influence correction coefficient of concrete: if the slump is larger than 50mm and not larger than 90mm, β is valued at 0.85; if the slump is larger than 90 mm and not larger than 130mm, β is valued at 0.90; if the slump is larger than 130 mm and not larger than 180mm, β is valuate at 1.0.

v——Placing speed, valued at the ratio of placement height (thickness) to placing time (m/h).

H——Total height from calculation position of concrete side pressure to the top surface of freshly placed concrete (m).

The diagram of the calculation distribution for concrete side pressure is shown as Figure 4-13 in which $h = H/\gamma_c$.

There in:

h——The effective pressure head height (m);

H——The total height of concrete in formwork;

F——The maximum side pressure.

(5) The characteristic value of the load Q_{1k} generated by construction personnel and construction equipment may be calculated according to actual conditions and may be 3.0kN/m².

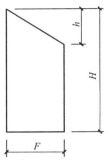

Figure 4-13 Concrete side presssure distribution

(6) The characteristic value of the horizontal load Q_{2k} generated by concrete pumping, dumping may be selected 2% of the weight of concrete and reinforcement, and acting along horizontal direction at the top of support by means of linear loads.

(7) Characteristics value of wind load Q_{3k}:

The anti-overturning stability should be considered for the formwork under larger wind pressure, and the formwork which is easy to fall down under the action of wind load.

Characteristic of wind pressure shall be adopted in accordance with the stipulations in "Load code for design of building structures GB 50009—2012". In this case, the reference wind pressure maybe valued in accordance with the decennial wind pressure, but shall not be less than 0.20kN/m².

Beside of adjusting the reference wind pressure according to different topography, the temporary structural adjustment coefficient of 0.8 can be multiplied as shown in Formula (4-3).

$$\omega_k = 0.8\beta_z\mu_s\mu_z\omega_0 \qquad (4-3)$$

There in:

ω_k——Characteristic value of wind load (kN/m²);

β_z——Dynamic effect factor of wind at a height of z;

μ_s——Shape factor of wind load;

μ_z——Exposure factor for wind pressure (variation coefficient of wind pressure height);

ω_0——Reference wind pressure (kN/m²).

3. Design value of load

Design value of load of forms and their supports can be obtained by multiply partial coefficient to those characteristic values calculated above.

4. Load combination

As for the bottom forms and their supports of horizontal concrete structure members, higher and bigger formwork support, side forms and its supports of vertical concrete structure members and horizontal concrete structure members, the most un-favorable effect combination of forms and their supports can be determined according to Table 4-3.

Most un-favorable combinatin of design load value Table 4-3

Type of forms and supports	Most un-favorable combination	
	Load-carrying capacity	Deformation checking
Bottom forms and their supports of horizontal concrete structure members	$G_1+G_2+G_3+Q_1$	$G_1+G_2+G_3$
Higher and bigger formwork support	$G_1+G_2+G_3+Q_1$	$G_1+G_2+G_3$
	$G_1+G_2+G_3+Q_2$	$G_1+G_2+G_3$
Side forms and its supports of vertical concrete structure members and horizontal concrete structure members	G_4+Q_3	G_4

5. Load-carrying capacity

In the design of formwork system, the calculation diagram should be determined according to its actual structure, but the emphasis is different for different structure components. The bending moment and deflection are mainly considered for typified form and, beam formwork, wooden girders; the stability under pressure is mainly considered for support column and support frame; the bending moment should be considered for upper chord of support truss; as for wooden structure members, shear force and pressure-bearing at supports should be considered.

Load-carrying capacity of structural members of forms and their supports shall be calculated according to the temporary design conditions. Calculation of load-carrying capacity shall meet the requirements of Formula (4-4).

$$\gamma_0 S \leqslant R/\gamma_R \qquad (4-4)$$

There in:

γ_0 ——Significance coefficient of structure, which should be $\gamma_0 \geqslant 1.0$ for important forms and their supports and shall be $\gamma_0 \geqslant 0.9$ for ordinary forms and their supports;

S ——Design value of effects calculated according to the fundamental combination of loads for forms and their supports;

R ——Design value of load-carrying capacity for structural members of forms and their supports, which shall be calculated according to relevant current national standards;

γ_R —— Adjustment coefficient of design value of load-carrying, which shall be determined according to the reuse conditions of forms and their supports and shall not be less than 1.0.

Design value of effects generated by the fundamental combination of loads on the forms and their supports may be calculated according to Formula (4-5).

$$S = 1.35\alpha \sum_{i \geqslant 1} S_{G_{ik}} + 1.4\psi_{cj} \sum_{j \geqslant 1} S_{Q_{jk}} \tag{4-5}$$

There in:

$S_{G_{ik}}$ ——Value of effect generated by the characteristic value of the ith permanent load;

$S_{Q_{jk}}$ ——Value of effect generated by the characteristic value of the jth variable load;

α ——Type coefficient of forms and their supports: 0.9 for the side forms, 1.0 for the bottom forms and their supports;

ψ_{cj} ——Combination value coefficient of the jth variable load, which should be $\psi_{cj} \geqslant 0.9$.

6. Deformation analysis

Height-width ration of supports should not be greater than 3, otherwise the overall stability measures shall be strengthened.

Slenderness ratio of steel members in supports shall not exceed the allowable value specified in Table 4-4.

Allowable slenderness ratio of steel members in supports　　　　Table 4-4

Member type	Allowable slenderness ratio
Shores and truss in supports for compression members	180
Diagonal braces and diagonal bracings for compression members	200
Steel bracing for tension members	350

Deformation analysis of forms and their supports shall meet the following requirements as shown in Formula (4-6).

$$\alpha_{fG} \leqslant \alpha_{f,lim} \tag{4-6}$$

There in:

α_{fG} ——Deformation value of members, calculated according to the characteristic value of permanent load;

$\alpha_{f,lim}$ ——Deformation limit value of members, determined according to the requirements as followings:

For the formwork with the exposed structure surface, the maximum deformation value should be 1/400 of the calculation span of form members;

For the formwork with concealed structure surface, the maximum deformation value should be 1/250 of the calculated span of form members;

For limit value of axial compression deformation or lateral deflection of supports should be 1/1000 of the calculated height or calculated span.

4.1.4　Installation of formwork

1. General stipulation

(1) Formwork shall be processed and fabricated according to the drawings. Universally used

formwork should be fabricated into typified forms.

(2) Ribs on back of form panels should be with uniform section height. During the fabrication and installation of formwork, the panels shall be spliced tightly. As for walls with water-proof requirements, water stop pieces, which shall be connected with split bolts by circumference welding, shall be arranged in the middle of split bolts.

(3) The special supports matching with the universal steel tube supports shall be processed and fabricated according to the drawings. The girder placed on the adjustable fork head on the top of supports may be made of batten, timber beam or section steel with symmetrical sections.

(4) Shore in support and vertical forms, when installed on the soil layer, shall meet the following requirements:

① Sub plates with sufficient strength and load-carrying area shall be arranged.

② The soil layer shall be solid and provided with drainage measures; as for the collapsible loess and expansive soil, waterproof measures shall be provided; as for the frost heaving soil, frost heaving prevention measures shall be provide.

③ As for soft soil foundation, the preloading method may be adopted, if necessary, to adjust the installation height of form panels.

(5) During the formwork installation, surveying and setting-out shall be carried out and the position measures shall be taken to ensure the accurate formwork position. As for the forms and their supports for vertical members, the measures shall be taken for sides way resistance, anti-floating and overturning resistance of vertical formwork according to the primary placing height and placing speed of concrete. As for the forms and their supports for horizontal members, the effective tying measures between supports, forms or between forms and their supports shall be taken in combination with the different types of support and form panel. Wind-proof measures shall be taken for the formwork that may carry relatively large wind load.

(6) As for beams and slabs with span not less than 4m, the camber height of formwork construction should be $1/1000 \sim 3/1000$ of the span thereof and the camber shall not reduce the section height of member.

(7) When steel tubes with couplers are used as the supports, the erection of supports shall meet the following requirements:

① The specification of steel tubes and couplers used for the erection of supports shall meet the design requirements; longitudinal spacing of upright tubes, transverse spacing of upright tubes, lift height of supports and detailing requirements shall meet the requirements of the special construction schemes.

② Longitudinal spacing of upright tubes and transverse spacing of upright tubes shall not be greater than 1.5m; lift height of supports shall not be greater than 2.0m. Bottom reinforcing tubes should be arranged on the longitudinal and transverse directions of upright tubes. The distance from the longitudinal bottom reinforcement tubes to the bottom of upright tubes should not be greater than 200mm. Transverse bottom reinforcing tubes should be arranged under longitudinal bottom reinforcing tubes. Base plate or sub plate should be arranged at the bottom of upright tubes.

③ As for the extension of upright tubes, except that the lift height on the top level may adopt the lap splicing, the joints of lift height of other levers shall be connected by butt couplers and the joints of the two adjacent upright tubes shall not be arranged in the same lift height.

④ Two-way horizontal tubes shall be arranged at both upper and lower ends of the lift height of upright tubes; the cross points of horizontal tubes and upright tubes shall be connected with couplers; the distance between the connecting couplers of two-way horizontal tubes and upright tubes shall not be greater than 150mm.

⑤ Vertical diagonal bracing shall be continually around the supports. When either the length or width of supports is greater than 6m, the longitudinal or transverse vertical diagonal bracing shall be arranged in the middle. The spacing between diagonal bracing and the width of single diagonal bracing should not be greater than 8m. The inclined angle between diagonal bracing and horizontal tubes should be 45°~60°. When the height of support is greater than 3 times of the lift height, one horizontal diagonal bracing should be arranged on the top of supports and the diagonal bracing shall be extended to the surroundings.

⑥ Splicing length of upright tubes, horizontal tubes and diagonal bracing shall not be less than 0.8m and they shall be connected with at least 3 couplers; the distance from the edge of coupler cover plates to the tube ends shall not be less than 100mm.

⑦ Tightening torque of coupler bolts shall not be less than 40kN/m and shall not greater than 65kN/m.

⑧ Vertical deviation of erection for upright tubes of supports should not be greater than 1/200.

(8) When steel tubes with couplers are used as the tall formwork supports, the erection of supports shall also comply with the following requirements:

① Adjustable fork head should be inserted on the top of upright tubes of supports; the out diameter of adjustable fork head screw shall not be less than 36mm; the length of screws inserted into steel tubes shall not be less than 150mm and the length of screws extended outside of steel tubes shall not be greater than 300mm; the cantilever length of adjustable for head extended outside of top horizontal tubes shall not be greater than 500mm.

② Longitudinal spacing of upright tubes and transverse spacing of upright tubes shall not be greater than 1.2m; lift height of supports shall not be greater than 1.8m.

③ When splicing is adopted within the top lift height of upright tubes, the splicing length shall not be less than 1m and at least 3 couplers shall be used.

④ Bottom reinforcing tubes should be arranged on the longitudinal and transverse directions of upright tubes; the distance from longitudinal bottom reinforcing tubes to the bottom of upright tubes should not be greater than 200mm.

⑤ The longitudinal or transverse vertical diagonal bracing should be arranged in the middle; the spacing between diagonal bracing should not be greater than 5m; the spacing between the horizontal diagonal bracing arranged along the height direction of supports should not be greater than 6m.

⑥ Vertical deviation of upright tubes of supports should not be greater than 1/200 and should not be greater than 100mm.

⑦ Overall stability of supports shall be strengthened by the effective connecting measures which are taken according to the conditions of peripheral structures.

(9) When steel tubes with bowl buckle, plate lock or plate pin couplers are used as supports, the erection of supports shall meet the following requirements:

① As for supports with bowl buckle, plate lock or plate pin couplers, the horizontal tubes and upright tubes shall be firmly fastened and shall not slip.

② Spacing between upper and lower horizontal tubes on upright tubes shall not be greater than 1.8m.

③ The cantilever length of adjustable fork form, inserted on top of upright tube, extended outside of top horizontal tube shall not be greater than 650mm and the length of screw inserted into the steel tube shall not be less than 150mm. the clearance between screws to inner wall of tube shall be not greater than 6mm. The lift height of the horizontal tube in the top level of supports shall be reduced by one node spacing from standard lift height.

④ Dedicated diagonal tubes or diagonal tubes with couplers shall be arranged between upright tubes to strengthen supports.

(10) If the supports are erected with frame steel tubes, the erection shall meet the relevant requirements of the current professional *Technical Code for Safety of Frame Scaffolding with Steel Tubes in Construction* JGJ 128—2010.

(11) Vertical diagonal braces and horizontal diagonal braces of supports shall be erected with supports at the same time and supports shall be tied with formed concrete structure. Erection of vertical diagonal braces and horizontal diagonal braces in steel tube supports shall meet the requirements of relevant current national standards related to the steel tubular scaffolds.

(12) As for the cast-in-situ multi-storey and high-rise concrete structures, the upright tubes of supports on the upper and lower storeys should be aligned. Components of forms and their supports shall be stacked in a scattered way.

(13) Formwork installation shall ensure the accuracy of the shapes, dimension and relative position of the concrete structural members, which shall also prevent the paste leakage.

(14) Formwork installation shall be carried out in coordination with the reinforcement installation. The formwork of joints of columns and beams should be installed after installation of reinforcement.

(15) The contact surface between forms and concrete shall be cleaned up and brushed with release agent that shall not contaminate the joints between reinforcement and concrete.

(16) Forms and their supports of post-cast strips shall be arranged separately.

(17) Embedded parts and preformed holes fixed on the forms shall not be missed and the installation shall be firm and position shall be accurate.

Formwork sides and soffit are shown as in Figure 4-14.

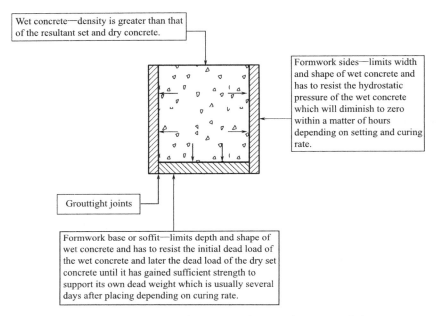

Figure 4-14　Formwork sides and formwork base (soffit)

2. Formwork for foundation

Characteristics: small height with big volume.

1) Formwork for stepped foundation

Each step of the formwork for stepped foundation is matched by four side forms. Two of themhave the same long as the according step side and another two should be 150～200mm longer than the according step side. Then nail the four side forms to a square frame with studs. The bottom match board of the two sides of the upper step should be lengthened as joist, which can be supported by the formwork of the first step. Around the four sides, diagonal and horizontal brace should be set up, one side is nailed to the stud of the side board, and another side is nailed to the wood pile.

Before erecting, we should mark center line on the inside face of the board and on the bottom of the foundation pit with chalk. Then aim at the two center line, calibrate the elevation, erect the wood pile and the brace. Illustrations of formwork for stepped foundation are shown in Figure 4-15 and Figure 4-16.

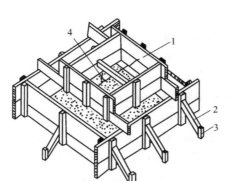

Figure 4-15 Formwork of stepped foundation (timber board)
1—Wooden panel and batten; 2—Diagonal bracing; 3—Wooden stake; 4—Iron wire

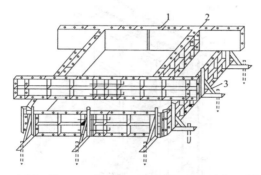

Figure 4-16 Formwork of stepped foundation (steel board)
1—Flat steel fitting; 2—T shape steel fitting; 3—Angle iron triangle bracket

2) Socket-connected footing

It is similar to the stepped foundation. It needs to put a cup-shape form into the top step. It also needs to nail two battens on the top of two sides of the cup-shape form as joists, so it can be shelved on the top of the formwork of the top step.

3) Strip foundation

Formwork for strip foundationis composed by side board, flat brace and diagonal brace. The side board should be composed by long strip board nailed to vertical stud or short strip board nailed to horizontal stud.

First, mark the border line of the foundation at the bottom of the foundation trench, then vertically erect the side board aim at the border line, after adjusted and calibrated, and then, fixed by brace. If the foundation is too long, we can erect the side board at each end, then tie two strings to the top of side board, maintain the width of foundation and erect the board along the strings. Some battens should nail on the top of the side boards in proper space to avoid the deformation of the form work.

3. Formwork of wall

Characteristics: big height, small thickness and main side pressure. So it needs to strengthen rigidity of formwork and set sufficient braces to ensure there is no deformation and displacement.

The formwork for walls is composed of side board, soldier or stud, waler and brace. The side board is long battens fixed to the vertical stud to make a big block. Horizontal walings are fixed to the stud from the bottom at an interval of 1~1.5m. The whole assembly is then strutted by the braces and piles.

Before erecting, mark the center line and border line on the floor, then erect one side first. Use plumb bob calibrate the uprightness, then fixed by brace. After the reinforcement is placed, erect the other side board. The two shutters are kept apart equal to the thickness of the wall, by providing spacers or φ12~16mm bolt fixed between the two sides. Details of formwork of wall are shown in Figure 4-17~Figure 4-20.

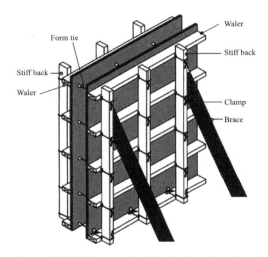

Figure 4-17 Formwork of wall (wooden form)

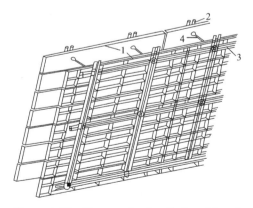

Figure 4-18 Formwork of wall (steel form)
1—Wall form; 2—Minor ridge; 3—Major ridge; 4—Split bolt

Some examples are show below with the alternative coil tie system. Spacing of wall form is shown as in Figure 4-21.

Chapter 4 Concrete Structure Engineering

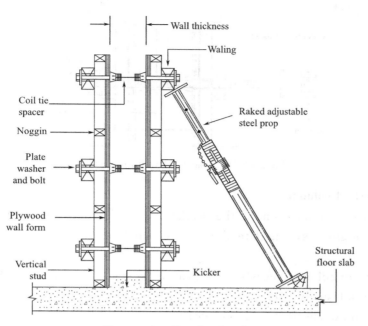

Figure 4-19 Details of tie bar

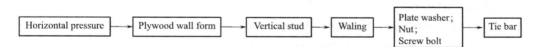

Figure 4-20 Force transmission routes of wall formwork

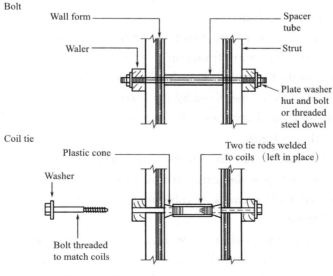

Figure 4-21 Spacing of wall form (one)

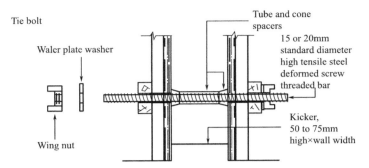

Figure 4-21 Spacing of wall form (two)

4. Formwork of column

Characteristics: dimension of the section is small while the height is big. So the problems of uprightness, big lateral pressure and lateral stability in the construction should be noticed.

Formwork for column consists of vertical or horizontal board called sheeting, cleats, wedge and yokes. Usually, the vertical board is 25~50mm thick and the horizontal board is 25~30mm thick. The width of the boards may vary depending on the section of the column.

The cleats used at the base of the box are larger so as to withstand the pressure which is exerted by the weight of concrete. The sides of the box are secured firmly together by means of wood or steel yokes (adjustable column ban made of angle or flat iron), the two side wood yokes can be connected together by two bolts of 16mm dia. Four wedges, one at each corner, are inserted between the bolt and the cleats. The sheeting is nailed to the wood yokes. The interval of yokes or column ban should be decided by the section of the column. The interval should be smaller at the bottom part of column. Generally, the interval should not be more than 1m.

Reinforcement should be placed before erecting the formwork. Mark the axes and border line on the floor, fix square frame on the floor, erect the side boards and fixed by temporary brace. Calibrate the uprightness by the plumb bob and check the elevation and position without the mistake, then fixed by the brace. When the height of the column is $\geqslant$ 4m, it should be fixed around the four sides. When the height is more than 6m, it should be supported with other columns.

For a row of columns, columns on the two sides should be erect first, then two strings can be tied on the top of the formwork maintaining the width of column to calibrate the columns in the middle. Formwork of column is as shown in Figure 4-22~Figure 4-24.

5. Formwork of beam

Characteristics: big span and small width, the height maybe varies from 20cm to 2m. Because it is built on stilt (aerial), there will be lateral and vertical pressure on the formwork. The formwork and support system should have good stability, enough strength and rigidity.

Chapter 4 Concrete Structure Engineering

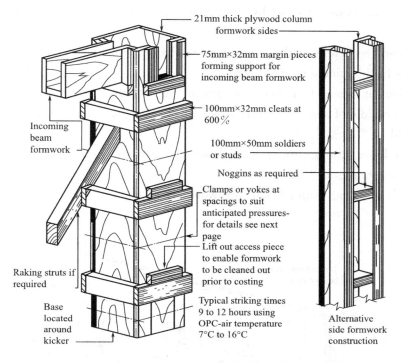

Figure 4-22 Typical column formwork details

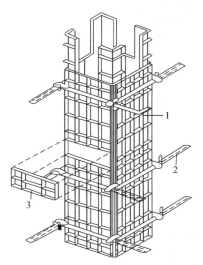

Figure 4-23 Steel formwork of column
1—Plain steel form; 2—Column hoop;
3—Board of concrete placing window

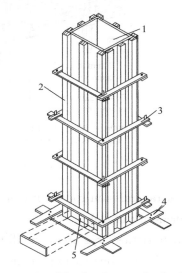

Figure 4-24 Wooden formwork of column
1—Poly wood; 2—Wooden minor ridge; 3—Column hoop;
4—Location wood frame; 5—Cleaning window

Formwork for beams consists of side board, soffit board, cleat, beam band and prop. The side board is 25mm thick, long batten and the soffit board can be 40~50mm thick.

161

(1) Space of props: decided by the dimension of the section of the beam, generally 0.8-1.2m.

(2) When the height of the beam is more than 0.7m, it should be strained by iron wire or bolts in the middle of side board.

(3) When the span of the beam is equal or more than 4m, the form work should be arched in the middle of span to prevent droop after placing concrete. The height of the arch (rise) should be 2‰~3‰ of the span.

(4) The brace among props: set a brace 0.5m above the ground or floor, upon it, the space of the braces is 2 m.

Details of a typical simple beam formwork are as shown as in Figure 4-25 and Figure 4-26.

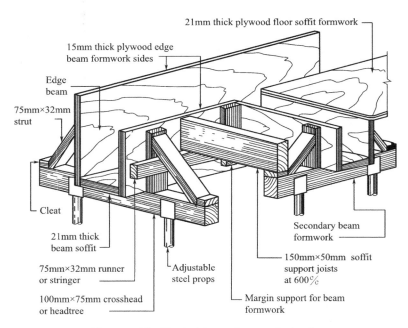

Figure 4-25 Typical beam formwork details

Forcet ransmission routes of construction load are shown as in Figure 4-27.

6. Formwork of slab

Characteristics: big area, small thickness. Side pressure is small, vertical pressure of concrete and the construction loads are the main pressure for formwork and sustain system.

The slab is continuous over a number of beams. The slab is supported on 2~2.5cm thick board. The board is supported on wooden battens which are laid between the beams at some suitable spacing and the batten is supported by props placed at about 60cm distance in order to reduce deflection.

Wooden formwork of beam and slab is as shown in Figure 4-28.

Chapter 4 Concrete Structure Engineering

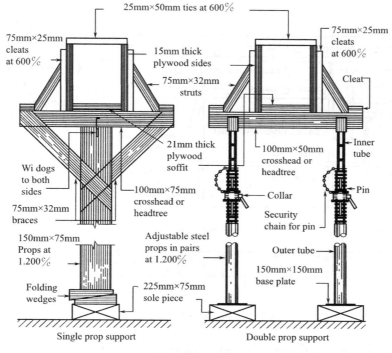

Figure 4-26 Typical simple beam formwork details

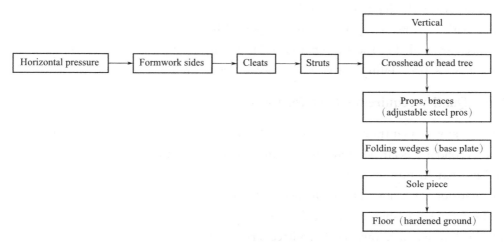

Figure 4-27 Force transmission routes of beam formwork

7. Formwork for stair

In constructing formwork for stairs, the landing is first set in position. The process is the same as that of floors. After landing has been set, two strings are tied to the landing and grounds (or upper) floor maintaining the width and the inclination of the flight. The soffit is most often prefabricated especially when the flight is short. The soffit boards are held underneath by timber measuring 100mm × 75mm placed at 300mm centers. The

163

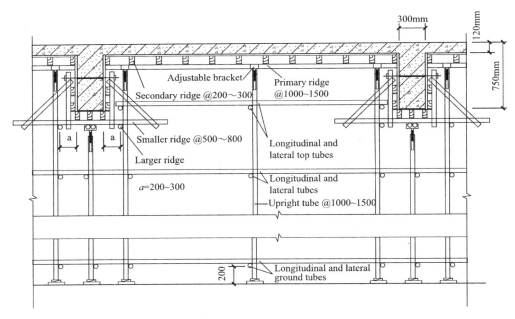

Figure 4-28 Formwork of beam and slab

prefabricated soffit is raised and its poison checked with strings.

The stringers are set in position, on these boards, the position for tread and risers are marked off with chalk. The face boards for risers are cut to the required height. They are then nailed to the stringers and are supported by cleats.

Braces are nailed to the side boards and to the protruding part of bearer. These prevent the side boards from falling apart when concrete is poured on due to vibration.

4.1.5 Quality requirements of formwork

For easily dismantling, the side board should cover the soffit board, the formwork for the secondary beam should not extent into main beam and the formwork for beam should not extent into column.

Dimension deviation shall be inspected for formwork after installation.

The formwork should be fixed tightly after erecting, settlement and deformation should be limited in allowance. Permitted errors of formwork for cast-in-place structure should accord to the following form, see Table 4-5.

Permitted errors of formwork for cast-in-place structure   Table 4-5

Number	Items		Permitted error(mm)
1	Position of axes		5
2	Level of the bottom board		±5
3	Internal dimension of the section	Foundation	±10
		Column, wall and beam	+4, −5

Continued

Number	Items		Permitted error(mm)
4	Uprightness	Height≤5m	6
		Height >5m	8
5	The difference of elevation of the surface of two adjacent boards		2

The embedded parts and preformed holes fixed on the formwork should not be forgotten, they should be fixed tightly and the position should be exact. The permitted errors should be in line with the following form, see Table 4-6.

Permitted errors for embedded parts and preformed holes   Table 4-6

Number	Items		Permitted error(mm)
1	Position of the embedded steel plate		3
2	Position of the embedded steel tube		3
3	Embedded bolt	Position of axes	2
		External length	+10
4	Central line of preformed holes		3
5	Preformed opening	Central line	10
		Interior dimension	+10

4.1.6 Removal of formwork

1. Dismantling

Forms and their supports should be removed as soon as possible to provide, but not until the concrete has reached sufficient strength to ensure structural stability and to carry both the permanent load and any construction loads that may be imposed on it.

Only when the concrete strength can ensure that the surface and edge angle are free from damage, the side formwork may be removed.

Bottom forms and their supports shall be removed after the concrete strength reaches the design requirements; where there are no specific requirements in the design, the compressive strength of concrete cubic specimens cured under the same condition shall be in according with those specified in Table 4-7.

Concrete strength requirments for bottom formwork removal   Table 4-7

Type of structure	Spans(m)	Percentage of the standard strength of concrete(%)
Slab	≤2	50
	>2≤8	75
	>8	100
Beam, arch	≤8	75
	>8	100

Continued

Type of structure	Spans(m)	Percentage of the standard strength of concrete(%)
Cantilever structure	≤2	75
	>2	100

The bottom supports of continuous formwork erecting among several storeys, the removal time shall be determined according to load distribution and concrete strength increase among the storeys of continuous formwork erecting.

For a multi-story building, when concrete is being cast on the upper floor, the props for under floor must not be removed; the props for the next under floor can be partly removed. If span of beam is ≥4m, some props should be reserved, the spaces should not more than 3m.

Spacing of upright tubes in the fast-removed support system shall not be greater than 2m. During the process of formwork removal, the upright tubes shall be kept and maintained jacking and supporting floor, and the concrete strength may be determined as thevalues for member span of 2m in Table 4-7.

2. Sequence

Removal sequence of formwork may be opposite to the erection sequence and may in the order from nonbearing formwork to load-carrying formwork and from top to bottom.

Dismantling formwork should be under guidance: which erected first should be removed last; which erected last should be removed first. Non-load-bearing structure should be removed first, load-bearing structure second.

When dismantling formwork for beam and slab, the side board of the beam can be dismantled first, then dismantle the formwork for slab; finally dismantle the soffit board of the beam. When dismantling the props for a large span beam, the sequence should be from midst to the two sides.

3. Notice

(1) When dismantling formwork, the worker should stand in a secure place.

(2) Don't overexert in dismantling formwork to avoid damage of the finish of the concrete and the formwork.

(3) The forms and supports removed shall not be thrown and shall be stacked in designated place in a scattered way and be cleared and transported in time.

(4) After formwork removal, the surface of members shall be cleaned, while the deformed and damaged parts shall be repaired.

4.2 Reinforcement engineering

In the reinforced concrete structure, the construction quality of the reinforced concrete project plays a key role in the quality of the structure, and the reinforced concrete project is a hidden project, after concrete is placed poured, the quality of the reinforcement cannot be checked. Therefore, from the acceptance of the raw materials, to a series of reinforcement processing and connection, till the end of the project, quality control must be strictly carried out to ensure the quality of the entire structure.

Illustration of reinforcements inside of a structure memberis as shown in Figure 4-29.

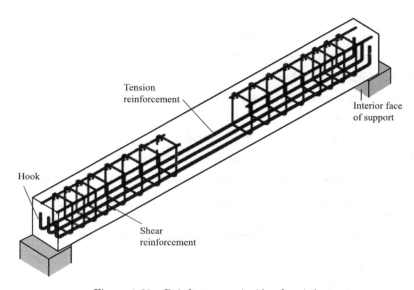

Figure 4-29　Reinforcement inside of a girder

4.2.1　Classification categories and acceptance of reinforcement

1. Categories

The reinforcements commonly used in civil engineering can be classified by the following aspects.

According to chemical component, they can be classified into carbon steel and ordinary low carbon alloy steel. According to the difference of carbon content, carbon steel can be further classified into low carbon content reinforcement (carbon content less than 0.25%), medium carbon content reinforcement (carbon content 0.25%~0.70%) and high carbon content reinforcement (carbon content 0.7%~1.4%). Ordinary low carbon alloy steel is made of low carbon steel and medium carbon steel with a small amount of alloying elements, such as titanium, vanadium, manganese, etc., the content of which generally doesn't exceed 3% of the total amount, and the steel with high strength and good comprehensive performance can be obtained.

According to mechanical properties, they can be classified into different grades of HPB300, HRB335, HRB400, HRB500, the higher the grade of reinforcement, the higher its strength and hardness, but the plasticity gradually reduced.

According to its rolling surface shape, they can be classified into smooth round reinforcement and deformed reinforcement (crescent rib, spiral rib and herringbone rib).

According to supply form, they can be classified into circular rebar (diameter not more than 10 mm) and straight bars (diameter of 12mm and above), the length of straight bars is generally 6~12m and can also be supplied according to the size required from buyer.

According to diameter, they can be classified intosteel wire (diameter 3~5mm), small diameter reinforced bar (6~10mm), medium diameter reinforced bar (12~20mm) and thick reinforced bar (diameter more than 20mm).

According to manufacturing technique, they can be classified into hot rolled rebar, cold-rolled ribbed wires and bars, cold-rolled and twisted bars, quenching and self-tempering ribbed steel bars, refined screw-thread steel bars.

1) Hot rolled rebar

Hot rolled rebar is formed by hot rolling and naturally cooled finished steel bar, it can be classified into hot rolled plain bars and hot rolled ribbed bars.

Mechanical properties of hot rolled rebar are as shown in Table 4-8.

Characteristic values of strength for ordinary steel bars (N/mm²) Table 4-8

Types	d (mm)	f_{yk}	f_{stk}
HPB300	6~22	300	420
HRB335, HRBF335	6~50	335	455
HRB400, HRBF400, RRB400	6~50	400	540
HRB500, HRBF500	6~50	500	630

There in:

f_{yk}——Characteristic value of yield strength of hot rolled rebar.

f_{stk}——Characteristic value of ultimate tensile strength of hot rolled rebar.

2) Cold-rolled ribbed wires and bars

Cold-rolled ribbed bars are formed from hot-rolled plain bars after being cold rolled, with three or two transverse ribbed evenly distributed along the length direction on the surface. They can be classified into CRB550, CRB650, CRB800, CRB970 and CRB1170, 5 grades.

CRB550 is used in ordinary reinforced concrete; other grades are used in pre-stressed concrete.

3) Cold-rolled and twisted bars

Cold-rolled and twisted bars is also called deformation steel bar. It is a continuous spiral steel bar with specified section shape and pitch which is straightened, rolled and twisted by special cold rolling and twisting machine.

It has high strength, sufficient plastic performance, and excellent bonding with concrete, generally, it can save more than 30% of steel consumption, has obvious economic benefits in engineering construction.

4) Quenching and self-tempering ribbed steel bars

Quenching and self-tempering ribbed steel bars, is the finished steel bar which is formed by hot rolled steel bar which then goes through water immediately, cooling down by surface control and tempering by using the residual heat in the steel bar core. The surface shape of the steel bar is crescent rib.

5) Refined screw-thread steel bar

The refined screw-thread steel bar is made by hot rolling method on the surface of the whole steel bar without longitudinal rib with connector for extension and nuts for end anchorage.

2. Acceptance of reinforcement

When rebar enters the construction site, there should be product qualification certificate, ex-factory inspection report and acceptance in batches according to variety, batch number and diameter.

The acceptance includes steel bar sign and appearance inspection, and rebar performance should be randomly sampling tested according the relevant stipulations.

The rebar performance test can be further classified into two categories of mechanical test and chemical composition test.

1) Appearanceinspection

All rebar shall be inspected for appearance.

Inspection content includes whether the rebar is flat and damage, whether the surface has cracks, oil and rust, etc..

The bent steel bar shall not reused as load-bearing rebar after knocking straight. There shall be no corrosion or contamination on the surface of the rebar which affects the strength and anchorage performance of the rebar.

The requirements for the appearance inspection of the commonly used rebar are as follows: the surface of hot rolled rebar shall not have cracks, scars and folds, the surface convex block shall not exceed the maximum height of the transverse rib, and the external dimensions shall meet the requirements; no visible cracks, scars or folds on the surface ofthe quenching and self-tempering ribbed steel bar, if there is a convex block, it shall not exceed the height of the transverse rib and the surface shall not be contaminated with oil. For cold-rolled and twisted bars, the surface shall be smooth, without cracks, folded interlayer, etc., and without indentation or pits with a depth of more than 0.2mm.

2) Performance inspection

The requirements stipulated in *Steel for the reinforcement of concrete-Part 1: Plain bars* GB/T 1499.1—2017, *Steel for the reinforcement of concrete-Part 2: Hot rolled ribbed bars* GB/T 1499.2—2018 and *Quenching and self-tempering ribbed steel bars for the reinforcement of concrete* GB/T 13014—2013 shall be complied with, take out the specimen for mechanical performance best, that is, re-inspection on site, its quality must meet the provisions of the relevant standard.

The inspection quantity is specified in the relevant standards, it shall be complied with.

As for structures with requirements for seismic resistance, the properties of longitudinal steel reinforcements shall meet the design requirements; where there are no specific requirements in the design, as for the frames and diagonal brace members (including stair flight) designed according to Aseismatic Grade Ⅰ, Ⅱ and Ⅲ, the longitudinal steel reinforcements shall adopt HRB335E, HRB400E, HRB500E, HRBF335E, HRBF400E or HRBF500E, while the strength and measured value of total elongation at maximum force shall meet the following requirements:

(1) The ratio of the measured value of tensile strength to that of yield strength of steel reinforcement shall not be less than 1.25.

(2) The ratio of the measured value of yield strength to the characteristics value of yield strength of steel reinforcement shall not be less than 1.30.

(3) The total elongation of steel reinforcement at maximum force shall not be less than 9%.

During construction, measures shall be taken to prevent mixing up, rusting or damage of steel reinforcements.

During construction, when brittle fracture, poor welding performance or obviously abnormal mechanical performance of steel reinforcement is observed, the batch of steel reinforcements shall be taken out of service and be examined by chemical composition inspection or other special inspections.

When it is impossible toaccurately judge the type and designation if steel reinforcements, such inspections as analysis on chemical composition and grain size shall be added to.

4.2.2 Reinforcement blanking

The steel reinforcement in the structure members shall be precisely cut off and reprocessed into various shapes according to the design drawings. For this reason, it is necessary to understand the concrete cover, relevant provisions such as steel bar bending, lap and hook, etc., and correct calculation method should be adopted to calculate the actual length according to the size in the drawing.

After the reinforcement being bended, its exterior surface have been extended and interior surface shrinkage, only the center line of the bent reinforcement has no changes. So the actual straight length of the reinforcement is exactly the length of the reinforcement center line.

Considering the exterior surface measurement methods used in construction site commonly and conveniently, the bending adjustment value should be calculated in advance for obtaining the length of its center line.

1. Length calculation

Length of rebar = sum of exterior dimensions of reinforcement-bending adjustments + hook length.

1) Exterior dimension

Exterior dimension = exterior dimension of structure member-thickness of concrete cover. The thickness of concrete cover shall comply with the stipulations in Table 4-9.

Minimum thickness of concrete cover for longitudinal streessed steel reinfrocement (mm)

Table 4-9

Environmental categories	Slab, wall, shell	Beam, column, rod
I	15	20
II a	20	25
II b	25	35
III a	30	40
III b	40	50

Note: (1) If the concrete grade is greater than C25, 5 mm should be added to the numbers in table above.
(2) Cushion should be set up for reinforcement concrete foundation, and the thickness of concrete cover of the reinforcement foundation should not be less than 40 mm measured from the bottom of the cushion top.

2) Bending adjustments

Exterior dimension:
$$2\left(\frac{D}{2}+d\right) \cdot \tan(\alpha/2) \tag{4-7}$$

Length of center line:
$$\left(\frac{D}{2}+\frac{d}{2}\right) \cdot \alpha \frac{2\pi}{360} = (D+d) \cdot \frac{\pi\alpha}{360} \tag{4-8}$$

Bending adjustment:
$$2\left(\frac{D}{2}+d\right) \cdot \tan(\alpha/2) - (D+d) \cdot \frac{\pi\alpha}{360} \tag{4-9}$$

There in:

D——Inner diameter of arc of bent steel reinforcement (Figure 4-30);

d——Diameter of the bent rebar (Figure 4-30);

α——Bent angle (Figure 4-30).

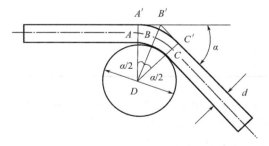

Figure 4-30 Diagram of bending adjustment

Inner diameter of arc of bent steel reinforcement shall meet the following requirements:

As for the plain steel bar, it shall not be less than 2.5 times of diameter of the steel bar.

As for Grade 335MPa and 400MPa ribbed steel bar, it shall not be less 4 times of diameter of the steel bar.

As for Grade 500MPa ribbed steel bar, it shall not be less than 6 times of diameter of the steel bar in diameter less than 28mm and shall not be less than 7 times of diameter of the steel bar in diameter of or above 28mm.

As for the upper longitudinal steel reinforcement of beam and outer longitudinal steel reinforcement of column at the joint on the top layer of frame structure, at the corner bending point of the joint, it should not be less than 12 times of diameter of the steel reinforcement in diameter less than 28mm and should not be less than 16 times of diameter of the steel reinforcement in diameter of or above 28mm.

At the bending point of stirrup, it shall not be less than diameter of the longitudinal steel reinforcement; when the longitudinal steel reinforcements are lapped bars or bundled bars at the bending point of stirrup, the inner diameter of stirrup shall be determined according to the actual configuration of steel reinforcement.

Bending adjustment calculation table is as shown in Table 4-10.

Bending adjustment (d)   Table 4-10

Inner diameter of arc of bent steel reinforcement $D(d)$ \ Bent angle	30°	45°	60°	90°
Plain round rebar: 2.5	0.29	0.49	0.77	1.75
Hot rolled ribbed steel bar of 335MPa and 400MPa: 5	0.30	0.54	0.90	2.29
Hot rolled ribbed steel bar of 500MPa with diameter<28mm: 6	0.31	0.56	0.95	2.50
Hot rolled ribbed steel bar of 500MPa with diameter≥28mm: 7	0.32	0.59	1.01	2.72
Rebar at joints of frame structure top layer with diameter <28mm: 12	0.35	0.69	1.28	3.79
Rebar at joints of frame structure top layer with diameter ≥28mm: 16	0.37	0.78	1.49	4.65

Exterior dimension: $2\left(\frac{D}{2}+d\right) \cdot \tan(\alpha/2)$; Length of center line: $(D+d) \cdot \frac{\pi\alpha}{360}$; Bending adjustment: $2\left(\frac{D}{2}+d\right) \cdot \tan(\alpha/2) - (D+d) \cdot \frac{\pi\alpha}{360}$.

There in:

D——Inner diameter of arc of bent steel reinforcement;

d——Diameter of the bent rebar.

The length of straight segment of bent longitudinal steel reinforcement shall meet the design requirements and relevant requirements of the current national standard *Code for Design of Concrete Structure* GB 50010—2010.

3) Hook length

When the end of the plain steel bar is made into 180° hook, the length of straight segment of bend hook shall not be less than 3 times of diameter of the steel bar.

End of stirrup and tie bar shall be made into hooks according to the design requirements and shall meet the following requirements:

Chapter 4 Concrete Structure Engineering

As for general structural members, the bend angle of stirrup hook shall not be less than 90° and the length of straight segment after bending shall not be less than 5 times of diameter of the stirrup.

As for structural members with requirements for seismic resistance or special design requirements, the bend angle of stirrup hook shall not be less than 135°and the length of straight segment after bending shall not be less than 10 times of the stirrup or 75mm, whichever is larger.

Lap length of circular stirrup shall not less than its tension anchorage length and both ends shall be made into the hooks not less than 135°. The length of straight segment after bending shall not be less than 5 times of diameter of the stirrup for general structural members, which shall not be less than 10 times of diameter of the stirrup or 75mm, whichever is larger, for structural members with requirements for seismic resistance.

When the tie bar is used as the one between single-leg stirrup or beam waist stirrup in the beam and column compound stirrup, the bend angle of hooks at both ends shall not be less than 135°, and the length of straight segment after bending shall meet the relevant requirements of above for stirrups. When the tie bar is used as the one in such as shear wall and floor slab, the hooks may be made into 135°at one end and 90°at the other end, and the length of straight segment after bending shall not be less than 5 times of diameter of the tie bar.

Hook length calculation table is as shown in Table 4-11.

Hook length (d) Table 4-11

Grade	Type	D (d)	l_0 (d, mm)	Bent angle of the hook		
				90°	135°	180°
HPB300	Plain round rebar hook	2.5	3	3.50	4.87	6.25
	Stirrup without earthquake		5	5.50	6.87	8.25
	Stirrup with earthquake(d=6mm)		75(mm)	—	86.24(mm)	94.49(mm)
	Stirrup with earthquake(d=6.5mm)		75(mm)	—	87.18(mm)	96.11(mm)
	Stirrup with earthquake($d \geqslant$8mm), tie rebar		10	—	11.87	13.25
	Column stirrup without earthquake		5	—	6.87	—
	Column stirrup earthquake		10	—	11.87	—
HRB335, HPB400	Stirrup without earthquake	5	5	6.21	8.57	10.92
	Stirrup with earthquake, d=6mm		75(mm)	—	96.41(mm)	63.42(mm)
	Stirrup with earthquake($d \geqslant$8mm), tie rebar		10	—	13.57	15.92
	Column stirrup without earthquake		5	—	8.57	—
	Column stirrup earthquake		10	—	13.57	—

Hook length: $\left[2\left(\dfrac{D}{2}+d\right) \cdot \tan(\alpha/2) - (D+d) \cdot \dfrac{\pi\alpha}{360}\right] + l_0$

There in:

l_0——Length of straight segment after bending.

2. Ingredients list and label board of reinforcing rebar

1) Ingredients list

Ingredients list of reinforcing rebar is numbered according to its variety, specification, shape size and quantity in the design drawing, the blanking length is calculated and expressed in the form of table.

Ingredients list is the basis of reinforcing processing; as well as that of proposing material plan, issuing task list and limited material requisition. Reasonable ingredients can not only save steel, but also simplify the construction operation.

When preparing the ingredients list of reinforcement rebar, firstly, calculate the blanking length according to the shape and specification of each numbered reinforcing rebar; and then calculate the total length of each numbered steel bar according to its number; at last, sum up the total length of each specification of steel bar and obtain the total weight.

When the steel bar need to be extended, the layout of the joints shall be considered according to the supply situation of raw materials and the joint form, required length of the joint shall be added to the blanking length.

The sequences of reinforcement ingredient list are as follow:

Familiar technical drawings→draw the reinforcement diagram→calculate the blanking length of each specification→fill in and prepare steel bar ingredient list→fill in the label board.

2) Label board

Besides of that ingredient list of steel bar, in the reinforcement engineering construction, label board is also needed to be the basis of reinforcement processing and binding. Label board has size of 100mm×70mm and made of thin wooden board or fiber board, and is transferred successively in each process of rebar processing and is finally attached to the processed rebar as a sign. The label board must be checked strictly in the construction so as to keep its accurate and avoid rework wastes.

Example 4-1: Calculation of reinforcement length

A reinforcement concrete frame structure with seismic precautionary category for building structure of Grade 3, frame column concrete strength of grade C40, frame beam concrete strength of grade C35. The longitudinal reinforcement of frame column is $\Phi 25$ and stirrup is $\phi 10$. Concrete cover of frame beam is 20mm. Frame beam longitudinal reinforcement should reach to the interior side of the frame column's exterior reinforcement; straight reinforcement supplying length is 12m; straight threaded sleeve method is adopted in reinforcement connections of column, beam. The anchorage length of the beam structure reinforcement at the support is $15d$ (d is diameter of the structure reinforcement). Diagram of reinforcement drawings of the frame beam is as show in Figure 4-31.

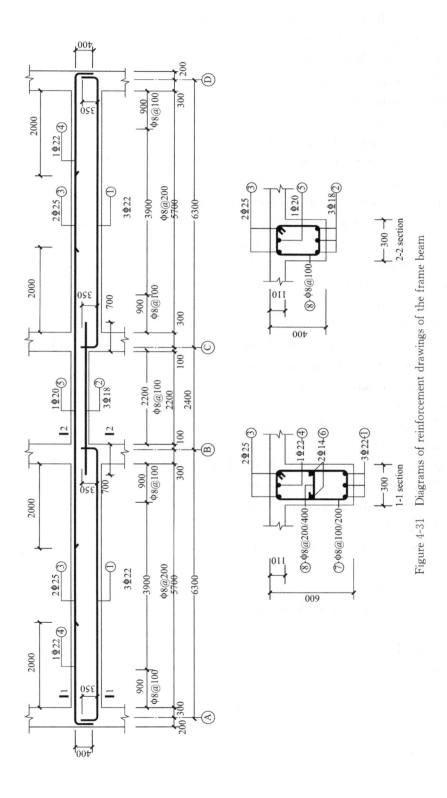

Figure 4-31 Diagrams of reinforcement drawings of the frame beam

Complete the following calculation, the result preserved as an integer.

(1) Blinking length of reinforcement ① (Φ22)
Horizontal exterior dimension: $(6300+100+200)-(20+10+25)\times 2=6490$mm
Vertical exterior dimension: 350mm
Bending adjustment: $2.29\times 22\times 2=100.76$mm
Blinking length: $6490+350\times 2-100.76=7089$mm

(2) Blinking length of reinforcement ② (Φ18)
Horizontal exterior dimension: $2400-100\times 2+700\times 2=3600$mm
Blinking length: 3600mm

(3) Blinking length of reinforcement ③ (Φ25)
Horizontal exterior dimension: $(6300+2400+6300)+200\times 2-(20+10+25)\times 2$
$=15290$mm
Vertical exterior dimension: 400mm
Bending adjustment: $2.29\times 25\times 2=114.50$mm
Blinking length: $15290+400\times 2-114.50=15976$mm

The blinking length is 15.976m, is greater than that of supply length of 12m, thus, the reinforcement need to be extended. One piece of it can be $15976/2=7988$mm, then the connection spot stays at the up middle of the span of B and C.

(4) Blinking length of reinforcement ④ (Φ22)
Horizontal exterior dimension: $(2000+300+200)-(20+10+25)=2445$mm
Vertical exterior dimension: $400-20-8=372$mm
Bending adjustment: $2.29\times 22=50.38$mm
Blinking length: $2445+372-50.38= 6490+350\times 2-100.76=2767$mm

(5) Blinking length of reinforcement ⑤ (Φ20)
Horizontal exterior dimension: $2000\times 2+300\times 2+2400=7000$mm
Blinking length: 7000mm

(6) Blinking length of reinforcement ⑥ (Φ14)
Horizontal exterior dimension: $5700+15\times 14\times 2=6120$mm
Blinking length: 6120mm

(7) Blinking length of reinforcement ⑦ (Φ8)
Exterior dimension: $[(300-20\times 2)+(600-20\times 2)]\times 2 =1640$mm
Hook length: $11.87\times 8\times 2=189.92$mm
Blinking length: $1640+189.92=1830$mm

(8) Number of reinforcement⑧ (Φ8) of span of AB
Dense part: $(900-50)/1+1=10$
Normal part: $3900/200+1-2=19$
Total: $10\times 2+19=39$

4.2.3 Reinforcement substitution

In the process of construction, the variety, grade or specification of the rebar must be adopted according to the design requirements. However, due to the shortage of supplying, the variety, grade or specification of the rebar cannot meet the design requirement, then, the substitution should be done in order to ensure the quality and progress of the construction.

When the reinforcement is required to be substituted, the design change documents shall be prepared.

1. Principle and method

(1) Equivalent strength substitution:

When the structure members are controlled by strength, the reinforcement can be substituted according to the strength equivalent principle, as shown in Formula (4-10).

$$\begin{cases} A_{s1} f_{y1} \leqslant A_{s2} f_{y2} \\ n_1 d_1^2 f_{y1} \leqslant n_2 d_2^2 f_{y2} \\ n_2 \geqslant \dfrac{n_1 d_1^2 f_{y1}}{d_2^2 f_{y2}} \end{cases} \quad (4\text{-}10)$$

There in:

d_1, n_1, f_{y1}——Diameter, number and design strength of the original design reinforcement;

d_2, n_2, f_{y2}——Diameter, number and design strength of the reinforcement to be replaced.

(2) Equivalent area substitution:

When the structure members are controlled by minimum reinforcement ratio, the reinforcement can be substituted according to the area equivalent principle, as shown in Formula (4-11).

$$A_{s1} = A_{s2} \quad (4\text{-}11)$$

There in:

A_{s1}——Calculated area of the original design reinforcement;

A_{s2}——Area of the reinforcement to be substituted.

(3) When structural members are controlled by crack width or deflection, the checking of crack width or deflection should be carried after substitution.

2. Notes in substitution

(1) The design change document shall be prepared when the variety, grade or specification of rebar need to be replaced.

(2) For some certain important structure components, such as crane beam, lower chord of truss, HRB335 and HRB400 might not be replaced with HP235 smooth round steel rebar.

(3) After replacement completed, the structural requirements of the reinforcement shall be satisfied, such as the minimum diameter, spacing, numbers and anchorage

length of the reinforcement.

(4) Within the same section, replacement rebar of different types and diameters can be used at the same time, but the tension difference of each rebar should not be too large (for the same type of rebar, the diameter difference is generally not greater than 5mm) if order to avoid uneven stress on components.

(5) In order to ensure the strength of normal section and oblique section, the longitudinal reinforcement and bending reinforcement should be replaced respectively.

(6) When the eccentric compression member (such as the frame column, column of workshop with crane, upper chord) or the eccentric tension member is replace with its reinforcement, it should be replaced separately according to the stress surface (compression or tension) rather than that of whole section.

(7) When the structure member is controlled by the crack width, if the small diameter rebar is substituted for the large diameter rebar, or the lower strength rebar is substituted for the higher strength rebar, the crack width may not be checked.

(8) After the replacement of rebar, sometimes, due to the increase in the diameter of the rebar or the numbers of the rebar, the effective height of the structure members section h_0 value will be reduced, then, the section strength must be checked in this situation.

(9) During the construction, if the longitudinal bearing force reinforcements in original design have to be replaced by those with higher strength grade, the conversion shall be made according to equal tensile capacity design values of such reinforcements, and shall also comply with requirements of minimum reinforcement ratio.

4.2.4 Reinforcement Fabrication

Rebar processing includes: derusting, straightening, cutting, connecting, bending and forming.

1. Derusting

Due to improper or long period storage of reinforcement, its surface will form a layer of rust, serious rust will affect the bond effect between rebar and concrete, and affect the effect of the use of components, so, they should be cleaned before use.

The derusting of rebar can be completed by means of cold drawing (rebar diameter less than 12mm), motor derusting machine, manual derusting (with wire brush), sandblasting and pickling derusting.

The surface of steel reinforcement shall be cleaned up before fabrication. Steel reinforcement with granular of flaky old rust damaged surface shall not be used.

2. Straightening

Manual straightening, mechanical straightening and cold drawing straightening methods can be adopted.

Manual straightening: a small hammer or winching can be used for the rebar with diameter less than 12mm. The large diameter rebar has a bit bending commonly, manual operation straightening methods may be selected by the helps of the fixed steel column on working platform.

Mechanical straightening: small diameter rebar are often straightened by machines, such as steel straightening machine, double end steel straightening machine or CNC steel straightening cutting machine.

The mechanical straightener has three functions of derusting, straightening and cutting. The three processes can be completed in one operation. Where CNC rebar straightening and cutting machine can control the length of rebar in millimeter accuracy and count the rebar number automatically by means of its photoelectric length measuring system and photoelectric counting device.

Cold drawing straighteningby winch can also be adopted for large diameter round rebar, the rust falls off due to the deformation of the rebar during tensioning process, but the elongation shall be trolled strictly in cold drawing straightening process.

Steel reinforcement should be straightened by mechanical equipment or with cold drawing method. In the case of straightening by mechanical equipment, the straightening equipment shall not be with extension function. In the case of straightening with cold-drawing method, the cold drawn rate of HPB300 plain steel bar should not be greater than 4% and the cold drawn rate of HRB335, HRB400, HRB500, HRBF335, HRBF400 and HRBF500 ribbed steel bar should not be greater than 1%. During the steel reinforcement straightening process, the transverse rib of ribbed steel bar shall not be damaged. Straightened steel reinforcement shall be straight and free from partial bending.

3. Cutting

Manual hydraulic cutter and rebar cutting machine are often used for steel bar cutting process. The former has the advantages of small in size, light weight and easy to be carried, it can cut off rebar with diameter less than 16mm; the latter can cut a variety of diameter of rebar with diameter of 6~40mm.

4. Bending

Steel reinforcement is often bent into a certain shape according to design requirement. The bending and forming of rebar generally adopts rebar bending machine and four-end rebar bending machine (mainly used for bending stirrups). Also, it is possible to use a hand wrench to bend the small diameter rebar as well as bend the large diameter rebar by the helps of chuck and wench head in the case of lack of bending machine.

For rebar with complex shape, each bending point should be marked according to the size marked on thelabel board before process.

4.2.5 Reinforcement connection

Rebar will be supplied in straight strips when diameter of rebar is greater than 12mm, the straight length is generally 6~12mm, which brings the inevitable rebar connecting in the construction of reinforced concrete structure.

Connection of steel reinforcements shall be selected according to the design requirements and construction conditions.

At present, there are three kinds of connecting methods: mechanical connection, welding connection and binding connection. Mechanical connecting method has been widely used in the connection of large diameter rebar due to its advantages of reliable connection, free from the influence of climate and fast in speed; weldingand bending methods are the traditional connecting methods, comparing with the binding connecting method, the welding connecting method can save steel, improve the mechanical performance of the structure, ensure the quality of the project and reduce the construction cost, should be adopted priority.

1. Welding Connection

Welding connection is a method to connect steel bars by welding technology, which is widely used. However, welding is a special technology requiring special training for welders, as well as holding a certificate on the post; welding construction is greatly affected by climate and electronic current stability. Its connection quality is not as reliable as mechanical connection. Flash butt welding, resistance spot welding, electric-arc welding, electro-slag welding and submerged arc pressure welding are commonly used in steel welding connection process.

1) Flash butt welding

Flash butt welding is a method of connecting two rebar along its axis so that the end face of the rebar is in connected. Flash butt welding shall be carried out on the butt-welding machine. When operation, make the end faces of two section of rebar in contact and convert electrical energy into heat energy through low voltage and strong current. After the rebar is heated to a certain temperature, the two rebar are welded together by pressing through axial direction and the butt welding connection formed after it cools.

Flash butt welding has the characteristics of low cost, good quality, high efficiency, and can be applied to all kinds of rebar, therefore, it is widely used. Schematic diagram of butt-welding is as shown in Figure 4-32.

2) Resistance spot welding

Resistance spot welding is that of the vertical overlapping of the two batches of rebar, placed between the two electrodes, pressured and clamped, then, electrify the connected point to produce resistance heat, steel materials melt and form a tight connection point by means of pressing, strong connection point obtained after condensation as shown in Figure 4-33.

Resistance spot welding methods is used for welding steel mesh or skeleton.

3) Electric-arc welding

Arc welding is to use arc welding machine to produce high temperature electric arc

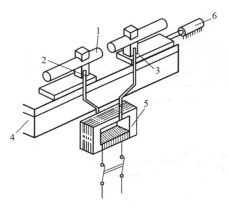

Figure 4-32 Schematic diagram of flash butt-welding

1—Rebar; 2—Fixed electrode; 3—Movable electrode; 4—Engine base; 5—Transformer; 6—Pressing machinery

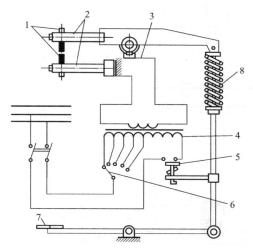

Figure 4-33 Schematic diagram of resistance spot welding machine

1—Electrode; 2—Electrode arm; 3—Secondary transformer winding; 4—Primary transformer winding;
5—Circuit breaker; 6—Adjusting switch of transformer; 7—Pedal; 8—Pressing mechanism

between the welding rod and the welding piece, so that the welding piece and welding rod within arc combustion range are melted, after solidification it will form a weld or joint, in which the electric arc refers to the strong and lasting discharging phenomenon between the welding rod and the welding piece metal in the air medium. There are two kinds of arc welding machines used in arc welding: AC arc welding machine and DC arc welding machine.

Arc welding is widely used inrebar welding, steel skeleton welding, steel and steel plate welding, assembled reinforced concrete structure joint welding and various steel structure welding.

As for the extension of the rebar, there are different types as shown in Figure 4-34.

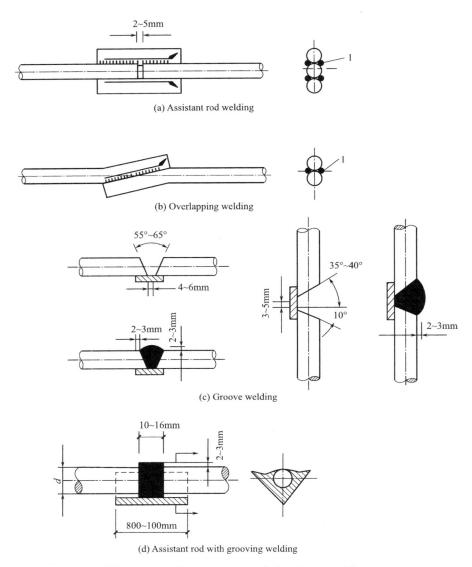

Figure 4-34　Connector type of electric arc welding

4) Electric-slag welding

Electro-slag welding is to use the resistance heat generated by the electrical current through the slag pool to melt the ends of the steel bar, and then apply pressuring through the axial direction of rebar. It is mainly used for extending grade HRB300, HRB335, HRB400 vertical or oblique (incline within 4∶1) rebar with diameter of 14～40mm in cast-in-place structure.

This welding method has the advantages of simple operation, good working conditions, high work efficiency and low cost. It can save more than 80% of the electric power than arc welding connection, save 30% of the rebar than binding connection, assistant rod welding and overlapping welding, and improve the work efficiency by 6～10 times. Schematic diagram of electrode welding is as shown in Figure 4-35.

Flash butt welding should be adopted for welding sealed stirrup, beside gas pressure welding or single welded lap splice may also be adopted. During the welding process, the special equipment should be adopted. Fabrication length for welding sealed stirrup and each socket fabrication shall be determined according to welding procedure. Welding points for welding sealed stirrup shall be arranged according to the following requirements:

(1) There shall be one welding point on each stirrup. The welding point should be in the middle of a certain side of polygon stirrup. The distance between the welding point and the stirrup bending point should not be less than 100mm.

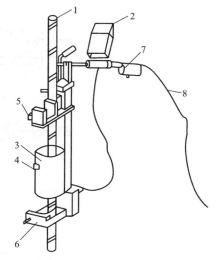

Figure 4-35 Schematic diagram of electrode welding
1—Rebar; 2—Monitor device; 3—Flux box;
4—Flux box snap ring; 5—Movable clamp;
6—Fixed clamp; 7—Operating handle;
8—Control cable

(2) Welding point of stirrup for rectangular column should be arranged on the short side of the column; welding point of stirrup for equilateral polygon column may be arranged on any side; welding point of inequilateral polygon column stirrup shall be arranged on different sides.

(3) Welding point of stirrup for beam shall be arranged on the top edge or the bottom edge.

2. Mechanical connection

Rebar mechanical connection has many advantages such as: simple equipment, easy operation technology, higher construction speed; its construction is not affected by weather conditions, especially in flammable, explosive, safe and reliable operation at high altitude and other construction conditions.

Although the cost of mechanical connection is higher, but its comprehensive economic benefits and technical effects are significant, it has been widely used in cast-in-situ long span structures, high-rise buildings, bridges, hydraulic structures and other projects for the connection of larger diameter rebar.

The mechanical connection methods of rebar include sleeve extrusion connection and thread sleeve connection as shown in Figure 4-36~Figure 4-39.

3. Binding lap splices

Rebar lap splice bind two rebar together by means of 20~22# galvanized iron wire or dedicated annealed wire, the process is simple, high efficiency and don't need connection equipment. However, it has high amount of rebar consumption because of that a certain

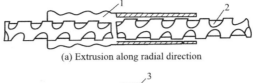

(a) Extrusion along radial direction

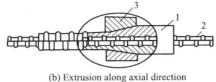

(b) Extrusion along axial direction

Figure 4-36　Sleeve extrusion connection
1—Steel sleeve; 2—Ribbed rebar; 3—Compression molding

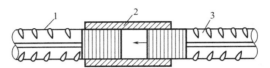

Figure 4-37　Straight thread sleeve connection
1—Rebar connected; 2—Straight thread sleeve; 3—Rebar to be connected

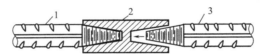

Figure 4-38　Tapered thread sleeve connection
1—Rebar connected; 2—Tapered thread sleeve; 3—Rebar to be connected

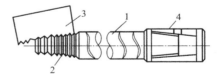

Figure 4-39　Check of tapered thread connection
1—Rebar; 2—Tapered thread on surface of rebar; 3—Tooth shape gauge; 4—Snap gauge

length is needed at lap splices section, also, its mechanical performance is lower than that of mechanical connection and welding connection.

The binding lap splices shall not be adopted for longitudinal load-carrying steel reinforcement in the axial tension and small eccentric tension members; the diameter of tensile steel reinforcement should not be larger than 25mm and the diameter of compressive steel reinforcement should not be larger than 28mm if the lap splices are adopted for steel reinforcement in other members.

The connection joint of load-carrying steel reinforcement in the concrete structure should be arranged at the place where the stressed force is relative small. The joints on a same load-carrying steel reinforcement should be arranged as less as possible. The connection joints should not be arranged for longitudinal load-carrying steel reinforcement in important members and key force transfer positions of the structure.

In the lap length of the joint, at least three points should be tied.

4.2.6 Reinforcement binding and installation

The processed rebar can be transported to the construction site for binding. The rebar grade, diameter, shape, size and quantity shall be checked to see whether they are consistent with the ingredients list and label board before binding, only if there are no any errors, can the binding process start. 20~22# galvanized iron wire are generally adopted in binding process.

For the steel mesh of foundation, slab and wall, all intersecting points of two row rebar in surroundings should be bound and fastened, where in the middle area, jumping and staging way may be adopted, both ways should ensure that there are no any displacement of rebar; all intersecting points should be bound and fastened for two-way forces of steel mesh.

For column, the stirrup joints (hooks overlapping spot) should be alternating binding in the longitudinal reinforcement along the four corner of the section. The intersection spot of stirrup corner and the longitudinal reinforcement should be bound tightly, the intersection spot of stirrup straight parts may be alternating binding between the longitudinal reinforcement.

For beam, the stirrup joints (hooks overlapping spot) should be staggered arrangement on the upper two longitudinal rebar.

In order to control the concrete cover, precast cement mortar pad block is often adopted between the reinforcement and formwork, the thickness of pad block is exactly the same as that of the concrete cover, plastic clamp could be adopted too for controlling the concrete cover.

Pad block or plastic clamp are generally arranged in a quincunx pattern with spacing not exceeding 1m. When there are double or multiple rows of rebar in the beam, short rebar with a diameter greater than 25mm should be padded between the two rows of rebar to maintain its spacing between. The steel bracing supports should be provided to maintain the position of the upper reinforcement rebar mesh with the spacing not more than 1 m apart in the case of that there are double layer rebar mesh of the foundation and slab. Special attentions should be paid for the cantilever slab such as canopy and balcony, the position of the upper negative bending moment reinforcement should be strictly controlled to avoid the cantilever slab fracture after formwork removal. When double layer rebar mesh is used in wall, bracing rods should also be set to maintain the distance between the two layers of reinforcing rebar, bracing rod can be made of steel with diameter of 6~10mm.

The binding process of the reinforcement shall be matched with the formwork erection. Rebar binding process should be completed before the formwork for column and wall. As for beam, generally, binding operation can start from the bottom formwork of the beam at first, then formwork of the one or two of its side formwork process starts; when the height of the beam is small, binding process can be completed at overhead of the beam by suspending and holding method, then falling down to the requiring position. Slab reinforcement binding can start after the completion of the slab formwork.

In order to accelerate the construction speed, making the single rebar into a steel mesh or frame by pre-binding, then transmitting and installation on-site.

4.2.7 Reinforcement splice and fixing

1. Splice of steel reinforcements should be arranged at the place where the bearing stress is smaller; as for structure with requirements for seismic resistance, in the densified portion of stirrup in beam end and column end, splices of steel reinforcements should not be arranged and lap splices shall not be adopted. The same longitudinal reinforcement should not be arranged with two or more splices. The distance from the splice end to the bending starting-point of the steel reinforcement shall not be less than 10 times of diameter of the steel reinforcement.

2. Mechanical splices of steel reinforcements shall meet the following requirements:

1) The operating personnel for fabrication joints of steel reinforcements shall receive the professional training and be qualified for their jobs; the fabrication of splices of steel reinforcements may be carried out after qualified in the process inspection.

2) Thickness of concrete cover at mechanical splices should meet the provisions on the minimum thickness of the concrete cover for longitudinal steel reinforcements as stated in the current national standard *Code for Design of Concrete Structure* GB 50010—2010 and also shall not be less than 15mm. The transverse clearance between splices should not be less than 25mm.

3) After installation of screwed splices, the torsional moment shall be checked and tightened with special torque wrench. Fluctuation range of impression diameter of squeezed splice shall be controlled within the allowable fluctuation range and shall be inspected with the special gauge.

4) Application scope, process requirements, sleeve materials and quality requirements of mechanical splices shall meet the relevant requirements of the current professional standard *Technical Specification for Mechanical Splicing of Steel Reinforcing Bar* JGJ 107—2016.

3. Welding of steel reinforcements shall meet the following requirements:

1) The welders engaging in welding of steel reinforcements shall possess the reinforcement welder qualification test certificates and operate according to the scope specified in the certificates.

2) Before welding of steel reinforcements, the welders involved in such welding shall perform welding test in site and may start to weld after qualification. During the process of

welding, in the case of change of destination and/or diameter of reinforcement, the welding test shall be carried out again. The materials, equipment, accessories and operational conditions applied in welding tests should be consistent with those in actual construction.

3) As for find grain hot-rolled steel bars and ordinary hot-rolled steel bars in diameter greater than 28mm, the welding parameters shall be determined according to tests; remained heat treatment ribbed steel bars should not be welded.

4) Electroslag pressure welding shall be only applicable to splices of vertical steel reinforcements in such members as column and wall.

5) Application scope, process requirements, electrodes and flux selection, welding operation as well as quality requirements of welded splices of steel reinforcements shall meet the relevant requirements of the current professional standard *Specification for Welding and Acceptance of Reinforcing Steel Bars* JGJ 18—2012.

4. When mechanical splices or welded splices are adopted for longitudinal steel reinforcements, the arrangement of splices shall meet the following requirements:

1) The splices set in the same member should bestaggered in batches.

2) Splice connecting region shall be $35d$ in length and shall not be less than 500mm, where d is the smaller diameter of the two interconnected steel bars. The splices whose midpoints are within the splice connecting segment shall belong to the same connecting segment.

3) Within the same connecting segment, the splice area percentage of longitudinal steel reinforcements is the ratio of the sectional area of longitudinal steel reinforcements with splices and that of all longitudinal steel reinforcements; the splice area percentage of longitudinal steel reinforcements shall meet the following requirements:

(1) As for the tensile splice, it should not be greater than 50%; as for the compression splice, it may be unrestricted.

(2) As for the mechanical tension splices in slabs, walls and columns as well as tension splices at connection of members in precast concrete structure, it may increases according to actual conditions.

(3) As for structural members directly bearing dynamic loads, welded splices should not be adopted; when mechanical splices are adopted, it shall not exceed 50%.

5. When binding lap splices are adopted for longitudinal steel reinforcements, the arrangement of splices shall meet the following requirement:

1) The splices set in the same member should be staggered in batches. Transverse clearance (s) of each splice shall not be less than the diameter of steel reinforcement or 25mm.

2) The length of splice connecting segment shall be 1.3 times of splicing length, the splices whose midpoints are within the splice connecting segment shall belong to the same connecting segment. The splicing length may be calculated according to the smaller

diameter of the two interconnected steel bars. The minimum splicing length of longitudinal steel reinforcements shall meet the requirements of Table 4-12.

Minimum splicing length of longitudinal tensile reforcement   Table 4-12

Reinforcement type		C20	C25	C30	C35	C40	C45	C50	C55	≥60
Plain reinforcement	Grade 300	48d	41d	37d	34d	31d	29d	28d	—	—
Ribbed reinforcement	Grade 335	46d	40d	36d	33d	30d	29d	27d	26d	25d
	Grade 400	—	48d	43d	39d	36d	34d	33d	31d	30d
	Grade 500	—	58d	52d	47d	43d	41d	39d	38d	36d

Note: d refers to the diameter of the lapped bar. As for two bars in different diameters, there splicing length shall be calculated according to the smaller diameter of the bar.

3) Within the same connecting segment, the splice area percentage of longitudinal steel reinforcement is the ratio of the sectional area of longitudinal steel reinforcement with splices and that of all longitudinal steel reinforcementsas shown in Figure 4-40; splice area percentage of longitudinal compressive reinforcement may be unrestricted; splice area percentage of longitudinal tensile reinforcement shall meet the following requirements:

(1) As for such members as beams, slabs and walls, it should not exceed 25%; as for the foundation raft slabs, it should not exceed 50%.

(2) As for column members, it should not exceed 50%.

(3) When increasing the splice area percentage is necessary and required; as for beam members, it shall not be larger than 50%; as for other members, it may increase properly according to the actual conditions.

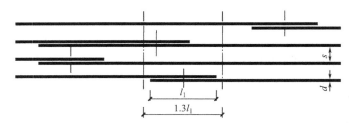

Figure 4-40 Connection area of reinforcing bar binding lap joints and the surface percentage of joints
Note: Lapping reinforcing bar at the same connecting area of lap joints indicated in the drawings are two reinforcing bar, when the diameter of each reinforcing bar is the same, the joint surface percentage is 50%.

6. As for such members as beams and columns, stirrup shall be configured within the splicing length range of the longitudinal steel reinforcements according to the design requirements and shall also meet the following requirements:

1) The diameter of stirrup shall not be less than 25% of the relevant large diameter of the lapped bars.

2) The spacing between stirrup in the tensile splice segments shall not be greater than 5 times of thesmaller diameter of lapped bars or 100mm.

3) The spacing between stirrup in the compressive splice segments shall not be greater than 10 times of the smaller diameter of lapped bars or 200mm.

4) When the diameter of longitudinal steel reinforcement in columns is greater than 25mm, two pieces of stirrup shall be arranged respectively within the range of 100mm beyond the two end faces of lap splice and the distance between stirrups in each region should be 50mm.

7. The binding of steel reinforcements shall meet the following requirements:

1) Binding lap splices of steel reinforcements shall be fastened tightly with wires at the center and both ends of the splices.

2) As for steel reinforcement cage in walls, columns and beams, all the intersection points of reinforcement mats on each vertical face shall be bound; as for slabs, all the intersection points of reinforcement mats of the upper part shall be bound and intersection points of reinforcement mats at bottom may be bound in a staggered way except for the peripheral part.

3) Stirrup hooks and welding points for weld closed stirrups in beams and columns shall be arranged in a staggered way along longitudinal steel reinforcements.

4) Longitudinal steel reinforcements of constructional columns should be bound synchronously with structural structures.

5) As for stirrup in beams and columns, steel reinforcements distributed horizontally in walls and steel reinforcements in slabs, the start distance to the edges of members should be 50mm.

8. Arrangement of steel reinforcement at the joints of members shall meet the design requirements. When there are no specific requirements in the design, position of steel reinforcements in major loaded members and those along main loaded direction in members shall be guaranteed. Longitudinal steel reinforcements of beams at joints of frame should be placed inside the longitudinal steel reinforcements of columns. When the elevation is the same at the bottom of girders and secondary beams, steel reinforcements of lower part of secondary beams shall be placed above those of girders. The steel reinforcements distributed horizontally in shear walls should be placed in the outside part and be bent and anchorage at the end of walls.

9. During the fixing process, the steel reinforcements shall be fixed with locating pieces and special locating pieces should beadopted. Locating pieces shall be possessed of sufficient load-carrying capacity, rigidity, stability and durability. The quantity, spacing and fixation method of locating pieces shall ensure that the position deviation of reinforcements meets the requirements of relevant current national standards. Metal locating pieces should not be placed within the concrete cover of frame beams and columns.

10. During the fixing process of reinforcement, when steel reinforcements are to be welded according to the requirements of construction operation, relevant requirements of the current relevant professional standard *Specification for Welding and Acceptance of*

Reinforcing Steel Bars JGJ 18—2012 shall be met.

11. When combined stirrups are adopted, the stirrup periphery shall be closed. Closed stirrups should be adopted for combined stirrups of beam members; single-leg stirrups may also be adopted in the case of odd legs; single-leg stirrups may be adopted partially in combined stirrups of column members.

12. Measures shall be taken in reinforcements fixing process to prevent steel reinforcements from being polluted by the release agent on the internal surface of forms and molds.

4.2.8 Quality control

1. After reinforcement straightening, the mechanical properties and weight deviation per unit length shall be checked. The check may be disregarded for steel reinforcements straightened by mechanical equipment without extension functions.

2. Dimension deviation shall be checked after reinforcement fabrication; type, grade, specification, quantity and position of steel reinforcements shall be checked after reinforcement fixing.

3. Quality control for splice of steel reinforcements shall meet the following requirements:

1) Process inspection shall be carried out before welding and mechanical splice of steel reinforcements. Effective reports for splice types shall be checked for mechanical splice.

2) Appearance inspection shall be carried out for all the welded splices and mechanical splices of steel reinforcements; the splicing length shall be randomly inspected for lap splice.

3) Tightening torque values shall be randomly inspected for screwed splices.

4) During welding of reinforcement, welders shall carry out self-checking timely. In the case of welding defects and abnormal phenomenon observed, the cause shall be found out and measures shall be taken to timely solve problems.

5) Splice percentage of steel reinforcements shall be checked.

6) Mechanical splices and welded splices of steel reinforcements shall be sampled for the mechanical properties inspection according to the relevant requirements of the current professionalstandard *Technical Specification for Mechanical Splicing of Steel Reinforcing Bars* JGJ 107—2016 and *Specification for Welding and Acceptance of Reinforcing Steel Bars* JGJ 18—2012.

The deviation of the installation position of the rebar should be in accordance with the provisions of the Table 4-13.

Deviation of the installation position of the rebar　　　　Table 4-13

Type		Deviation(mm)	Checking methods
Binding steel net	Length, width	±10	Steel gauge
	Size of net	±20	Maximum value from the continuous three measurements by steel gauge

Continued

Type			Deviation(mm)	Checking methods
Binding steel frame	Length		±10	Steel gauge
	Width, height		±5	Steel gauge
Force-carrying reinforcement	Spacing		±10	Maximum value from ends and middle part by steel gauge
	Distance between rows		±5	
	Concrete cover	Foundation	±10	Steel gauge
		Column, beam	±5	Steel gauge
		Slab, wall and shell	±3	Steel gauge
Binding stirrup, spacing of transverse reinforcement			±20	Maximum value from the continuous three measurements by steel gauge
Position of bending spot			20	Steel gauge
Embedded parts	Location of central line		5	Steel gauge
	Elevation difference		+3,0	Steel gauge and wedge gauge

4.3 Concrete engineering

Concrete engineering includes: mixing, transportation, placing, curing and other process. Every process is closely related to each other and affects each other. If any process is not handled property, the final quality of concrete project will be affected.

The quality of concrete requires not only the correct shape and size, but also enough strength, compactness, uniformity and integrity. Therefore, reasonable measures should be taken in the construction of each process to ensure the quality of concrete works.

4.3.1 Concrete preparation

In order to meet the strength grade required by the design, durability such as impervious and frost resistance, and the requirements of the construction operation on the workability of concrete mixture, the design mix ratio of concrete must be implemented in the construction.

Because the raw materials of concrete affect the quality of concrete directly, the raw materials must be controlled. Temperature, humidity and volume of all sorts of material keep in changing, weight changes a lot even with the same volume, so the mix proportion of concrete should be measured by weight in order to ensure the mix proportion accurately and reasonability so as to meet the quality requirements.

1. Raw materials

The raw materials that make up concrete include cement, sand, stone, water, admixtures, etc.

1) Cement

Commonly used types of cement are: Portland cement, Ordinary Portland cement, Portland slag cement, Portland pozzolana cement, Portland fly-ash cement, 5 in total; other kinds of cement may also be used under some special conditions, but the performance index of cement must conform to the provisions of the current national standards. Setting time, early strength, hydration heat, water absorption and performance of anti-erosion differs as variety and composition changes, so the varieties of cement should be selected reasonability.

The variety, grade, package or bulk warehouse number, date of production shall be checked when cement entering construction site, its strength, stability and other necessary performance should be retested, its quality must meet the provisions of the current national standards. When the cement in use is subject to unfavorable ambient influences or the cement has been delivered over three months (a month for rapid hardening Portland cement), re-inspection shall be carried out and cement shall be used according to the re-inspection results. Chloride cement is strictly prohibited to be used in reinforced concrete structures and prestressed concrete structures.

The stored cement shall be stacked up separately according to the variety, strength grade and date of production, as well as putting up signs and under the rules of first coming first serving. Commixture usages of different varieties are forbidden. In order to prevent the cement from moisture, the no-site warehouse should be as close as possible. When bagged cement is stored, it should be raised about 30cm from the ground and the distance from the wall should be more than 30cm. Generally, the stacking height shall not exceed more than 10 packages.

The temporary storage in open air should also be covered with rainproof tarpaulin, the baseboard should be higher from ground, and moisture-proof measures should be adopted.

2) Fine aggregates

The fine aggregate used in concrete is generally sand. Coarse sand or medium sand with good gradation should be selected.

For better technical performance of concrete, the solidity, slit content and harmful substance must meet the requirement of relevant national standards, in addition, the content of activate silica in sand should be controlled to avoid the alkali-aggregate reaction of concrete.

3) Coarse aggregates

The coarse aggregate commonly used in concrete are gravel or pebble.

Coarse aggregates should adopt clean crushed stone or pebble with good particle shape and sufficient soundness.

Aggregate grading and maximum particle size of gravel or pebble have great influence on the quality of concrete. In reinforced concrete structure, the maximum particle size of the coarse aggregates shall not exceed 1/4 of the minimum size of the cross-section or 3/4 of the minimum clearance of the reinforcements; for solid concrete slab, the maximum particle size of coarse aggregate should not exceed 1/3 of the slab thickness and it shall not exceed 40mm.

4) Water

When drinking water is used for concrete mixing, it may not be inspected. For other water like reclaimed water, cleaning water in mixing plants or circulating water in construction site, composition shall be analyzed.

Water for mixing and curing concrete shall meet the relevant requirements of current professional standard *Standard of Water for Concrete* JGJ 63—2006.

Untreated seawater must not be used for mixing and curing concrete in reinforced concrete structures and pre-stressed concrete structures.

5) Mineral admixtures

Part of cement can be replaced and the physical and mechanical properties and durability of concrete improved by adding mineral admixtures into concrete.

Commonly used mineral admixtures are: fly ash, ground slag, zeolite powder, silica powder, among which fly ash is the most widely used.

Mineral admixtures shall be selected according to the design and construction requirements and the environment conditions as well as the related national standard, the amount of mineral admixtures used shall be determined by test.

6) Admixtures

Varieties of admixtures can be classified into four categories according to its main functions: (1) admixtures for improving the rheological properties of concrete, such as water reducer, air-entraining agent and pumping agent. (2) admixture that adjusting concrete setting and hardening duration, such as early strength agent, acceleration agent and set retarder agent. (3) admixture for improving durability of concrete, such as air-entraining agent, antifreeze and antirust agent. (4) admixtures for improving other properties of concrete, such as expanding agent. Commonly, commercial admixtures are compound admixtures with several functions.

The storage and measuring of row materials also need to be complied with following requirements:

The materials after acceptance in site, shall be stored and stacked by type and batch and shall be marked clearly and shall meet the following requirements:

(1) Powder materials like bulk cement and mineral admixtures shall be stored in bulking tanks respectively; bagged cement, mineral admixtures and admixtures shall be stacked by type and batch respectively, applied with rain-proof and damp-prof measures. Sun-screen measures shall also be taken in high temperature seasons.

(2) Aggregate shall be stacked by type and specification respectively, prevented from impurities mixing and kept clean and uniform size grading. The ground for aggregate

stocking shall be applied with hardening treatment and measures of drainage, dust control and rain-proof.

(3) Liquid admixtures shall be kept in dry, shady and cool place, prevent from insolation, pollution and immersion and should be stirred uniformly before use; the liquid admixtures with any segregation or color changed shall be used after qualified by test:

The material amount shall be accurately measured in concrete mixing and the following requirements shall be met:

The precision of the measuring instruments shall meet the relevant requirements of current national standard Building Construction Machinery and Equipment-Concrete *Mixing* Plant (Tower) GB/T 10171—2016. The measuring instruments shall be applied with periodic calibration and shall be adjusted to zero before use.

The raw materials shall be measured by weight, water and admixture solutions may be measured by volume. The allowable measuring deviation shall meet the requirements of Table 4-14.

Allowable measuring deviation of raw materials for concrete (%)   Table 4-14

Type of raw materials	Cement	Fine aggregate	Coarse aggregate	Water	Mineral admixture	Admixture
Allowable measuring deviation for each batch	±2	±3	±3	±1	±2	±1
Accumulative allowable measuring deviation	±1	±2	±2	±1	±2	±1

Note: 1. The allowable measuring deviation of raw materials in-situ shall meet the requirements in allowable measuring deviation for each batch.
2. Accumulated allowable measuring deviation refers to the deviation accumulated in measuring each material in each concrete batch, and this index is only applicable to the mixing plant with computer control measuring.

2. Mix Proportioning

The design of the concrete mixture ratio shall comply with relevant provisions of the current national standard "design code for the mix proportion of ordinary concrete" and determined in accordance with the strength grade, durability and workability. As for the concrete with special requirements of anti-freezing and anti-seepage, the mixture ratio design shall meet the special provisions of relevant standards, the principle of reasonable use of materials and economy should also be considered in the mixture ratio design, and the design will be determined after test.

The concrete mixture ratio is determined in laboratory according to dry sand and dry stone as shown in Formula (4-12), but in construction site, it contains some water and the moisture content changes with the climate. Therefore, the actual moisture content of sand and stone aggregate should be measured before mixing concrete, the laboratory mixture ratio shall be converted into construction mixture ration according the measured result of moisture contents as shown in Formula (4-13).

The aggregate moisture content shall be tested frequently and test times shall be increased in rainy or snowy days. The amount of coarse / fine aggregates and water shall be adjusted in time when the actual water content in coarse /fine aggregates changes.

(1) Laboratory mixture ratio of concrete:
$$1:S:G:W \tag{4-12}$$

(2) Construction mixture ratio of concrete:
$$1:S(1+\omega_s):G(1+\omega_g):(W-S\cdot\omega_s-G\cdot\omega_g) \tag{4-13}$$

There in:

S——Proportion of sand in concrete mixture ratio;

G——Proportion of gravel (or pebble) in concrete mixture ratio;

ω_s——Water content of sand in construction site;

ω_g——Water content of Gravel (or pebble) in construction site;

W——Proportion of water in construction site (also known as water binder ration).

Example 4-2: Calculation construction mixture ratio of concrete

The known laboratory concrete ratio is $1:1.30:2.75:0.45$, weight of cement is 310 kg per cubic meters concrete, moisture content of sand and gravel measured in construction site is 3% and 1% respectively. Calculate the actual consumption of all materials in per cubic meter of concrete.

Answer:

(1) Construction concrete ratio:

$1:1.30\times(1+3\%):2.75\times(1+1\%):(0.45-1.30\times3\%-2.75\times1\%)=1:1.339:2.778:0.384$

(2) Actual consumption of all materials in per cubic meter of concrete:

Cement: 310kg

Sand: $310\times1.339=415.09$kg

Gravel: $310\times2.778=861.18$kg

Water: $310\times0.384=119.04$kg

3. Concrete mixing

1) Selection of concrete mixer

Concrete mixers can be classified into self-falling mixers and force action mixers according to the mixing principle.

Self-falling mixer: the inner wall of the mixing cylinder is equipped with blades, when the mixing cylinder rotates vertically, the blades lift the materials to a certain height, and then the material falls freely and being mixed. This is the principle of gravity mixing which is suitable for mixing plastic concrete and low-fluidity concrete.

Force action mixer: the shaft is equipped with blades, which rotate horizontally to force the materials in the mixing cylinder. It is the shear mixing principle which is suitable for mixing dry hard concrete and light aggregate concrete.

The discharge volume of concrete mixer is used for express its type generally. The type of the mixer should be determined according to the engineering volume, concrete slump requirement and aggregate size, which should not only meet the technical requirement, but also consider the economicbenefits and energy conservation.

2) Concrete mixing regulations

When mixing methods of feeding in batch is adopted, such process parameters as feeding sequence, quantity and mixing time of materials shall be determined by test. Mineral admixtures should be fed with cement, while liquid admixtures should be fed after water and cement; power admixtures should be fed after being dissolved.

(1) The loading volume of the mixer refers to the sum of the loose volumes of various raw materials required to mix one cylinder (also known as one plate) of concrete. In order to ensure sufficient mixing of concrete, the loading volume is usually only $1/3 \sim 1/2$ of geometric volume of the mixer. The volume of a single batch of concrete mixture is called the discharge volume (i.e., the standard specification of the mixer), approximately $0.5 \sim 0.75$ of the loading volume.

(2) Feeding sequence:

One time feeding method is often used at present, that is the material is put into the mixing cylinder in the order of sand, cement and stone, and then mixed with water. The advantage of this feeding sequence is that the cement stays between the sand and gravel, which can reduce the cement flying when entering the mixing cylinder. At the same time, sand and cement forms mortar which can shorten the time of wrapping stones and improve the mixing quality.

(3) Duration of concrete mixing:

The duration of concrete mixing refers to time quantum from all materials being fed into the mixing cylinder to starting discharging.

If the mixing duration is too short, the uneven mixing of concrete will affect its strength; but if the mixing duration is too long, it will not only reduce the production efficiency, but also will make the concrete workability reduced or segregation phenomenon.

The determination of mixing duration is related to the type of mixer and the fluidity requirement of concrete.

The concrete shall be mixed uniformly and forced action concrete mixer should be adopted. The minimum duration of concrete mixing may be selected from Table 4-15 and the mixing duration may be reduced appropriately if uniformly mixing is ensured. The mixing duration for concrete with strength C60 and above shall be increased appropriately.

Minimum duration of concrete mixing (s)   Table 4-15

Concrete slum(mm)	Type of mixer	Discharge of mixer(L)		
		>250	250~500	>500
≤40	Forced action	60	90	120

Continued

Concrete slump(mm)	Type of mixer	Discharge of mixer(L)		
		>250	250~500	>500
>40 and ≤100	Forced action	60	60	90
≥100	Forced action	60		

Note: 1. The mixing duration shall be appropriately increased for concrete with admixtures and mineral admixtures;
  2. When gravity type of concrete mixer is adopted, the mixing duration should be extended 30s;
  3. When mixers of other type are adopted, the minimum mixing duration may be determined according to the requirements in the equipment instruction manual or determined by test.

The mix proportion used first shall be applied with first-batch inspection, which should include the following items:

(1) Conformity of raw materials in concrete to one used in mix design.
(2) Conformity of discharged concrete workability to mix design requirements.
(3) Concrete strength.
(4) Setting time of concrete.
(5) Concrete durability, if required in the engineering.

4.3.2 Concrete transportation

After the concrete is unloaded from the mixer, it shall be transported to the placing position in time.

1. Requirement for transportation

(1) Good uniformity, no stratification, no segregation, no grout leakage shall be ensured in the process of transportation.
(2) The specified slump should be ensured after being transported to the placing site.
(3) Minimize the transport time and transmit times in order to ensure that the concrete can be transported to the site, placing and compacting before its initial setting time.
(4) Ensure that the concrete placing work can be carried out continuously.
(5) When transferring concrete, pay attention to make sure that the mixture can be directly placed into the center of the loading vehicle, so as to avoid aggregate segregation.

2. Transport machine

Concrete transportation can be divided into ground horizontal transportation, vertical transportation and high altitude horizontal transportation.

The diesel dumper and the wheelbarrow can be adopted for the short range ground horizontal transportation. Concrete mixing truck and dump truck are adopted when the concrete quantity demanded is large or distance is longer. The concrete mixing truck is shown in Figure 4-41. A conical tilting discharging mixer is mounted on the chassis of a truck, which can continue to stir during the transportation of concrete to prevent concrete segregation; in the case of very long transport distance, only put dry materials into the cylinder at first, then add water in transportation process or in construction site so as to reduce the concrete slump loss cause by long-distance transport.

Concrete pump, tower crane, derrick or gantry, etc. are commonly adopted in concrete vertical transportation. The loading hopper is needed when a tower crane is adopted.

As for horizontal transportation of concrete at high altitude, concrete distribution spreader machine should be equipped for distributing materials if concrete pump is adopted; if tower crane is adopted, it will complete both horizontal and vertical transportation simultaneously; wheelbarrow is needed if the derrick or gantry is adopted.

Necessary measures shall be taken to ensure the formwork and reinforcement no displacement in transportation of high altitude.

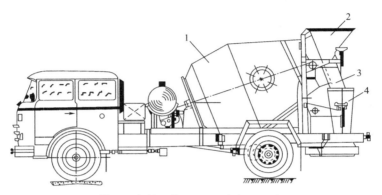

Figure 4-41 Concrete mixing truck
1—Concrete stirring cylinder; 2—Feed hopper; 3—Fixed discharge downspoutings; 4—Movable discharge hopper

The following requirements in transportation should be meet in order to ensure the project quality:

(1) Mixer trucks used in concrete transportation, shall meet the following requirements:

① Before charging, accumulated water in mixer truck tank shall be removed.

② When in transportation and waiting for discharging, the mixer truck tank shall keep normal revolving speed and shall not stop revolving.

③ The mixer truck tank should revolve at fast periodic for 20s prior to discharging.

(2) When mixer truck is used to transport concrete, traffic safety personnel shall be arranged at the site entry and exit, roads in site shall be unobstructed; cycle lanes should be arranged if permitted; warning signs shall be arranged in danger zones; good illumination shall be set for construction work at night.

(3) When mixer truck is used to transport concrete and the concrete slump loss is too large to satisfy the construction requirements, an amount of water reducer same to the one used in original proportioning may be added into the truck tank. The addition of the water reducer shall be determined by test and shall be recorded. After adding of the water reducer, the mixer truck tank shall revolve at fast speed to achieve uniform concrete mixtures and concrete shall be pumped or placed after reaching the required performance.

(4) When dumper is used to transport concrete, roads shall be unobstructed, and the road surface shall be lever and compact; temporary ramp or supports shall be firm and the planking joints shall be smooth.

4.3.3 Concrete conveying

1. Pumping should be adopted for concrete conveying. Schematic diagram of fixed concrete pump is as shown in Figure 4-42.

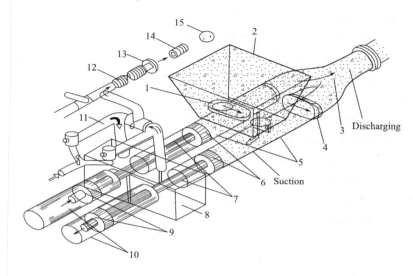

Figure 4-42 Schematic diagram of fixed concrete pump
1—Suction end horizontal valve; 2—Receiving hopper; 3—Y transportation pipe; 4—Push end horizontal valve; 5—The concrete cylinder; 6—Concrete piston; 7—Piston rod; 8—Water tank; 9—hydraulic piston; 10—hydraulic cylinder; 11—Reverse valve of washing device; 12—Heavy pressure hose of washing device; 13—Frange of washing device; 14—Piston of washing device; 15—Sponge ball

2. The selection and layout of concrete pumps shall meet the following requirements:

(1) The type selection of concrete pumps shall be determined by engineering features, concrete conveying height and distance, and concrete workability.

(2) The quantity of concrete pumps shall be determined according to amount of concrete and placing condition; if necessary, standby pumps shall be arranged.

(3) Locations of concrete pumps shall satisfy the construction requirements and the site shall be level and compact; the road shall be unobstructed.

(4) There shall be no obstruct within the operating range of concrete pumps; facilities for preventing overhead falling shall be arranged in locations of concrete pumps.

3. Setting of concrete pump pipe and supports shall meet the following requirements:

(1) Concrete pump pipe shall be selected according to concrete pump model, mixture performance, output quantity, unit output, conveying distance and particle size of coarse aggregate.

(2) When maximum particle size of concrete coarse aggregate is not larger than

25mm, concrete pump pipe with inner diameter not less than 125mm may be used; when maximum particle size of concrete coarse aggregate is not larger than 40mm, concrete pump pipe inner diameter not less than 150mm may be used.

(3) Concrete pump pipe shall be connected tightly, and the turns of pipe should be rounded.

(4) Concrete pump pipe shall adopted supports for fixing, the supports shall be connected with the structure firmly and the supports on pump pipe turns shall be increased; supports shall be determined by calculation; the locating location on structure shall be analyzed and reinforcement members shall be taken if necessary.

(5) When concrete is pumping upwards, the total converted length of straight and bent concrete pump pipe on ground should not be less than 20% of the vertical pumping height or 15m.

(6) When concrete is pumping inclined or vertically downwards and the vertical difference is larger than 20m, straight pipes or bent pipes shall be arranged at the lower end of inclined or vertical pipes; the total converted length of straight and bent pipes should not be less than 1.5 times of the vertical difference.

(7) When the pumping height is larger than 100m, stop valve shall be arranged at the discharge port of concrete pump pipe.

4. Concrete pump pipe and supports shall be inspected and maintained frequently.

(1) Placing equipment shall be matched with concrete pumps; the inner diameter of concrete conveying pipe of placing equipment should be same to that of concrete pump pipe.

(2) The quantity and location of placing equipment shall be determined according to placing equipment working radius, construction areas and construction requirements.

(3) Placing equipment shall be installed firmly and applied with antidumping measures; structure or special devices where placing equipment are installed shall be applied with calculation and reinforcement measures shall be taken if necessary.

(4) Wall thickness of bent pipes for placing equipment shall be inspected frequently and the one with serious abrasion shall be replaced in time.

(5) There shall be no obstruct within the operating range of placing equipment; facilities for preventing overhead falling shall be arranged.

5. Pipes, vessels and downspoutings for concrete conveying shall not absorb water or leak paste and they shall guarantee smooth conveying. Thermal insulation, head insulation and rainproof measures shall be applied for conveying concrete according to environment conditions.

6. Conveying concrete with pump shall meet the following requirements:

(1) Inspection for pumping water shall be performed first and such parts as hopper and piston in contact with concrete directly shall be fully wet; accumulated water in concrete pump shall be removed after the inspection.

(2) Before conveying concrete, cement mortar should be conveyed first to lubricate

concrete pump and conveying pipes.

(3) Concrete conveying shall be slow at first and accelerate gradually and the concrete shall be conveyed at normal speed after the system runs smoothly.

(4) In concrete conveying process, mesh enclosures shall be arranged for collecting hopper of concrete pump and enough concrete shall be ensured in the collection hopper.

7. Conveying concrete with cranes and hoppers shall meet the following requirements:

(1) Different hoppers shall be selected according to different structure type and concreting method.

(2) The volume of hoppers shall be determined according to crane capacity.

(3) Concrete delivered to the site should be directly loaded into the hopper for conveying.

(4) Concrete should be placed on the designated placing spot directly by hoppers.

8. Conveying concrete with lifting equipment and dollies shall meet the following requirements:

(1) Quantity of lifting equipment and dollies used, travel route and unloading spot of dollies shall be able to satisfy the concrete placement demand.

(2) Concrete delivered to the site should be directly loaded into the dollies for conveying and dollies should be loaded near the lifting equipment.

4.3.4 Concrete placing

1. Preparation works before concrete placing

(1) The position, elevation, size, strength, stiffness and other aspects of the formwork shall be checked to see whether they meet the specified requirements, as well as whether the joints of the formworks are firmly and closely.

(2) The rebar and embedded parts shall meet the requirements in terms of varieties, specifications, quantity, placement and concrete cover, record of acceptance of concealed work should be done.

(3) Any residues on internal surface of forms shall be removed, and the wooden formwork should be watered and moistened but not allowed accumulated water.

(4) Arrangement should be completed about supply of materials and water and electricity, preparation of machinery, road, water and materials, labor organization, etc. , security technology disclose should becarried out.

2. Technical requirement

1) General requirement

(1) The concrete shall be placed into the form immediately after it arrives at the placing site. If layering and segregation are found, the mixture should be stirred to ensure its uniformity. In the case of large loses of concrete slump, cement mortar should be added in stirring process.

(2) Concrete should becasted before the initial setting, if there is a phenomenon of

initial setting, it should be adopted a strong stirring to restore its liquidity before placing.

(3) In order to prevent the concrete from layering or segregation during concrete placing, the free falling height of concrete shall not exceed 2m, in vertical structure members (such as wall, column), the free falling height should not exceed 3m, otherwise, downspouting, tumbling barrel, sliding tube or vibrating sliding tube should be adopted. Methods of dropping concrete from a height are shown in Figure 4-43.

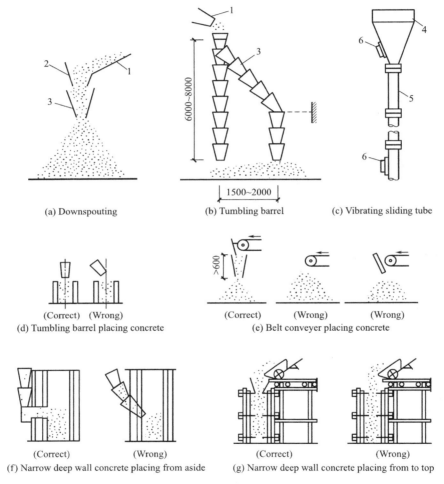

Figure 4-43 Methods of dropping concrete from a height
1—downspouting; 2—Baffle plate; 3—Tumbling barrel; 4—Funnel; 5—Sliding tube; 6—Vibrator

(4) Before concrete placing of vertical structure (such as wall and column), cement mortarwhich has the same composition as it in the concrete should be first placed with a thickness of 50~100mm, so as to avoid the quality defects such as honeycomb, pitted surface and exposed stone in the bottom of components due to the lower content of mortar.

(5) Concrete in and after placing process, measures should be taken to prevent cracks; non-structural surface cracks of concrete due to settlement and shrinkage shall be repaired before final setting.

2) Thickness of concrete placing

In order to ensure the compactness of concrete, concrete must be placing and tamping in layers, and the thickness of the placing layer shall comply with the provisions of Table 4-16.

Thickness of the concrete placing (mm)　　Table 4-16

Methods of concrete tamping		Thickness of the concrete placing
Immersion vibrator		1.5 times of length of the acting part of the vibrator
Plain vibration		200
Artificial tamping	In foundation, unreinforced concrete or structure with sparse reinforcement	250
	In beam, wallboard and column structure	200
	Structure with density rebar	150
Light-weight aggregate concrete	Immersion vibrator	300
	Plain vibration (load added while vibration)	200

3. Interval duration

To ensure the integrity of concrete, placing should be carried out continuously. Where the interval is necessary, the interval duration should be shortened as far as possible.

The placement of upper concrete layer should be completed prior to the initial setting of the lower layer.

The concrete transportation, conveying and placing processes shall be guaranteed with continuous operation. The duration from transportation to placing should not exceed those specified in Table 4-17 and shall not exceed those specified in Table 4-18. For concrete mixed with hardening accelerating and water reducing admixtures and hardening accelerating agents and concrete with special requirements, the allowable duration shall be determined according to the design and construction requirements.

Construction joints must be arranged if the initial setting time is exceeded.

Duration from transportatin to placing (min)　　Table 4-17

Condition	Temperature	
	$\leqslant 25℃$	$>25℃$
Without admixture	90	60
With admixture	150	120

Total time limit of transportatin, placing and intervals (min)　　Table 4-18

Condition	Temperature	
	$\leqslant 25℃$	$>25℃$
Without admixture	180	150
With admixture	240	210

4. Construction joint and post-cast strip

Due to technical or construction organization reason, if the whole structure concrete cannot be placed continuously and the interval duration exceeds the time specified, the construction joints shall be reserved in the appropriate position.

Construction joint refers to the connecting surface between the later placing concrete and the pre-placed concrete that has been solidified and hardened. They are weak parts of the structure and must be considered seriously.

Position of construction joints and post-cast strips shall be determined before concrete placing. Construction joints and post-cast strips should be set in the positions convenient for construction where the structures carry the relatively small shear force. As for structural members with complex stresses or those with water-proof and permeability requirements, the setting positions of construction joints shall be confirmed by the design organization.

1) The setting position of horizontal construction joints shall meet the following requirements:

As for columns and walls, construction joints may be set on the top surface of foundation and floors; the distance between construction joints for column and top surface of structures should be 0~100mm; the distance between construction joints for wall and top surface of structures should be 0~300mm.

As for columns and walls, construction joints may also be set on the bottom surface of floors; the distance between construction joints and bottom surface of structures should be 0~50mm; when there are beam studs under the slab, construction joints may be set 0~20mm below the beam studs.

As for high columns, walls or beams, or thick foundation, horizontal construction joints may be set in middle parts according to construction requirements; when the reinforcement of members shall be adjusted due to stress state changing caused by construction joints settings, the adjustment shall be confirmed by the design organization.

Horizontal construction joints set in the special structure positions shall be confirmed by the design organization. Diagram of column and wall concrete joint are as shown in Figure 4-44 and Figure 4-45, as for beam-slab casted as a entirety, concrete joint position is as shown in Figure 4-46.

2) Setting position of vertical construction joints and post-cast shall meet the following requirements:

As for floors with girders and secondary beams, the construction joins shall be set within the range of 1/3 in middle of secondary beam span as shown in Figure 4-47.

As for one-way slabs, the construction joints shall be set in any position parallel to the direction by which forces transfer, the one-way slab concrete joint position is as shown in Figure 4-48.

As for stairs, the construction joints should be arranged within the range of 1/3 at the

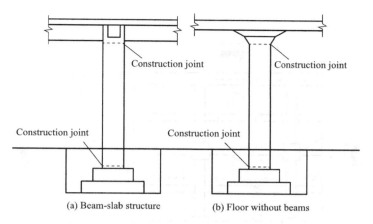

Figure 4-44 Concrete joint position of column

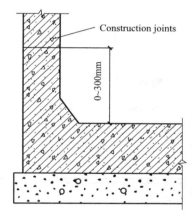

Figure 4-45 Horizontal construction joint of wall

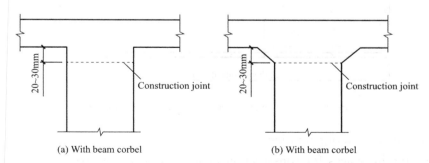

Figure 4-46 Concrete joint position in beam-slab casted as a entirety

end of stair slab span.

As for walls, the construction joints should be arranged within the range of 1/3 of lintelmidspan at the floor opening, or the joints of vertical and cross walls.

Setting position of post-cast strips shall meet the design requirements:

(1) Vertical construction joints set in the special structure positions shall be

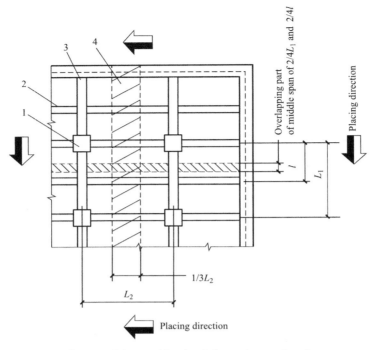

Figure 4-47 Concrete joint position in girder and secondary beam structure
1—Column; 2—Secondary beam; 3—Girder; 4—Slab;
L_1—Span of girder; L_2—Span of secondary beam; l—Span of slab

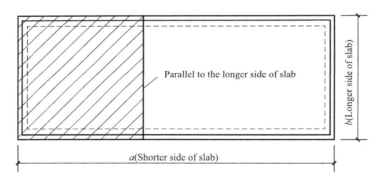

Figure 4-48 Concrete joint position in one-way slab

confirmed by the design organization.

(2) Setting interfaces for construction joints and post-cast strips shall be vertical to structural members and longitudinal steel reinforcements. For relatively thick or high structural members, the interface for construction joints or post-cast strips should be sealed with special materials.

During concrete placing, if the construction joints are required to be arranged temporarily due to special causes, the construction joins should be set orderly and should be vertical to the surface of members. If necessary, the technical measures such as adding joint bars and subsequent chiseling may be taken.

Such protective measures as rust prevention or inhibition for steel bars shall be taken for the construction joint and post-cast strip.

3) Treatments of construction joints:

Concrete placementat the construction joints shall not start until the compressive strength of the casted concrete reaches $1.2kN/mm^2$, and the necessary treatments must be carried out on the construction joints to strengthen the connection between the old and the new concrete, and minimize the adverse impact of the construction joints on the structural integrity, the methods are listed as following:

(1) Binding face shall be rough and removed from laitance, loosened gravels and weak concrete.

(2) Binding face shall be fully wet by water but without any accumulated water.

(3) The thickness of receiving layer for cement mortar in horizontal construction joints for columns and walls shall not be larger than 30mm and cement mortar for receiving layer shall be same to concrete paste in composition.

(4) Strength grade and performance of concrete for post-cast strips shall meet the design requirement; if there is no specific requirement in the design, the strength grade of concrete for post-cast strips should be on grade higher than that of concrete on both sides and technical measures to reduce shrinkage should be adopted.

(5) Concrete placed at the construction joints shall be carefully vibrated and compacted to make the old and new concrete closely combined.

5. Placing of cast-in-situ concrete main structure

The main components of the main structure are column, wall, beam and slab, etc., in multi-storey and high-rise building structures, these components are repeated along the height direction, thus, layered construction methods are adopted according to the building storeys; each floor could also be segmented construction in the case of large horizontal area in order to make the process into flow construction.

In the construction of each layer and each section, theplacing sequence is to placing column and wall at first, then the beam and slab.

1) Placing of column

(1) Column concrete placing should be carried out before formwork erecting and rebar binding of beam and slab, so that the beam and slab formwork can be used to stabilize the column formwork and serve as the operation platform for placing column concrete.

(2) The sequence ofplacing a row of columns should go to middle from both ends at the same time, rather than from one end to the other end, so as to avoid the transverse pressure forces generated by the expansion of formwork after placing concrete caused by water absorption, those pressure forces will lead to column bending and deforming.

(3) Placing height of concrete in formwork of column and walls should comply with the regulations of Table 4-19, otherwise, tumbling barrel and downspouting might be

adopted.

Limit of concrete placing height in column and wall formwrok (in)   Table 4-19

Situation	Limit of concrete placing height
Diameter of coarse aggregate more than 25mm	≤ 3
Diameter of coarse aggregate equal or less than 25mm	≤ 6

Note: concrete placing height might not be limited by Table 4-19 when there are reliable measures which can prevent concrete from segregation.

Some placing and vibrating methods often used are listed in Figure 4-49 and Figure 4-50.

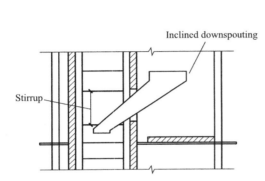
Figure 4-49 Casing concrete from opening

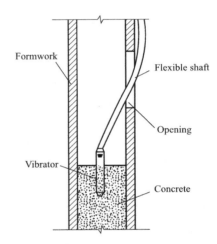
Figure 4-50 Vibrating through opening

2) Placing of wall

(1) The wall shall be casted in segmenting way along the height direction.

(2) While placing shear wall, flow construction should be adopted along the length direction of wall, placing and vibrating should be carried out in different segment along horizontal direction and layers along vertical direction, rising evenly, and the placing thickness of each layer should be controlled around 60 mm.

(3) Whenplacing the opening of doors and windows in the wall, both sides of the openings should be casted at the same time, and the height of the concrete on both sides should be roughly the same in order to prevent the moving of the formwork of the openings of doors and windows; the part under the windows sill should be casted at first, a short while after, the wall between the windows will be casted.

(4) In order to avoid uneven strength of concrete caused from accumulation of a large amount of slurry water while placing to a certain height of wall, for placing shear wall, thin wall and other narrow and deep structures, amount of water should be reduced when placing reaches an appropriate height.

3) Placing of beam and slab

(1) Concrete for beams and slabs is generally casted simultaneously. The beam is

firstly casted in layers and form a stepped shape, then casted together with the concrete of the slab when it reaches the bottom of the slab. The placing of the slab goes forward constantly along with the continuous extensions of the stepped ladder.

The direction of dumping concrete shall be opposite to that of placing concrete, as shown in Figure 4-51.

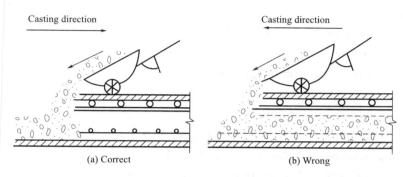

Figure 4-51　Dumping direction of concrete

(2) In the case of beam height is greater than 1m, it can be casted independently, the horizontal construction joints can be reserved at 2~3cm below the bottom of slab.

(3) Generally, at the intersection part of the column and girders, main beam and secondary beam, the rebar are relatively dense, especially the upper negative bending moment rebar with bigger diameter and dense arrangements, will result in difficulties of concrete dumping into. If necessary, this part can be converted to fine stone concrete, while the vibrator head can be changed to slice type and assisted by manual tamping.

(4) Where placing girder and slab which are integrated with column and column, the placing of the girder and slab should not continue until 1~1.5 hours after the completion of the placing of wall and column, in which the initial settlement and excreting water of wall and column complete.

6. Placing of mass concrete

Mass concrete refers to the thickness of more than or equal to 1m, and larger length and width of structure, such as high-rise buildings in the reinforced concrete box foundation of the bottom plate, industrial building equipment foundation, bridge and pier.

The construction characteristics of mass concrete structure: first, the integrity requirements are high, generally requires continuous placing, and construction joint is not allowed; second, due to the large volume of the structure, the hydration heat generated by concrete placing is large, and the accumulation is not easy to be distributed inside, therefore, form a larger temperature difference between inner and outer side, causes larger temperature difference stress, result in temperature cracks in the concrete.

The key of mass concrete construction is:

In order to ensure the integrity of the structure, a reasonable concrete placing schemes should be determined; the effective measures should be taken in order to avoid temperature cracks caused by temperature difference between inner and outer of concrete.

For ensuring the continuous placing, the upper layer of concrete should be casted before the initial setting of the lower layer concrete. Therefore, the number of concrete to be casted per hour shall be calculated according to, namely, the placing strength.

$$V = BLH(t_1 - t_2) \qquad (4\text{-}14)$$

There in:

V ——The amount of concrete casted per hour (m^3/h).

B, L, H ——The width, length and the thickness of the casted layer (m).

t_1 ——Initial setting time of concrete (h).

t_2 ——Transport time of concrete (h).

According to the amount of concrete casted, calculate the number of mixers, conveyors and vibrators required, determine the concrete placing scheme and labor organization.

The placing scheme of mass concrete shall be determined according to the actual situation of structure size and concrete supply, etc. Generally, there are three schemes of overall layering, segmented layering and inclined plane layering, as shown in Figure 4-52.

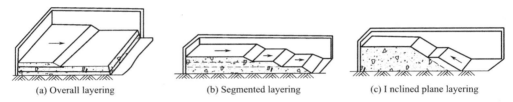

(a) Overall layering (b) Segmented layering (c) Inclined plane layering

Figure 4-52 Placing scheme of mass concrete

1) Overall layering [Figure 4-52 (a)]

The concrete is casted in layers in the whole structure. It is required that the concrete placing of each layer must be completed before the initial setting of the lower layer.

This method is suitable for the structure with a small plane size.

The construction should start from the short side and advance along the long size. If necessary, it can also be pushed from middle to both ends or from both ends to the middle of the structure surface.

2) Segmented layering [Figure 4-52 (b)]

The segmented layering method can be adopted if the concrete placing strength is too high to meet the construction requirement when the overall layering placing was adopted.

It is to divide the structure into several segments (sections) on the plane and several layers on the thickness. Concrete is casted from the bottom layer and then back to the second layer after a certain distance (length of the segment), so that the above layers are casted forward successively.

It is required to start the construction of the second segment before the initial setting of the first segment end concrete of the first layer, so as to ensure the better combination of concrete.

This method is suitable for the structure with small thickness but large area or length.

3) Inclined plane layering [Figure 4-52 (c)]

The inclined plane layering method should be adopted when the length of the structure is more than three times of the thickness.

During construction, the vibrator should start from the bottom of the placing concrete layer, move upward gradually to ensure the construction quality of concrete.

7. Concrete placing underwater

In foundation engineering and water conservancy engineering construction, such as pore-forming pile and underground continuous wall, concrete is often casted underwater directly, or, even in mud.

Placing concrete underwater or in mud generally uses the ducting method which has the follow characteristics:

The pipe is used to transport concrete down and isolate it from surrounding water or mud. The concrete inside the pipe falls down by its self-weight and squeeze the concrete around the pipe mouth at the lower part of the pipe, so that it flows and diffuses in the casted concrete, then keep lifting and placing till finished.

The ducting method can not only avoid the contact between concrete and water or mud, but also ensure that the aggregate and cement in concrete are not separated, so as to ensure the quality of underwater concrete placing. Schematic diagram of ducting method is shown as in Figure 4-53.

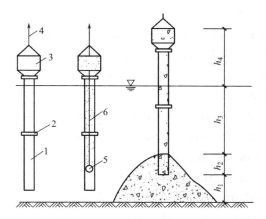

Figure 4-53 Schematic diagram of ducting method
1—Duct; 2—Joint; 3—Hopper; 4—Lifting rope; 5—Bulb stopper; 6—Wire

During construction, first, the duct pipe shall be sunken into the water about 100mm above the water body bottom, the ball stopper is suspended with wire or hemp rope with a height of 0.2m above the water body bottom, then, the concrete is placed into duct pipe. After the pipe and hopper are filled with concrete, the rope can be cut off. At this time, concrete pushes the bulb stopper down by its self-weight, rush out of the bottom of the duct pipe and spread around, form a concrete heap, and the duct pipe bottom is buried in the concrete heap. When the concrete is continuously placed into the duct pipe from the hopper and diffused in bottom, the concrete surface outside the duct pipe rises continuously, the duct pipe is also lifted upward correspondingly.

The height of each lifting should be controlled with the range of 150~200mm to ensure that the lower part of the duct pipe is always buried in concrete. The minimum embedding depth is shown in Table 4-20, and the maximum embedding depth should not exceed 5m in order to ensure the concrete placing continuously.

Minumum depth of duct pipe embedding (mm)　　　　Table 4-20

Placing depth for underwater concrete	Minimum depth of duct pipe embedding
≤10	0.8
10~15	1.1
15~20	1.3
>20	1.5

When the concrete diffuses from the bottom of the duct pipe to all sides, the concrete near the duct pipe mouth has good uniformly and high strength, while the concrete far from the duct pipe mouth is easy to isolate and its strength decreases.

In order to ensure the quality of concrete, the feeding radius of the duct pipe should not be greater than 4m, when multiple duct pipes are casting together, the spacing between duct pipes should not be greater than 6m, and distribution area of each duct pipe should not be greater than $30m^2$。

When multiple duct pipe placing concrete method is adopted, they should be placed from the deepest place, and the concrete surface should rise horizontally and uniformly. The elevation difference of adjacent duct pipe bottom should not exceed 1/15~1/20 of the distance between duct pipes.

Concrete placing shall be carried out continuously without interruption. The concrete supply shall be guaranteed to be greater than the amount of concrete needed to maintain the height of the concrete inside the duct pipe.

Due to its lower strength, the concrete structure on the surface in contact with water or mud should be removed after concrete placing. The taken value should be at least 0.2m in water surrounding environments, and 0.4m in mud, therefore, the elevation control of placing concrete should exceed this value.

4.3.5　Concrete compacting

After concrete is casted into the formwork, it cannot be filled and densified by itself due to the friction resistance between the aggregate and the viscous force of cement slurry, the certain volume of voids and bubbles inside make concrete cannot meet the requirement density, it will affect the strength and durability of concrete members. Therefore, the concrete must be compacted after filling into formwork to ensure that the shape and size of the components are correct, the surface is smooth, as well as the strength and other properties meet the design and use requirements.

There are three waysfor concrete compactness:

(1) By means of mechanical external force (such as mechanical vibration) to overcome the friction resistance inside the mixing material and make it liquefied and compactness.

(2) Increase the moisture in the mixing material to improve its fluidity, so that it is easy to form, then, centrifugal method, vacuum suction method can be selected to remove the excessive water and air.

(3) Efficient water reducer might be mixed into the mixture, so that its slump greatly increased, forming its shape by artesian pouring. It is a promising method.

Among many different compacting methods, mechanical vibration compacting method is often adopted at present which can be further divided into immersion vibrators, plate vibrator, clamping vibrator and vibration table (Figure 4-54). If necessary, additional compacting by hand may be adopted.

Concrete compacting shall be able to make concrete compacted and uniform in all positions and shall be free from any missed, lack or excess vibration.

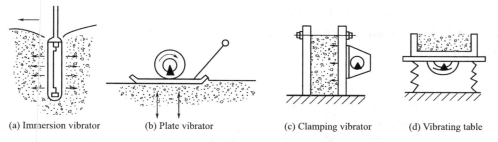

(a) Immersion vibrator   (b) Plate vibrator   (c) Clamping vibrator   (d) Vibrating table

Figure 4-54 Dumping direction of concrete

1. Immersion vibrator

An immersion vibrator is also called a plug-in vibrator, as shown in Figure 4-55. It is widely used in construction cause of its better vibration effect, simple structure, easy to

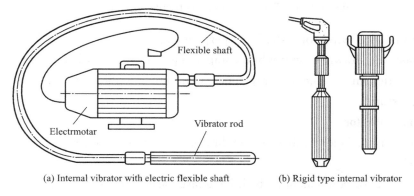

(a) Internal vibrator with electric flexible shaft   (b) Rigid type internal vibrator

Figure 4-55 Internal vibrating machine

use and long service life. Immersion vibrator is mainly used for vibrating large volume concrete, foundation, column, girder, wall, thick slab and other structure members.

Compacting concrete with internal vibrators shall meet the following requirements:

(1) Compacting shall be carried out respectively according to depth of placement layer, the front end of immersion vibrators shall be immersed into the last concrete layer and the immersion depth shall not be less than 50mm as shown in Figure 4-56.

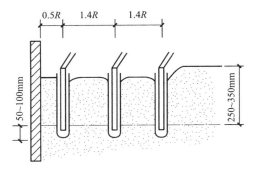

Figure 4-56　Depth immersed of immersion vibrator

(2) The distance between immersion vibrator and forms shall not be larger than 50% of the action radius of immersion vibrators; the spacing of compacting immersion spots should not be larger than 1.4 times of the action radius of immersion vibrators as shown in Figure 4-57.

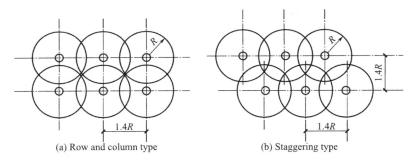

Figure 4-57　The spacing of compacting immersion spots

(3) Immersion vibrators shall be perpendicular to concrete surface and shall be immersed fast and drawn slowly while compacting uniformly; when concrete surface has no obvious settlement, cement slurry appears and no air bubble appears, the vibrating at this spot can stop.

2. Surface vibrator

Surface vibrator is also called plate vibrator, it is fixed itself to a base plate as shown in Figure 4-58.

It is suitable for components which have large area, flat surface and small thickness, such as floor slab, ground, road surface and thin shell. Compacting concrete with plate

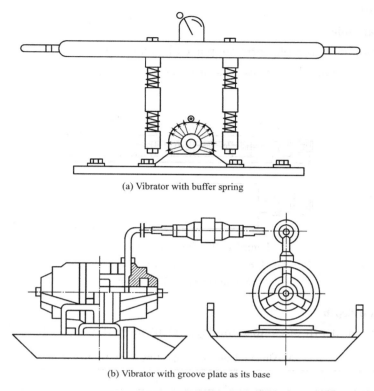

(a) Vibrator with buffer spring

(b) Vibrator with groove plate as its base

Figure 4-58 The surface vibrator

vibrator shall meet the following requirements:

(1) Surface vibrator shall cover all edges and corners of compacting area.

(2) The displacement spacing of plate vibrators shall cover the edge of earlier compacted concrete.

(3) Compacting of inclined surface of concrete shall run from lower part to upper part.

3. External vibrator

External vibrator can be also called attached vibrator. It is suitable for column, beam, plate, wall and other structure members which have denser reinforcement arrangement and has width not great then 300mm, as well as the structure which is not suitable to use the insert vibrator. Compacting concrete with external vibrators shall meet the following requirements:

(1) Clamping vibrators shall be clamped to the formwork and the spacing of location spot shall be determined by test.

(2) Clamping vibrators shall compact from lower to upper in sequence according to the height and speed of the concrete placement.

(3) When multiple clamping vibrators are used in a formwork simultaneously, the frequency shall be consistent and the vibrators shall be arranged in staggered way on forms

opposite each other.

4. Vibration table

The vibration table is a platform supported by an elastic support, under which there is a vibration machine and the formworks are fixed on the platform, as shown in Figure 4-59.

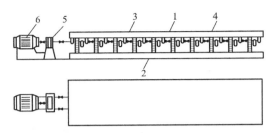

Figure 4-59　The vibration table

1—Vibration table; 2—Fixed frame; 3—Eccentricity vibrator units; 4—Supporting spring;
5—Synchronizer; 6—Electromotor

5. Maximum depth

The maximum depth of compacting layer shall meet the requirement in Table 4-21.

Maximum depth of compacting layer (min)　　　　Table 4-21

Compacting methods	Maximum depth of compacting layer
Immersion vibrator	1.25 times of the acting part length of the vibrator
Plate vibrator	200mm
Clamping vibrator	Determined by test according to location mode

4.3.6　Concrete curing

The setting and hardening of concrete is mainly due to the hydration of cement, which requires proper humidity and temperature. After concrete being poured, such as the hot weather, dry air and small humidity, the moisture in concrete will evaporate rapidly and dehydration occurs, which will cause the cement particles cannot be adequately hydrated, cannot be formed into the stable crystallization, lack of adequate bond, therefore, there will be flake or powdery spalling on the surface of concrete, affecting the strength of concrete. At the same time, premature evaporation of water will cause concrete shrinkage deformation, dry shrinkage cracks, affecting the integrity and durability of concrete. If the temperature is too low, the concrete strength growth is slow, it will affect the concrete structure and members put into use earlier.

The curing of concrete is to provide the necessary temperature and humidity conditions for the hardening of concrete to ensure that it meets the design requirements of strength within the specified age and to prevent shrinkage cracks.

Before the concrete strength reaches 1.2MPa, there shall be no stamping on, no materials stack, and no forms and supports erection on the concrete surface.

Curing condition for curing specimens under the same curing conditions shall be the same as those of the actual structure positions and shall be stored properly.

The construction site shall be possessed of the concrete standard specimens fabrication conditions, besides the standard specimen curing rooms or curing boxes shall be provided. Standard specimen curing shall meet the requirements of relevant current national standards.

1. Nature curing

The nature curing of concrete means to keep the concrete wet for a certain period of time under the natural condition of the average temperature higher than 5 celsius degree.

It can be further categorized into cover watering and curing, plastic film curing.

1) Cover watering and curing

Water absorbing and moisturizing materials are adopted in cover watering and curing methods, such as grass curtain, sacks, sawdust, etc., which cover the exposed surface of concrete with water spraying to keep its moisture state. This method is often used to cure the foundation, beam, plate and other structural components.

Curing time shall meet the following requirements:

(1) Concrete should be covered and watered within 8~12 hours after pouring.

(2) As for the concrete prepared by Portland cement and ordinary Portland cement or Portland slag cement, the curing time shall not be less than 7d.

(3) As for the concrete mixed with set retarding admixtures or large amount of mineral admixtures, the curing time shall not be less than 14d; as for the impermeable concrete and the concrete of C60 or above strength grade, the curing time shall not be less than 14d; as for post-cast strips, the curing time shall not be less than 14d.

(4) As for the concrete prepared by other types of cement, the curing time shall be determined according to the properties of cement; as for walls and columns on the ground floor of basement and those on the first floor of superstructure, the curing time should be properly increased; as for mass concrete, the curing time shall be determined according to construction scheme.

(5) Watering times should be able to keep the concrete in the state of wetting. If the daily minimum temperature is below 5 celsius degree, providing water curing shall not be adopted.

(6) Curing water should be same as that of used in concrete mixing, shall meet the requirements of current professional standard *Standard of Water for Concrete* JGJ 63.

2) Plastic film curing

Plastic film curing is to use plastic film to close the concrete surface tightly, prevent the evaporation of water inside of the concrete, ensuring its hardening under hydration. It can be further categorized into covering curing and curing compounds.

(1) Covering curing shall meet the following requirements:

Plastic membrane shall cling to the exposed surface of concrete; condensation water shall be maintained in plastic membrane.

Cover materials shall be enclosing, the member of layers of which shall be determined according to construction scheme.

Its advantages are no watering, easy operation, can improve the early strength of concrete. It is widely used in column, floor and ground.

(2) Curing compounds:

The method is to spray the plastic solution on the concrete surface, the solvent volatilized and condensed into a layer of plastic film, so as to insulate the concrete from air.

This method is used for walls that are not easily watered and covered for curing, such as tall structures, large areas of concrete structures and areas lack of water.

Curing compounds method shall meet the following requirements:

Concrete shall be cured by spraying dense curing compounds on its exposed surface.

Curing compounds shall be uniformly sprayed on the surface of structural members and it shall be ensured that the entire area of surface is sprayed. The curing compounds shall be possessed of reliable moisturizing effects that can be examined by tests.

Use of curing compounds shall meet the relevant requirements of product instruction.

2. Steam curing

Steam curingis to place concrete members into steam curing rooms witch are full of saturated steam or mixture of steam and air. Curing process is at higher temperature and relative humidity so as to speed up the process of concrete hardening by a relatively short time duration to achieve the required strength.

Steam curing process could be further categorized into 4 phases, standing, warming up, constant temperature and cooling.

1) Standing phase

After the concreteplacing completed, the concrete members should be stored at normal temperature for a certain time to improve its resistance ability against damaging at warming up stage. For ordinary Portland cement production component, the time should be 2~6h; for Portland cement of volcanic ash or slag Portland cement, there is no need for standing process.

2) Warming up phase

The heat absorption phase of the component. The heating rate should not be too fast, so as to avoid excessive temperature difference between the surface and internal of the component which might cause cracks. The heating rate shall not exceed 25 celsius degree per hour for thin-walled members (such as multi-ribbed floor, perforated porous floor, etc), 20 celsius degree for other members; structural members that made of hard drying concrete shall not exceed 40 celsius degree.

3) Constant temperature phase

At this phase, the strength of concrete increases the fastest, the relative humidity of 90%~100% should be guaranteed, the constant temperature, should not be more than 80 celsius degree for Portland cement concrete, increase to 90 celsius degree for slag cement, pozzolanic cement concrete.

Generally, the duration of keeping constant temperature is 5~8 hours.

4) Cooling phase

It is the heat dissipation phase of the component. Cooling speed should not be too fast, otherwise, the concrete surface cracks might come appear.

In general, with the thickness of components around 10 cm, the cooling speed should not exceed 20~30 celsius degree per hour. In addition, the temperature difference between component staying outside and the outdoor space should not exceed 40 celsius degree; and not more than 20 celsius degree whenever the outdoor temperature is negative.

4.3.7 Concrete quality control

Quality control for concrete slump and Vebe consistency shall meet the requirements:

(1) Inspection methods for slump and Vebe consistency shall meet the following requirements of current national standard *Standard for Test Method of Performance on Ordinary Fresh Concrete* GB/T 50080—2016.

(2) Allowable deviation of concrete slumps and Vebe consistency shall meet the requirements of Table 4-22.

Allowable deviation of concrete slump and Vebe consistency   Table 4-22

\multicolumn{4}{c}{Slump(mm)}			
Design value (mm)	≤40	50~90	≥100
Allowable deviation (mm)	±10	±20	±30
Vebe consistency			
Design value (s)	≥11	10~6	≤5
Allowable deviation (s)	±3	±2	±1

(3) Slump verification for ready-mixed concrete shall be carried out on the delivery point.

(4) For concrete with slump larger than 220mm, its slump flow may be tested according to requirements and the allowable deviation of the slump is 30mm.

The air content of concrete mixed with air-entraining agents or admixtures of air entraining type shall be tested according to the relevant requirements of current national standard *Standard for Test Method of Performance on Ordinary Fresh Concrete* GB/T 50080—2016 and it should meet the requirements of Table 4-23.

Air content limit of concrete	Table 4-23
Maximum nominal particle size of coarse aggregate (mm)	Air content in concrete (%)
20	≤5.5
25	≤5.0
40	≤4.5

4.3.8 Crack control of mass concrete

1. Long-term strength of mass concrete should be adopted as the basis of mix proportion design, strength evaluation and acceptance. As for foundations, the age of concrete may take 60d (56d) or 90d for determining the concrete strength; as for columns and walls with concrete strength grade is not lower than C80, the age may take 60d (56d) for determining the concrete strength. If the age of more than 28d is adopted for determining the concrete strength, the age shall be confirmed by design organization.

2. Mix design for mass concrete construction shall meet the requirements of related national codes and the concrete curing shall be strengthened.

3. During construction of mass concrete, temperature of concrete shall be controlled, which shall meet the following requirements:

(1) The temperature of mixture placing to forms should not be higher than 30℃; maximum temperature rise of concrete should not be greater than 50℃.

(2) During the covering curing or curing with forms, the difference between temperature of the position at 40～100mm below concrete surface and temperature at concrete surface shall not be greater than 25℃. After the covering curing or forms removal, the difference between temperature of position at 40～100mm below concrete surface and ambient temperature shall not be greater than 25℃.

(3) The temperature difference between the two adjacent temperature measurement points in the concrete shall not be greater than 25℃.

(4) Descending speed of temperature for concrete should not be greater than 2.0℃/d; when there are reliable experiences, the requirements of descending speed of temperature may be broadened properly.

4. The temperature measurement points of mass concrete foundation shall meet the following requirements:

(1) Two typical intersected vertical profiles should be selected for temperature measuring and the intersection of the vertical profile should pass through the center of foundation.

(2) Temperature measurement points shall be set around and within each vertical profile as well as at the intersection of the two vertical profiles; the temperature measurement points on surface of concrete shall be set at the bottom of the heat insulation covering or on the inner surface of forms, while the position and quantity shall be

corresponding to those around two profiles; the measuring points for ambient temperature shall not be less than 2.

(3) Temperature measurement points around each profile shall be set at 40~100mm below the concrete surface; temperature measurement points of each profile should be aligned vertically and transversely; temperature measurement points vertically set for each profile shall not be less than 3 in position with spacing not less than 0.4m and not greater than 1.0m; temperature measurement points transversely set for each profile shall not be less than 4 in position with spacing not less than 0.4m and not greater than 1.0m.

(4) When the thickness of foundation is not larger than 1.6m and there are reliable technical measures for crack control, temperature measurement may not be necessary.

5. The temperature measurement points for mass concrete columns, walls and beams shall meet the following requirements:

(1) If the minimum size of column, wall and beam structure is greater than 2m and the concrete strength grade is greater than or equal to C60, the temperature measuring shall be carried out.

(2) Two transverse profiles shall be selected longitudinally along members for temperature measuring and temperature measurement points shall be set around and in middle of each transverse profile; temperature measurement point on the concrete surface shall be set on the inner surface of forms, while the positions and quantity shall be corresponding to those around two profiles; the measurement points for ambient temperature shall not be less than 1 in position.

(3) Temperature measurement points around each transverse profile shall be set at 40~100mm below the concrete surface; temperature measurement points of each transverse profile should be aligned; temperature measurement points for each profile shall not be less than 2 in position and the spacing shall not less than 0.4m and should not greater than 1.0m.

(4) The technical measures for differential temperature control may be improved in accordance with the first temperature measuring result and temperature measuring may not be carried out for follow-up constructions.

6. The temperature measuring of mass concrete shall meet the following requirements:

(1) The temperature of mixture placing to forms at each measuring point should be determined in accordance with the temperature at temperature measurement points covered by concrete for the first time.

(2) The temperature at the temperature measurement points within the surface around concrete, on concrete surface and for ambient temperature shall be carried out synchronously with the placing and curing processes.

(3) The temperature measuring report shall be provided in time in accordance with the requirement of temperature measuring frequency. The temperature measurement report shall include details such as the temperature data of each temperature measurement point,

differential temperature date, temperature variation curve of representative points, temperature variation trend analysis.

(4) If the difference between temperature of the position at 40～100mm below concrete surface and the ambient temperature is less than 20℃, temperature measuring may stop.

7. The temperature measuring frequency of mass concrete shall meet the following requirements:

(1) The 1st day to 4th day: at least once for every 4h;

(2) The 5th day to 7th day: at least once for every 8th;

(3) The 7th day to the end of temperature measuring: at least once for every 12h.

4.3.9 Repair of concrete defects

1. The defects of concrete structures may be classified into the dimension deviation defects and appearance defects. The dimension deviation defects and appearance defects may be classified into the common defects and serious defects. When the dimension deviation of concrete structures exceeds those specified in the code but has no influence on performance and use functions of structures, it shall belong to the common defects; otherwise, when it has influence on performance and use functions of structure, it shall belong to the serious defects. The classification of appearance defects shall be compliance with those specified in Table 4-24.

Classification for appearance defects of concrete structures   Table 4-24

Name	Phenomenon	Serious defects	Common defects
Reveal of reinforcement	Steel bar in structural members are exposed as not covered by concrete	Exposed longitudinal steel reinforcement	Few reveal of other kinds of steel bar
Honeycomb	Exposed gravels due to lack of cement mortar on concrete surface	Honeycomb exist in the main load-carrying positions of members	Few honeycombs in other positions
Hole	Both the depth and length of space in concrete exceed the thickness of concrete cover	Hole in the main load-carrying positions of members	Few holes in other positions
Slag inclusion	Foreign matter included in concrete with depth exceeding the thickness of concrete cover	Slag inclusion exists in the main load-carrying positions of members	Few slag inclusions in other positions
Looseness	Part of concrete with poor compaction	Looseness in the main load-carrying positions of members	A small amount of looseness in other positions
Crack	Gap extends from concrete surface to concrete inner	Cracks which have influence on the performance or use function of structures exist in the main load-carrying positions of members	Few cracks not influencing the performance or use functions of structures in other positions

Continued

Name	Phenomenon	Serious defects	Common defects
Defect at joint	Concrete defects at member joints and loose splice of reinforcements and loose couplers	Defects influence the structural force transmission at the joint	Defects almost not influencing the structural force transmission exist at the joint
Profile defect	Edge missing, corner losing, out-of-straight edge angle, uneven warping and convex rib on flying edge, etc.	Profile defects of off-form concrete members, influencing use functions or decorative effects	Profile defects of other concrete members, not influencing use functions
Surface defect	Pitted surface, flaking, sand rising and contamination, etc. on surface of members	Surface defects of off-form concrete members having significant decorative effects	Surface defects of other concrete members, not influencing use functions

2. When defects of concrete structures are found during construction, their cause shall be seriously analyzed. For serious defects, a special repair scheme shall be developed by construction organization and the scheme shall be implemented after approval through justification and shall not be handed without authorization.

3. When repairing the common appearance defects of concrete structures, the following requirements shall be complied with:

(1) As for defects including reveal of reinforcement, honeycomb, hole, slag inclusion, looseness and surface defect, the concrete in the position where are of poor cementing shall be chiseled away; the surface shall be cleaned and finished with $1:2 \sim 1:3$ cement mortar after moistening with water.

(2) Cracks shall be sealed.

(3) Defects at joint and profile defect may be handled along with the decorative finish operation.

4. When repairing the serious appearance defect of concrete structures, the following requirements shall be complied with:

(1) Members of civil building structures which are in contact with aqueous medium, like the basement, lavatory and roof shall all be treated to seal by grouting. Members of civil building structures which are not in contact with aqueous medium may adopt sealing treatment by grouting, painting of polymer mortar or sealing treatment by other surface sealants.

(2) Members of corrodent-free industrial building structures which are in contact with aqueous medium, like the basement, lavatory and roof, as well as all members of industrial building structures with corrodent shall all be treated to seal by grouting. Members of corrodent-free industrial building structures which are not in contact with aqueous medium may adopted sealing treatment by grouting, painting of polymer mortar or sealing treatment by other surface sealants.

(3) As for the serious profile and surface defects of off-form concrete, a polishing

machine should be adopted for finishing after repairing with cement mortar or fine aggregate concrete.

5. As for common defects of dimension deviation of concrete structures, repairing may be carried out alongwith the decoration construction.

6. As for serious defects of dimension deviation of concrete structures, a special repair scheme shall be prepared jointly with the designer. Structures shall be undergo inspection for acceptance after being repaired.

Questions

1. Briefly describe the technical requirement of formwork engineering.
2. Briefly describecategories of formwork according to materials used.
3. How many types does the steel form have in composite steel-form formwork?
4. Briefly introduce the design principle of formwork.
5. What are the characteristics of formwork of wall?
6. Briefly describe the characteristics of formwork of column and beams.
7. Write out the force transmission routes of beam formwork.
8. What are the characteristics of formwork of slab?
9. Briefly describe the concrete strength requirements of removal of side formwork and bottom formwork.
10. Write out the sequence of the formwork removal.
11. What are the reinforcement category contents according to reinforcement mechanical properties?
12. Why does the reinforcement substitution exist? What are the principles of the reinforcement substitution?
13. Write out the contents of rebar processing.
14. How many connection methods in reinforcement connection?
15. What are the raw materials in concrete?
16. Why should the concrete laboratory mixture ration be converted into construction site mixture ration?
17. What are the requirements of the concrete transportation?
18. Briefly introduce the preparation works before concrete placing.
19. Write out the concept of concrete construction joint.
20. What are the requirements of setting position of horizontal construction joints?
21. What are the requirements of setting position of vertical horizontal construction joints?
22. Write out the three schemes of mass concrete placing.
23. Briefly introduce the three types of concrete compactness.
24. What are the requirements for compacting concrete with internal vibrators?
25. Briefly introduce the necessary of concrete curing.

Chapter 4 Concrete Structure Engineering

Exercise

1. Determine the actual length of the steel bar ①, ② and ③, as shown as following. Curving angle of ③ steel bar is 45° at middle parts.

Measurement difference as for $D=2.5d$ with typical angles   Table 4-25

Dimension of core Red(D)	2.5d			
Bending angle(α)	30°	45°	60°	90°
Calculation value	0.2923d	0.4937d	0.7710d	1.7595d
Value taken	0.3d	0.5d	0.8d	2d

Hook increment as for $S=3d$, $D=2.5d$ with typical angles   Table 4-26

Dimension of core Red(D)	2.5d		
Bending angle(α)	90°	135°	180°
Calculation value	3.49053d	4.8608d	6.2310d
Value taken	3.5d	4.9d	6.25d

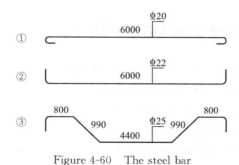

Figure 4-60 The steel bar

2. Laboratory mix ratio of concrete C20, 1 : 2.55 : 5.12, water cement ration 0.65, amount of the cements per cube meters 310kg. Water content ration of sand in site is 3%, water content ration of crushed stone is 1%. Try to calculate the site mix proportion and amounts of the aggregates per cube meters.

Chapter 5　Prestressed Concrete Structure Engineering

Mastery of contents: Construction technology of pre-tensioned and post-tensioned concrete.

Familiarity of contents: Various construction methods of prestressed concrete, construction of unbounded prestressed concrete (post-tensioned).

Understanding of contents: Construction equipment and accessories.

Key points: Construction technology of pre-tensioned and post-tensioned concrete, as well as construction methods of unbounded prestressed concrete (post-tensioned).

Difficult points: Key points of construction of pre-tensioned and post-tensioned concrete, construction methods of unbounded prestressed concrete (post-tensioned).

5.1　Overview of prestressed concrete

Ordinary reinforced concrete structure has many advantages, but it also has some disadvantages, such as premature cracking and smaller stiffness which cannot make full use of high strength materials, etc., thus affecting the application of reinforced concrete structure in construction engineering.

In order to overcome the shortcomings above, the prestressing of concrete is the best way: before the componentsbearing load, the reinforcement in component tensile area will be tensioned at first, applying its resilience forces on concrete by releasing the tensioned reinforcement, then, the pre-compression stress established at the tensile area of the concrete components. In this way, after the structural member bears the load, this pre-compression can offset most or all of the tensile stress generated by the load in the concrete tensile area, delaying the generation of cracks and inhibiting the development of cracks.

Comparing with the ordinary concrete, prestressed concrete has following advantages:

(1) Higher crack resistance, stiffness and good durability of components.

(2) Make full use high strength steel bar and high strength concrete.

(3) Reduce the section size of the component, reduce self-weight and decrease material consumption.

(4) The utility function of concrete structure could be expanded, and the comprehensive economic benefit is better.

Meanwhile, the disadvantages exist:

Chapter 5 Prestressed Concrete Structure Engineering

(1) Specialized mechanical equipment are necessary.

(2) Complex process, high technical lever is required.

(3) Strictly requirement for materials used in prestressed concrete engineering.

With the improvement of the construction technology, the application of pre-stressed concrete will be more and more extensive.

Prestressed concrete is widely used in civil engineering constructions: (1) single structure member, such as, roof truss, crane beam, large roof slab, large-span hollow floor slab. (2) structure system, such as cast-in-place frame structure system, cast-in-place floor structure system and the whole prestressed prefabricated slab-column structure system.

Prestressed concrete can be categorized into pre-tensioning and post-tensioning according to different methods of applying prestressing.

(1) Pre-tensioning

Reinforcement is tensioned before the concrete is poured, and the prestress in the reinforcement is transferred to the concrete by the bonding forces between the reinforcement and the concrete near around.

(2) Post-tensioning

Reinforcement is tensioned after the concrete is poured and reaching a certain strength, the prestress transferred by the anchors that fixed at the ends of the structure member.

Post-tensioningmethods can be further categorized into bonded post-tensioning and unbounded post-tensioning.

The former is a working cooperation type by means of grouting in the gap between tendons surface and the concrete inner surface in the reserved duct.

The later has no bonding effect due to the grease applied to the prestressed tendons, in this case, anchorages function is more important in this case.

5.2 Pre-tensioning

Tensioning of prestressed tendons has been completed before pouring concrete, prestressed tendons are temporarily anchored to the pedestal or steel formworks, after concrete pouring and reaches a certain strength, release the prestressed tendons so that its retraction will squeezing the concrete to establish the pre-stress.

Pre-tensioning is a process of tensioning the prestressed tendons before pouring concrete, temporarily anchoring them to a fixture on the pedestal or steel formwork, and then pouring concrete. After the concrete reaches certain strength, the prestressed tendons are released till cutting off, the prestressed tendons are applied pre-pressure stress to concrete when they shrink back. This pre-tensioning method is shown in Figure 5-1. which is widely used in the production of small and medium-sized prefabricated concrete

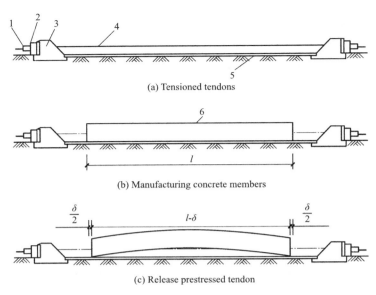

Figure 5-1 Production diagram of pre-tensioning method
1—Clamps; 2—Transverse beam; 3—Abutment; 4—Prestressed tendon;
5—Surface of tension field; 6—Concrete member

components.

Pre-tensioning methods can be further categorized into long line bench methods and short line bench methods.

Flowing construction by conveyor belt is adapted in short line bench method, every working process will be executed in different operation spot, it is often used in fabrication of concrete members in manufacture factory.

While for long line bench methods, all working process will be executed at the same pedestal. Long line bench methods will be introduced in this section.

5.2.1 Tendons, abutment, clamp, equipment and tools

1. Tendons

Steel wire and strand are commonly used in the production of prestressed concrete member production.

Steel wire should be stress relief steel wire, and spiral rib steel wire and scratch steel wire should be adopted in advantage to ensure the bond between steel wire and concrete.

Steel strand is divided into 1×2 steel strand and 1×7 steel strand according to its twist structure; standard steel strand and engraved steel strand according to its different processing methods. The latter is made of engraved steel wire twisted in order to increase the bonding force between steel strand and concrete.

The replacement of prestressing tendons shall be calculated specially and confirmed by the original design organization.

2. Abutment

The abutment is one of the main equipment in prestressing production process which bears the total tension of the prestressed tendons.

Therefore, the abutment should have sufficient strength, stiffness and stability to avoid the loss of prestress caused by the deformation, overturning and sliding of the abutment.

Abutment can be divided into pier abutment and groove abutment according to their different structure as shown in Figure 5-2, Figure 5-3.

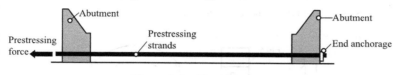

Figure 5-2　Pier abutment

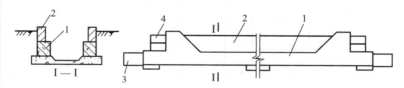

Figure 5-3　Groove abutment

1—Column; 2—Brick wall; 3—Below transverse beam; 4—Upper transverse beam

3. Clamp

In the prestressed concrete construction, the clamp is the tool for tensioning and temporary anchoring of prestressed tendons. Clamp should be reliable, simple structure and convenient in construction. Clamp can be either used in tensioning end or anchoring end.

Some clamp for single steel wire or single strand are show as in Figure 5-4~Figure 5-7.

1) Single steel wire clamp

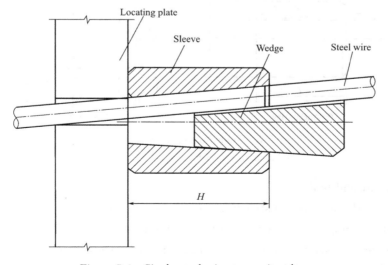

Figure 5-4　Single steel wire taper pin calmp

Civil Engineering Construction Technology and Organization

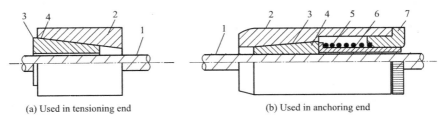

(a) Used in tensioning end (b) Used in anchoring end

Figure 5-5 Single steel wire clamping piece clamp

1—Steel wire; 2—Sleeve; 3—Clamping piece; 4—Steel wire ring; 5—Elastic ring; 6—Push rod; 7—Cover

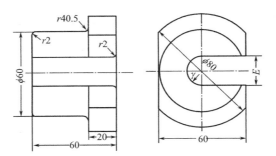

Figure 5-6 Single header clamp (mm)

2) Single strand clamp

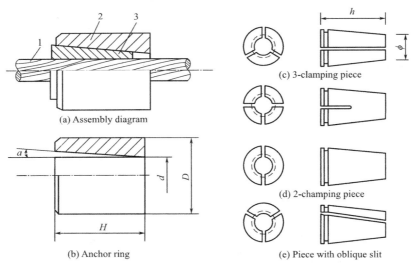

(a) Assembly diagram

(b) Anchor ring

(c) 3-clamping piece

(d) 2-champing piece

(e) Piece with oblique slit

Figure 5-7 Single hole clamping piece anchor

1—Steel strand; 2—Anchor ring; 3—Clamping piece

4. Equipment and tools

In prestressed concrete construction, prestressed steel wire can be tensioned either single or multiple. Single tensioning is often adopted at abutment production process, in this case, the electric screw tensioning machine or small electric winch tensioning machine is generally adopted cause of their small tension force. While, core jack is often selected for the prestressed strand tensioning.

5.2.2 Construction technology of pre-tensioning pre-stressed concrete

When pre-tensioning prestressed concrete members are produced on the abutment, the general process flow is shown in Figure 5-8.

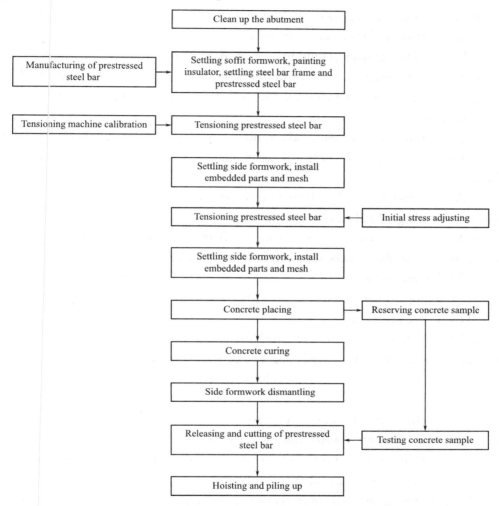

Figure 5-8 Construction technology of pre-tensioning

1. Tensioning

The tensioning of prestressed tendons is a key process in construction. The tensioning stress of prestressed tendons should be controlled strictly according to the design requirements.

1) Tensioning method

Single steel bar tensioning is often adopted in prestressed concrete construction, while group tensioning method can be selected when there are too much number with densely distributed, in this case, the initial stress of each single steel bar should be adjusted to make its length and tightness consistent to ensure the stress of the prestressed steel bars keep consistent after tensioning.

2) Tensioning sequence

Tensioning sequence of prestressing tendons shall meet the design requirements as well as the following requirements:

(1) The tensioning sequence shall be determined according to such factors as load-carrying characteristics of structures, convenience for construction and safety in operation.

(2) Prestressing tendons should be tensioned uniformly and symmetrically.

(3) As for prefabricated beam, the tension sequence should be symmetrical in left and right, if the pretension zone on the top of the beam is equipped with prestressed steel bar, it should be tensioned priority.

(4) The overturning moment and eccentric force affecting on the abutment should be considered as much as possible when determining the tensioning sequence of prestressed steel bar.

(5) As for prefabricated hollow beam, tensioning the middle one at first, then tension to two side gradually and symmetrically.

(6) As for cast-in-situ prestressed concrete floors, prestressing tendons of floors and secondary beams should be tensioned before those of girders.

(7) As for members placed in a horizontal and spliced way, such as precast roof trusses, tensioning sequence shall be from top to bottom.

3) Tensioning procedure

For the tensioning of prestressed steel wire, one time tensioning should be adopted because of the large amount of tensioning work, see Formula (5-1).

$$0 \rightarrow 1.03 \sim 1.05 \sigma_{con} \tag{5-1}$$

There in:

σ_{con} ——Designed tension control stress.

1.03, 1.05——Coefficient (considering the error of spring dynamometer, the insufficient stiffness of abutment beam or locating plate, the length of abutment doesn't meet the design value, and the influence of workers' operation).

As for prestressed strand, over tensioning should be selected when ordinary slack strand is adopted, while one time tensioning should be selected when low slack strand is adopted. Their procedures are shown as in Formula (5-2) and Formula (5-3) respectively.

$$0 \rightarrow 1.05 \sigma_{con} \xrightarrow{\text{Sustaining 2min}} \sigma_{con} \tag{5-2}$$

$$0 \rightarrow \sigma_{con} \tag{5-3}$$

2. Placing and curing

After the prestressed tendons are tensioned, the concrete pouring should be carried out as soon as possible.

(1) Low water-cement ration should be adopted when determining concrete mix ratio, and the cement amount should be controlled, also, the aggregate with good gradation

should be used to minimize concrete shrinkage and creep, so as to reduce the loss of prestress caused by it.

(2) The pouring of concrete must be completed at one time, construction joints are not allowed.

(3) In placing, the vibrator shall not collide with the prestressed tendons and ensure the compacting of concrete, especially the end and corner of the members should ensure the pouring quality, so as to make the concrete and the prestressed tendons have an effective bonding force and prestress transferring.

3. Releasing of steel bar

1) Requirement of releasing

Before releasing of prestressed rebar, concrete strength shall meet the design requirements and compressive strength of cubic concrete specimens cured under the same conditions shall meet the following requirements:

(1) It shall not be less than 75% of designed concrete strength grade.

(2) It shall not be less than 30 MPa when stress-relieved steel wires or steel strands are adopted as prestressing tendons for pre-tensioned members.

(3) It shall not less than the minimum concrete strength required in the product technical manual provided by anchorage supplier.

(4) As for post-tensioned prestressed beams and slabs, the age of concrete for cast-in-situ structures should not be less than 7d and 5d respectively.

Before releasing, the side formwork should be removed, the structure members can freely retract to avoid damage of formwork or component cracking.

2) Sequence of releasing

The releasing sequence of prestressed tendons should meet the design requirements, if there is no specifications in design, it can be carried out in the following order:

(1) For members bearing axial pressure, all prestressed tendons shall be released simultaneously (such as bar).

(2) For members bearing eccentric pressure (such as beam), the prestressed tendons in the area with small pressure should be released simultaneously, and then the prestressed tendons in the area with large pressure should be released simultaneously.

(3) In the case that the releasing can't be carried out according to the orders above, it should be released in stages, symmetrically and staggered in order to prevent member of bending and cracking, the fracture of prestressed tendons in the process of tension releasing.

3) Methods of releasing

In general, in the releasing process, the anchorage device should be released slowly so that all the prestressed tendons which in a high stress are released slowly at the same time.

For the releasing of multiple steel wire or steel strand, the sand box or wedge block equipment can be adopted as shown in Figure 5-9.

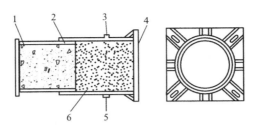

Figure 5-9　Sand box structure
1—Piston； 2—Steel sleeve box； 3—Sand inlet； 4—Base of steel sleeve box； 5—Sand outlet； 6—Sand

Both of these equipment are placed between the abutment and the transverse beam to control the releasing speed, which can provide reliable operation and convenient construction (Figure 5-10).

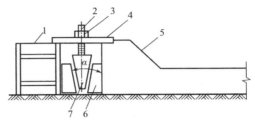

Figure 5-10　Wedge releasing equipment
1—Transverse beam； 2—Screw bolt； 3—Screw nut； 4—Load bearing plate；
5—Abutment； 6—Steel block； 7—Steel sedge block

5.3　Construction technology of bonded post-tensioning prestressed concrete

This methodsreserves channel prior the manufacturing of structure member or structure, the prestressed tendons are threaded through the channel and stretched on member or structure, anchored with anchorages, prestress is applied to the concrete by means of anchorages, at last, channel grouting is carried out. The production process is shown in Figure 5-11.

In the process, prestressed tendons are bonded to the concrete by grouting in tunnel, which reduces the effect of transferring prestress to the anchorage, and improves the reliability and durability of the anchorage.

The bonded post-tensioning prestressed concrete doesn't need the platform equipment, has more flexibility and widely used at construction site for producing large prefabricated prestressed concrete components and cast-in-site prestressed concrete structures.

5.3.1　Materials, equipment and tool in bonded post-tensioning process

1. Prestressed tendons

Bundle of steel wires, bundle of strand and single large diameter reinforcement can be adopted.

Chapter 5　Prestressed Concrete Structure Engineering

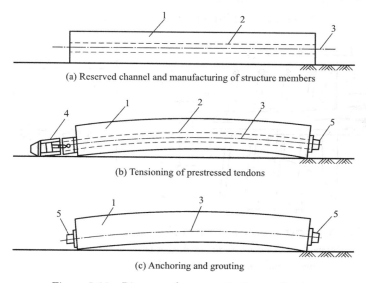

Figure 5-11　Diagram of post-tensioning production
1—Structure member; 2—Reserved tunnel; 3—Prestressed tendons; 4—Jack; 5—Anchorages

2. Anchorage

In post-tensioning structure member or structure, anchor is a permanent anchoring device that maintains the tension of prestressed tendons and transfers force to concrete.

As anchors, they should have reliable anchoring ability, simple in construction, convenient in operation, small in size and low in cost.

1) Anchor for bundle of steel wires

There are two types of equipment, heading anchorage and steel cone anchorage which are shown in Figure 5-12 and Figure 5-13.

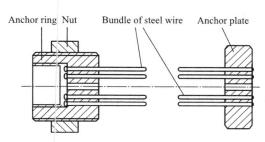 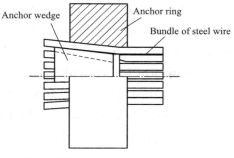

Figure 5-12　Heading anchorage for bundle of steel wires

Figure 5-13　Steel cone anchorage for bundle of steel wires

2) Anchor for bundle of strand

The multi-channels anchorage system which clamps each strand by means of each tapered channel fitted with a pair of clips is reliable and its advantage is that the failure of anchoring of any single strand will not lead to the failure of all anchoring. Anchoring system used in tensioning end and fixing end are shown in Figure 5-14 and Figure 5-15.

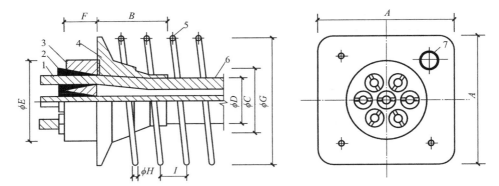

Figure 5-14 Anchorage system of clamping piece with multi-channels (Tensioning end)
1—Strand; 2—Clamping piece; 3—Anchor plate; 4—Base of anchor plate;
5—Spiral reinforcement; 6—Metal bellows; 7—Grouting entrance

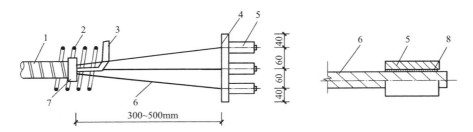

Figure 5-15 Squeezing anchorage (Fixing end)
1—Metal bellows; 2—Spiral reinforcement; 3—Went pipe; 4—Base of anchor plate;
5—Squeezing anchorage; 6—Strand; 7—Constraint circle; 8—Profiled steel wire ring

3) Anchor for single large diameter reinforcement

The finely-rolled deformed rebar with large diameter is often adopted as prestressed steel bar, which can be anchored by a special nut matching the screw-thread of the finely-rolled deformed rebar.

3. Equipment for tensioning

In the construction of post-tensioning prestressed concrete, the tensioning of prestressed tendons is carried out by hydraulic jack, equipped with electric oil pump and external oil pipe, as well as force measuring instrument.

5.3.2 Construction technology of bonded post-tensioning prestressed concrete

General process flow of post-tensioning prestressed concrete is shown in Figure 5-16.

1. Preformed duct

(1) Embedding metal spiral pipe is the method that most widely used in nowadays construction of bonded prestressed concrete, especially in cast-in-site concrete structures.

Chapter 5 Prestressed Concrete Structure Engineering

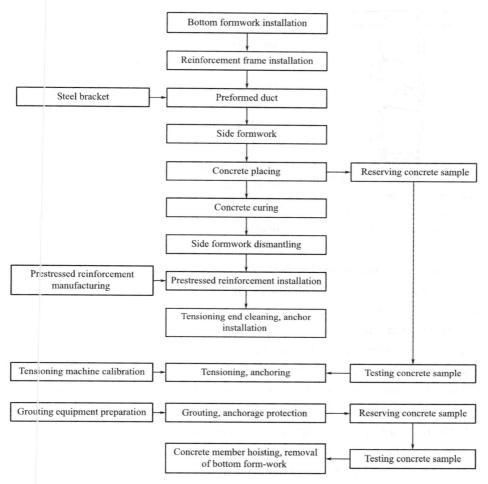

Figure 5-16 Construction technology of post-tensioning

Metal spiral pipe are also called corrugated pipe as shown in Figure 5-17 and Figure 5-18, made of cold rolled steel strip or galvanized steel strip after pressuring waves and spiral biting.

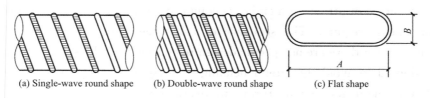

Figure 5-17 Metal spiral pipe

It has the advantages of light weight, better stiffness, simple connection, low friction coefficient and good bonding with concrete, it can runs in various shapes such as straight line, curve, polyline and other shapes of tunnel as an ideal materials for performed duct of concrete structure members.

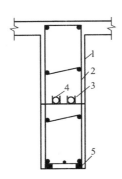

Figure 5-18 Fixing of metal spiral pipe
1—Side formwork of beam; 2—Stirrup;
3—Reinforcement supporter;
4—Metal spiral pipe; 5—Cushion block

(2) Extracting pipe methods can also be adopted in processes of preformed duct in concrete members.

Steel extraction pipe can be only used in straight shape tunnel, while rubber extraction can be used in both straight and curve shape tunnels.

2. Tensioning

1) Tensioning requirement

The tensioning of prestressed tendons is the key in the post-tensioning prestressed construction. Before tensioning, the concrete strength of the members or structures shall meet the design requirements, when there is no specific requirement in the design, it shall not less than 75% of the designed strength.

Before tensioning, the welding slag, burr, concrete residue at the end contact surface between the embedded steel plate and anchor should be cleaned.

2) Tensioning method

Post-tensioning tendons shall adopt one end or two ends tensioning according to the requirements of design and special construction schemes. When both ends of prestressing tendons are tensioned, they should be tensioned simultaneously. However, the way that one end is tensioned after the other end has been tensioned and anchored may be accepted. When there are no specific requirements in the design, the following requirements shall be met:

Bonded prestressing tendons in length not greater than 20m may be tensioned at one end, while bonded prestressing tendons in length greater than 20m should be tensioned at both ends; threshold length of tensioning at one end for straight pre-stressing tendons may be extended to 35m.

3) Tensioning sequence

When structure or structure member is equipped with multiple pre-stressing tendons, batch tensioning is required. The sequence of batch tensioning should meet the design requirements; when there are no specific design requirements, the symmetrical tensioning should be adopted to avoid the lateral bending or torsion of components under eccentric pressure. For components manufacturing in overlapping way at construction site, the tensioning sequence should be carried out layer by layer and from top to bottom.

4) Tensioning procedure

The tensioning procedure of prestressed tendons is mainly determined according to the component type, tensioning anchorage system, relaxation loss and other factors.

For ordinary relaxation prestressed tendons, in order to reduce the stress relaxation loss of prestressed tendons, the over-tensioning procedure should be used.

Chapter 5 Prestressed Concrete Structure Engineering

For heading anchorage and loadable anchorages, procedure is shown in Formula (5-4). While for clamping piece anchorage and un-loadable anchorages, procedure is shown in Formula (5-5).

$$0 \rightarrow 1.05\sigma_{con} \xrightarrow{\text{Standing by 2mins}} \sigma_{con} \qquad (5\text{-}4)$$

$$0 \rightarrow 1.03\sigma_{con} \qquad (5\text{-}5)$$

When low relaxation steel wire or strand are adopted, one time tensioning may be adopted, its procedure is shown in Formula (5-6).

$$0 \rightarrow \sigma_{con} \qquad (5\text{-}6)$$

3. Grouting

Grouting can both protect prestressed tendons from corrosion and bond the prestressed tendons and the structure member together effectively.

Ducts shall be grouted as early as possible after bonded post-tensioning tendons being tensioned and qualified. The cement grout in ducts shall be full and compact.

The bonded pre-stressing can control the crack development of the structure and reduce the load of the beam end anchors, therefore, attentions must be paid to the quality of grouting.

The following preparations shall be carried out prior to ducts grouting:

(1) Smoothness of ducts, air exhaust and bleeding pipes and grout holes shall be confirmed; ducts preformed by embedded pipes may be cleaned with compressed air.

(2) Seams of anchorage devices at ends shall be sealed with such materials as cement grout and cement mortar. Exposed anchorage devices may be sealed by using protection cover.

(3) The tightness of ducts shall be confirmed when using vacuum grouting.

The cement, water and admixtures for preparing cement grout shall meet requirements of relevant current standards and the following requirements:

(1) Ordinary Portland cement or Portland cement should be adopted.

(2) The water and admixtures mixed shall have no component adverse to prestressing tendons or cement.

(3) Mix proportioning test shall be carried out for admixtures and cement and mix proportion shall be determined.

Cement grout for grouting shall meet the following requirements:

(1) The consistency of cement grout shall be controlled within $12 \sim 20s$ for basic grouting process and should be controlled with $18 \sim 25s$ for vacuum grouting process.

(2) The water / cement rate shall not be larger than 0.45.

(3) The free bleeding rate in 3h should be 0 and shall not be larger than 1%. Amount of bleeding, if any, shall be absorbed by cement grout completely within 25h.

(4) The free expansion rate in 24h shall not be larger than 6% for basic grouting process and shall not be larger than 3% for vacuum grouting process.

(5) The content of chloride ion in cement grout shall not exceed 0.03% of cement mass.

(6) The compressive strength of cubic cement grout with side length of 70.7 mm after 28 da standard curing shall not be lower than 30 MPa.

(7) The test methods for consistency, bleeding rate and free expansion rate shall meet the requirements of the current national standard *Grouting Admixture for Prestressed Structure* GB/T 25182—2010.

Note: A set of specimens contains 6 single cement grout specimens. The compressive strength is the average value for a set of test specimens; when the difference of the maximum or minimum value of compressive strength to the average value in a set of test specimens exceeds 20%, the average value of the compressive strength of the other 4 test specimens shall be taken.

The preparation and use of cement grout for grouting shall meet the following requirements:

(1) The cement grout should be mixed by high speed mixer and the mixing period shall not exceed 5 minutes.

(2) The cement grout shall be filtered by sieve mesh with a mesh size not larger than 1.2mm×1.2mm before use.

(3) The cement grout that cannot be grouted into ducts in a short time after mixing shall be kept mixing slowly.

(4) The cement gout shall be grouted into ducts before initial set and the duration from stopping mixing to finishing grouting should not exceed 30 minutes.

The grouting shall meet the following requirements:

(1) The ducts in upper layer should be grouted after ones in lower layers.

(2) Grouting shall be continuous till the grout from air exhaust holes has the same consistency of that filled into grout holes and appears no bubble. Air exhaust holes are then sealed in sequence along the grout flowing direction. After air exhaust holes are all sealed, pressure of 0.5~0.7 MPa should continue being applied and grout holes shall be sealed after stabilizing pressure for 1~2min.

(3) When serious bleeding occurs, a second time grouting should be carried out and gravity grout replenishing for bleeding holes shall be performed.

(4) When grouting halt due to some reason, cement grout in ducts where grouting is not completed shall be cleaned out with pressure water.

4. Anchorage protection

Exposed anchorage devices and pre-stressing tendons shall be applied with reliable protection measures according to the design requirements.

The exposed excess length of post-tensioning tendons after anchoring should be cut off by mechanical method, or may be oxy-acetylene flame. The exposed residual length should not be less than 1.5 times of the diameter of pre-stressing tendons and shall not be less than 30mm.

Anchorage protection should be done as soon as the grouting finished. Concrete used for anchorage protection should be one grade higher than that of design strength of structure member, and its size should be greater than that of embedded steel plate, the concrete cover of the anchorage should not be less than 50mm. there shall be no cracks between anchor and the surrounding concrete after sealing.

5.4 Construction technology of un-bonded post-tensioning prestressed concrete

Un-bonded prestressed concrete is the development and important branch of post-tensioning technique. It is a kind of concrete structure equipped with un-bonded prestressed tendons, transferring prestress completely relying on anchorages.

The construction process of un-bonded prestressed is as follows:

The un-bonded prestressed tendons are laid in the formwork like ordinary steel bars, and then the concrete is poured. After the concrete reaches the strength specified in the design, the prestressed tendons are tensioned and anchored. The characteristics of this prestressed technology are: (1) No preformed duct and grouting, simple in construction. (2) Less friction force in tensioning. (3) Prestressed tendons are easy to be tent into the different of shapes, such as curves, it is most suitable for the structure with curve reinforcement. (4) Because the prestress can only permanently rely on the anchor to transfer to the concrete, so there are higher requirements for the anchor. At present, the technology of un-bonded prestressed concrete is widely used in one-way and two-way continuous slab with large span and multi-ribbed floor slab, which is economical and reasonable. There are also prospects for the development of multi-span continuous beams.

5.4.1 Un-bonded prestressed tendons, equipment and tools

1. Un-bonded tendons

Un-bonded prestressed tendons are composed of steel strand, coating layer and outer layer, as show in Figure 5-19.

Steel strand have the structure of 1×7 with diameter of 9.5mm, 12.7mm, 15.2mm and 15.7mm. The purpose of the coating layer is to isolate the prestressed tendons strand from the concrete, reduce the friction during tensioning, and prevent the corrosion of the prestressed stands.

The coating layer adopts anti-corrosion lubricating grease, the performance requirements are:

(1) Good chemical stability and has no erosion effect on surrounding materials.

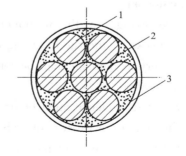

Figure 5-19 Un-bonded prestressed tendons
1—Steel strand; 2—Coating layer; 3—Outer layer

(2) Water proofing and non-hygroscopic. (3) Good corrosion resistance. (4) Good lubrication and low friction resistance. (5) Neither of flowing at high temperature and nor brittle at low temperature in the temperature range of -20 celsius degree ~ +70 celsius degree, and has some certain toughness. Because of its adequate tensile strength, toughness and abrasion resistance, the outer layer of un-bonded prestressed tendons is often made of high density Polyethylene plastic which can ensure that the prestressed stand are not easily damaged in the process of transportation, storage, laying and concrete pouring.

Un-bounded prestressed strand are supplied in plate after factory fabrication.

2. Anchorage

Single hole clamping piece anchorage is often adopted at tensioning end, as shown in Figure 5-7. In post-tensioning construction, it can form a single-hole clamping system together with pressure plate and spiral reinforcement as shown in Figure 5-20. It is commonly used for anchoring strand with diameter of $\phi^s 12.7$ and $\phi^s 15.2$.

Squeezing anchorage can be used for anchoring the fixed end of un-bonded prestress single strand, as shown in Figure 5-15.

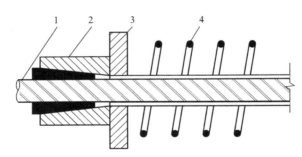

Figure 5-20 Single hole clamping system
1—Strand; 2—Single clamping piece anchorages; 3—Supporting steel plate; 4—Spiral reinforcement

3. Tensioning equipment

Hydraulic jack are often adopted at tensioning process.

5.4.2 Construction technology of un-bonded post-tensioning prestressed concrete

1. laying out

Before laying out the un-bonded prestressed tendons, the outer layer should be carefully checked, and the local slight damage should be repaired by wrapping waterproof tape, the serious damage shall be removed.

1) Sequence

Laying out sequence must be determined especially in two way slab constructions.

In the laying out process of slab, bottom ordinary reinforcement should be laid prior un-bonded prestressed tendons, pipes of water supply and electricity should be laid out after the prestressed tendons, non-prestressed negative moment bars at the support are laid at last.

2) Locating and fixing

After the vertical and horizontal positions of the un-bonded bars are checked correctly, they should be bound firmly with the non-prestressed steel bars and supporting steel bars with lead steel wire to avoid displacement and deformation in the process of pouring concrete.

3) Fixing at end part

The un-bonded bars of the tensioning end shall be perpendicular to the pressure plate, and the extrusion anchors of the fixing end shall be tightly attached to the pressure plate.

2. Tensioning

The general tensioning order of un-bonded prestressed concrete floor structure is to tension floor slab first and then tension floor beam.

The un-bonded bars in the slab can be tensioned one by one, tensioning symmetrically should be adopted in beam.

Un-bonded pre-stressing tendons in length not greater than 40m may be tensioned at one end, while un-bonded pre-stressing tendons in length greater than 40m tensioned at both ends.

3. Anchorage protection

In the un-bonded prestressed structure, the maintenance and transferring of the tension of the prestressed tendons to the concrete are completely dependent on the anchorage at the tendons end. There must be more strict sealing and anti-corrosion measures than that of bonded prestressed structure at the anchorage area to prevent moisture corrosion of prestressed tendons and anchors.

The exposed excess length of un-bounded tendons shall not be less than 30mm, the excess parts should be cut by hand grinding wheel saw.

The sealing structure of the anchorage area is shown in Figure 5-21.

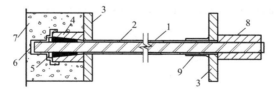

Figure 5-21 Full sealing structuue of un-bonded tendons

1—Outer layer; 2—Strand; 3—Supporting steel plate; 4—Anchor ring; 5—Clamping piece; 6—Plastic caps; 7—Sealing concrete; 8—Squeezing anchorage; 9—Plastic sleeve or paste tape

The requirements for cover of concrete or mortar in the anchorage are: no less than 25mm in the beam and no less than 20mm in the slab.

Questions

1. Comparing with the ordinary concrete, what are the advantages and disadvantages of prestressed concrete construction?

2. Write out the concept of prestressed pre-tensioning and post-tensioning prestressed concrete construction.

3. What's clamp and what's its function in pre-tensioning prestressed concrete construction?

4. Briefly describe the general process flow of pre-tensioning prestressed concrete construction.

5. Write out the tensioning method of pre-tensioning prestressed concrete construction.

6. Write out the tensioning sequence and tensioning procedure of pre-tensioning prestressed concrete construction.

7. What are the requirements in placing and curing of pre-tensioning prestressed concrete construction?

8. Write out the requirement of releasing in pre-tensioning prestressed concrete construction.

9. What's the sequence of releasing in pre-tensioning prestressed concrete construction?

10. Briefly describe the general process flow of post-tensioning prestressed concrete construction.

11. What are the advantages of preformed duct in post-tensioning prestressed concrete construction?

12. Write out the tensioning requirement and method in post-tensioning prestressed concrete.

13. Write out the tensioning sequence and procedure.

14. Describe briefly the requirements of preparation and use of cement grout for grouting in post-tensioning prestressed concrete construction.

15. Describe briefly the requirements of grouting in post-tensioning prestressed concrete construction.

16. How to do anchorage protection in post-tensioning prestressed concrete construction?

Chapter 6　Structure Installation Engineering

Mastery of Contents: Construction technology and methods of concrete and steel construction, as well selection of lifting machinery.

Understanding of contents: Type and application scope of lifting machinery.

Key points: Installation technology of reinforced concrete single layer industrial construe, selection of lifting machinery.

Difficult points: Installation technology of steel construction, selection of machinery.

6.1　Rigging equipment

6.1.1　Pulley blocks

The pulley blocks is composed of a certain number of fixed and movable pulleys, it can both lower the pulling force and change the direction of the force.

The number of ropes in pulley block that together sharing the gravity of the weight is called working line number which is numerically equals the number of ropes that running around the movable pulley.

The efficiency of lowering the pulling force depends on both the working line numbers and the friction resistance of sliding bearing.

The rope running head of the pulley block can be divided into two types: from the fixed pulley blocks and from the movable pulley (Figure 6-1).

The pulling force of the rope running head can be determined by Formula (6-1):

$$N = KQ \quad (6\text{-}1)$$

There in:

N——Pulling force of the rope running head.

Q——Calculated load, which equals the multiplying of weight and dynamic coefficient.

K——Efficiency coefficient of pulley block.

(1) When the rope head is drawn from the fixed pulley block, K is determined by Formula (6-2):

$$K = \frac{f^n \times (f-1)}{f^n - 1} \quad (6\text{-}2)$$

(2) When the rope head is drawn from the movable pulley block, K is determined by Formula (6-3):

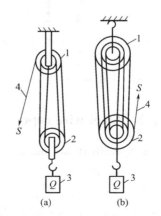

Figure 6-1　Pulley blocks
1—Fixed pulley; 2—Movable pulley;
3—Weight; 4—Steel wire rope

$$K = \frac{f^{n-1} \times (f-1)}{f^n - 1} \tag{6-3}$$

There in:

f ——Resistance coefficient of single pulley block (1.02 for rolling bearing, 1.04 for bronze axle sleeve bearing, 1.06 for bearing without axle sleeve).

n ——Working line number.

6.1.2 Winch

Electric winch which is widely used in building construction includes slow speed winch (JJM) and fast speed winch (JJK). Slow speed winch is mainly used for cold tensing steel bar and the prestressed steel bar; fast speed winch is mainly used for vertical and horizontal transport and piling operations.

Electric winch must be anchored by ground anchor to prevent its sliding or overturning during operation. There are four methods for fixing it as shown in Figure 6-2: bolt anchoring, horizontal anchoring, vertical piling anchoring and weight balancing anchoring.

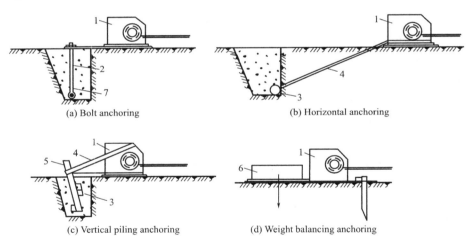

(a) Bolt anchoring (b) Horizontal anchoring

(c) Vertical piling anchoring (d) Weight balancing anchoring

Figure 6-2 Fixing methods

1—Winch; 2—Ground bolt; 3—Crossbar; 4—Pulling cable; 5—Wooden pile; 6—Balancing weight; 7—Pressing plate

6.1.3 Steel wire rope

1. Specifications and types of steel wire rope

1) Specifications

The steel wire rope is made of smooth steel wires with same diameter which are twisted into wire strand; the steel wire rope is accomplished by twisting the six wire strands around one core rope.

The steel wire rope can be divided into three specifications according to the number of wires in each strand.

6×19+1: Six strands, 19 steel wires for each strand, one core rope in the center. Wires are bold, harden and wear-resisting, difficult to bend, normally used for wind rope.

6×37+1: Six strands, 37 steel wires for each strand, one core rope in the center. Wires are slender, softer, normally used for pulley blocks and slings.

6×61+1: Six strands, 61 steel wires for each strand, one core rope in the center. Wires are softer, normally used for heavy lifting machinery.

2) Types

The steel wire rope can be divided into two types according to the directions between the directions that wires twisting into strands and that of the strands running around upon the core rope (same directions, opposite directions).

Same direction steel wire rope: the directions of the wires twisting into strands and that of the strands running around upon the core rope are same. It has good flexibility, flat surface, wear-resisting, but easy to be loose and kink curling, easy making the weights rotating while lifting it. Thus, it is normally used for dragging or traction equipment.

Opposite direction steel wire rope: the directions of the wires twisting into strands and that of the strands running around upon the core rope are different. It is stiff, not easy to loose, no rotating while lifting the weights. It is normally used for lifting works.

2. Allowable tensioning force

Allowable tensioning force can be determinedby the maximum working pulling forces, determined by Formula (6-4).

$$[F_g] \leqslant \frac{\alpha F_g}{K} \qquad (6-4)$$

Therein:

$[F_g]$ ——The maximum working pulling force (kN).

F_g ——Sum total breaking pulling forces of the steel wires (kN).

α ——Conversion coefficient, as shown in Table 6-1.

K ——Security coefficient, as shown in Table 6-2.

Conversion coefficient of sum total breraking pulling forces Table 6-1

Specifications of the steel wire rope	Conversion coefficient
6×19	0.85
6×37	0.82
6×61	0.80

Security coefficient of the steel wire rope Table 6-2

Application	Security coefficient	Application	Security coefficient
Wind rope	3.5	Sling without bending	6~7
Manually lifting equipment	4.5	Binding sling	8~10
Machinery lifting equipment	5~6	Manned elevator	14

6.2 Lifting machinery and equipment

6.2.1 Derrick crane

Major performance characteristic of Derrick crane:

Easy to manufacture, simple assembling and disassembling, large lifting weight (more than 1000 kN), less restricted by terrain, can be used for special structure and equipment while other lifting machineries are not available.

Small service radius, hard to move, more wind ropes are needed, used only in the case of the amount of works relatively concentrated.

By its different structures, Derrick crane can be classified into single pole, herringbone shifter lever, cantilever pole and guyed mast crane, as shown in Figure 6-3.

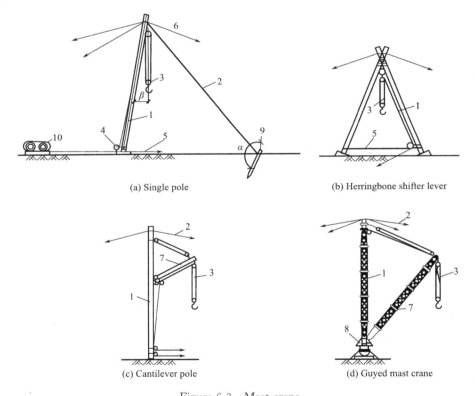

(a) Single pole (b) Herringbone shifter lever

(c) Cantilever pole (d) Guyed mast crane

Figure 6-3 Mast crane

1—Pole; 2—Wind rope; 3—lifting pulley blocks; 4—Guiding devices; 5—Pulling rope; 6—Main wind rope; 7—Lifting boom; 8—Rotating plate; 9—Anchorage; 10—Winch

1. Single pole

Components included: poles, lifting pulley block, winch, wind rope and anchorage, etc. Its pole can be made of round-wood, steel tube or metal latticed columns [Figure 6-3 (a)].

2. Herringbone shifter lever

Components included: two of the poles, or steel tubes, or single pole which has the metal latticed column sections, they are tied or kinky at the top part by using steel wire rope or iron accessories [Figure 6-3 (b)].

It has better lateral stability, less wind ropes, it mainly used for lifting heavy structure members or equipment for its small activity space under the poles after lifting the weights.

3. Cantilever pole

It is formed by placing a lifting boom at the middle or 2/3 of the height of the single pole, so as to increase its hoisting height as well as its serving radius [Figure 6-3 (c)]. It is convenient to use but has small less lifting weight, mainly used for lifting of light weight structure members.

4. Guyed mast crane

It is formed by placing a lifting boom at the bottom of the single pole, and the lifting boom can rotate around its axis both in horizontal and vertical plane [Figure 6-3 (d)].

The machine can turn round in 360 degree in horizontal plane which can provide better working flexibility; it has larger lifting weight and lifting radius which gives larger serving area that ensuring its ability of lifting the structure members to the required place.

6.2.2 Self-propelled derrick crane

Major performance characteristic of derrick crane:

It has more flexibility, easy transportation, wide range of application. It can be divided into crawling crane, truck crane and wheel crane.

1. Crawling crane

It is a kind of universal lifting machinery, formed by: running gear, rotating part, body parts and lifting boom, etc. Chain crawler is adopted for running gear in order to decrease the pressure on ground; rotating part is a rotating plate installed on the base that ensuring the body parts rotating in 360 degree; powering devices, winch and operation system are inside of the body part; lifting boom is made of angle steel lattice members which hinged at the bottom front of the body part, rotating together with the body part. The lifting boom length can changed by either adding sections to it or changing the different length lifting booms, there are two set of pulley blocks at the top of it, the steel wire ropes connect the pulley blocks and the winches inside of the body part (Figure 6-4).

Crawling crane has lager lifting capacity and working speed, can even walk on smooth solid ground while lifting loads but with lower walking speed, its chain crawlers have destructive effects to the road surface, so platform truck is needed while in a long distance transport.

Crawling cranes are widely used in installation construction of single story industrial workshop, dry land bridge, as well as other installation constructions.

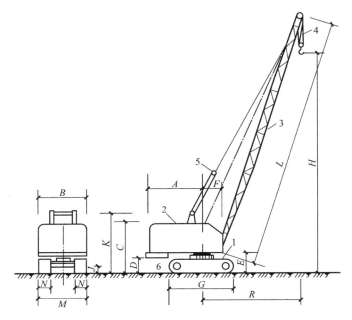

Figure 6-4 Crawling crane outside view
1—Walking gear; 2—Body parts; 3—Lifting boom; 4—Lifting pulley block;
5—Pulley blocks for changing angle of depression and elevation; 6—Crawler

1) Main technical performance

The main technical performance of crawler crane includes three parameters: Lifting load Q, lifting radius R and lifting height H.

Q refers to the maximum lifting load allowed for the crane to work safely, normally the weight, the weight of its hook is generally excluded.

R refers to the horizontal distance between the crane rotating center and the hook.

H refers to the vertical distance between the hook center and the stopping surface.

The three performance parameters have constraining relations for each other, as well as the lifting boom length (L) and the elevation angle (α).

When the length of the boom (L) is constant, as the elevation angle (α) of the boom increases, the lifting load Q increases, the lifting radius R decreases, and the lifting height H increases.

When the elevation angel (α) is constant, as the length of the boom (L) increases, the lifting load Q decreases, the lifting radius R increases, and lifting height H increases.

Length of the boom (L) is constant, as the elevation angle (α) of the boom increases, the lifting load Q decreases, the lifting radius R decreases, and the lifting height H increases.

Technical performance of the crawling crane used commonly is shown in Table 6-3.

Technical performance of the crawling crane   Table 6-3

Parameter		Unit	Model										
			W1-50		W1-100		W200A、WD200A			NE 78D(80D)			
Length of the lifting boom		m	10	18	18 (with bird mouth)	13	23	15	30	40	18.3	24.4	30.25	37
Maximum lifting radius		m	10.0	17.0	10.0	12.0	17.0	14.0	22.0	30.0	18.0	18.0	17.0	17.0
Minimum lifting radius		m	3.7	4.5	6.0	4.5	6.5	4.5	8.0	10.0	4.7	7.5	8.0	10.0
Lifting load	While in minimum radius	t	10.0	7.5	2.0	15.0	8.0	50.0	20.0	8.0	20.0	10.0	9.0	3.0
	While in maximum radius	t	2.6	1.0	1.0	3.7	1.7	9.4	4.8	1.5	3.3	2.9	3.5	1.0
Lifting height	While in minimum radius	m	9.2	17.2	17.2	11.0	19.0	12.1	26.5	36.0	18.0	23.0	29.1	36.0
	While in maximum radius	m	3.7	7.6	14.0	6.5	16.0	5.0	19.8	25.0	7.0	16.4	24.3	34.0

2) Operation cautions of the crawling crane

The following points should be noted in order to ensure the safety working of crawling crane:

(1) The certain safety distance between the lifting hook center and the fixed pulley of the pulley blocks at top of the lifting boom, normally is 2.5~3.5m.

(2) Crane must stay on the solid ground when it is fully loaded, firstly, lift the weight 20~30cm off the ground, then, check and confirm the stability, brake reliability and binding firmness of the lifting components before lifting can continue. The lifting action should be stable and two or more actions simultaneously are forbidden.

(3) For the crane which has no lifting limit devices, its maximum angle of elevation must not exceed 87° in the case of none any lifting limit devices.

(4) The road shall be smooth and solid, and the maximum allowable slope shall not exceed 3°.

(5) When double cranes lifting method adopted, the weight of the member shall not exceed 75% of the total allowable lifting weights of the two cranes.

3) Stability checking calculation of crawling crane

The stability of crane refers to the degree of stability of the whole body during lifting operation.

In normal working conditions, the crane can guarantee the stability of its body, but when carrying out overload lifting or changing to longer lifting booms, the stability checking calculation should be carried out to ensure that the crane will not overturn in the lifting process.

As shown in Figure 6-5, the worst stability situation, stability degree of crawling crane should be carried out by picking the center point of crawler (A at Figure 6-5) as the center of the overturning.

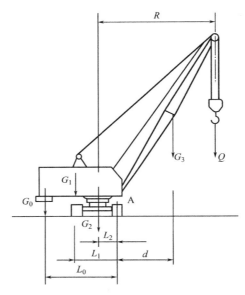

Figure 6-5 Stability checking calculatin of crawling crane

G_0—Blance gravity; G_1—Weight of the turning parts of the crane; G_2—Weight of the none turning parts of the crane; G_3—Weight of the lifting boom; Q—loads (including weight of the lifting components and the riggings); L_0, L_1, L_2, d, R—Distance between the overturning center and the weights above respectively

(1) Considering the lifting load and additional loads (wind load, brake force and centrifugal force), the stability degree of the crawling crane should meet the follow equation shown in Formula (6-5).

$$K_1 = \frac{M_S}{M_O} \geq 1.15 \qquad (6-5)$$

There in:

M_S ——Stability moment (kN·m).

M_O ——Overturning moment (kN·m).

K_1 ——Coefficient of stability with additional loads considered.

(2) Considering only the lifting without regard to the additional loads, the stability

degree of the crawling crane should meet the follow equation shown in Formula (6-6).

$$K_2 = \frac{M_S}{M_O} \geqslant 1.40 \qquad (6\text{-}6)$$

There in:

K_2——Coefficient of stability without additional loads considered.

(3) Stability checking calculation, as shown in Formula (6-7) considering case (2) above.

$$K_2 = \frac{G_0 \cdot L_0 + G_1 \cdot L_1 + G_2 \cdot L_2 - G_3 \cdot d}{Q \cdot (R - L_2)} \geqslant 1.40 \qquad (6\text{-}7)$$

For convenience of calculation, overturning moment M_O takes only the value of the lifting weight; while the stability moments M_S is the difference between the whole stability moment and other overturning moment.

2. Truck crane

It is a type of crane that install its lifting part on normal heavy truck or special designed vehicle base. The driving cab and the lifting control room are separated as shown in Figure 6-6.

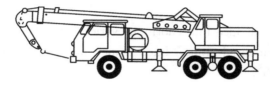

Figure 6-6 Truck crane

Truss or expansion boom are usually adopted as structure form for lifting boom, hydraulic expansion boom crane is widely used currently.

The advantages of truck crane are high speed, convenient transfer and little damage to theroad, it is much suitable for the works of large mobility and frequently changing locations.

Retractable supporting legs must landing on the solid surface, therefor sleepers are needed between the legs and the ground surface, which also enlarges the supporting area and ensures its stability. The truck crane can't carry loads while traveling or working on soft or muddy ground.

3. Wheel crane

Wheel crane is a type of full turning crane that installs its lifting part on special designed chassis frame consisting of heavy tires and axles. Its upper structure is same as that of crawling crane, but its traveling gears are tires, as shown in Figure 6-7.

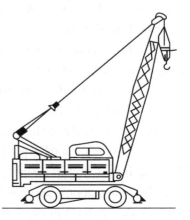

Figure 6-7 Wheel crane

The crane is equipped with four retractable supporting legs. When hoisting with small lifting weight on flat ground, the supporting legsmight not be used while travelling at low speed. However, the supporting legs are generally used to increase the stability of the body and protect the tires.

Comparing with the truck crane:

(1) The wheel crane has larger lateral dimensions, better stability, shorter body, shorter turning radius.

(2) It has lower speed while traveling, not suitable for a long distance driving and working on the soft or muddy ground.

6.2.3 Tower crane

1. Category and performance characteristic of tower crane

Tower crane is a kind of lifting machine which mounts its lifting boom on the top of the vertical tower body, and the lifting boom can rotate horizontally up to 360°.

It can not only used in structure installations but also widely used in multi-story and high rise vertical transportation.

1) Categories

Tower crane can be divided into enormous varieties according its differences of usage and assembling: rail mounted crane, fixed crane, attached crane and internal-climbing tower crane (Figure 6-8).

(1) Rail mounted crane

The rail mounted crane runs on straight or curved steel rail track and lifts loads while running, has high production efficiency, large operating surface, and cover a rectangular space. It is suitable for strip buildings or other structures. The crane has good mechanical condition, low cost, the advantages of quick assembly and disassembly, convenient transfer, and unnecessary to tie with the structure.

But it takes up more construction site area, and the work of laying steel rail track is large, so its shift cost is higher.

(2) Fixed crane

The fixed crane fixes its tower body on the concrete foundation. It is easy to install and occupies a small construction site, but the lifting height is not large (generally within 50m). it is suitable for multistory construction.

(3) Attached crane

The tower body of the attached crane is attached on the foundation of the near building or structure, and anchored to near structures with tie bars every 20mm height of building. It has better stability, higher lifting height. The crane can be jacked up along with the construction process by its jacking system. It takes up very small construction site, especially used in narrower construction site. But it has limited service range cause of its fixed tower body onto the building structure.

Chapter 6 Structure Installation Engineering

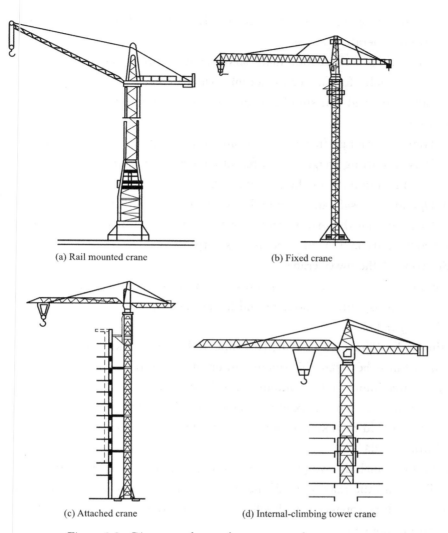

Figure 6-8 Diagrams of several common used tower crane

(4) Internal-climbing tower crane

Internal-climbing tower crane is mounted on the internal structure of building (such as elevator or stairwell, etc.). With the help of climbing mechanism, it will climb up while the building raising every 1~2 floors.

It has smaller steel consumption and lower cost cause of its shorter tower body, doesn't occupy construction sites and doesn't require rail and attachments.

The attached structure need to be reinforced; its assembling and disassembling are inconvenience.

The internal-climbing tower crane is much suitable for narrow construction site of high rise buildings; also enlarging the lifting serving range in the case of a larger construction site.

2) Major performance characteristic of tower crane

(1) Cover larger space by means of its higher tower body, longer lifting boom and larger horizontal operation range.

(2) Lift various construction materials, manufactured goods, prefabricated parts and equipment, especially for supper longer and wider items.

(3) Lift, turn and run simultaneously so as to operate vertically and horizontally at the same time.

(4) Higher working efficiency by means of its multiple operation speed.

(5) Raise its lifting weight for different construction requirements by means changing the "working line number" of the pulley block.

(6) Operation is safe and reliable by means of its complete safety devices.

(7) By mounting the driver cab on its tower top, the driver has better view so as to improve productivity and ensure operation safety.

2. Selection of the tower crane

Selection of the tower crane is mainly depended on its characteristics required: lifting radius, lifting loads, lifting moment and lifting height.

1) Lifting radius

Lifting radius can also be called turning radius or working radius which means the horizontal distance between the swing center of tower crane and the hook center line, including the maximum and the minimum of lifting radius.

Whether or not the maximum range of the tower crane meet the needs of construction, should be taken account at first while in selection.

2) Lifting loads

Lifting loads include two parameters: the maximum lifting weight while in maximum lifting radius, and the maximum lifting weight. Lifting loads include the weights and the accessories (such as ropes, iron shoulder poles and container).

3) Lifting moment

Multiplying the lifting radius by its corresponding lifting loads, is called lifting moment. Rated lifting moment is one of the leading characteristic that reflecting the tower crane's lifting capacity.

After the lifting radius and lifting loads are determined preliminarily in the progress of selecting tower crane, the inspection should be done to check and ensure the lifting moment shall not exceed the rated lifting moment according to the data of the tower crane technical specifications.

4) Lifting height

Lifting height is the vertical distance from the surface of steel rail upon the tower crane's foundation and the center of the hook. It mainly depends on the height of the tower body and its brachial structural forms.

The determination of the lifting height in construction site should be decided by the items as follows: total height of the building, maximum height of the prefabricated parts or components, structure size of the scaffold, etc.

New tower cranes have emerged at home and abroad. At home there are QT4-10、QT16、QT25、QT45、QT60、QT80、QT100、QTZ200 and QT250. Lifting performance of QT4-10 tower crane is shown in the Table 6-4.

Lifting performance of type of QT4-10 tower crane Table 6-4

Length of lifting boom (m)	Type of installation	Lifting radius (m)	Pulley block rate	Lifting height (m)	Lifting weight (t)	Length of lifting boom (m)	Type of installation	Lifting radius (m)	Pulley block rate	Lifting height (m)	Lifting weight (t)
30	Fixed or movable type	3~16	2	40	5	35	Fixed or movable type	3~16	2	40	4
			4	40	10				4	40	8
		20	2	40	5			25	2	40	5
			4	40	8				4	40	5
		30	2	40	5			35	2	40	3
			4	45	5				4	45	4
			4	50	4				4	50	3,4
	Adhesion type or climbing type	3~16	2	160	5		Adhesion type or climbing type	3~16	2	160	4
			4	80	10				4	80	8
		20	2	160	5			25	2	160	4
			4	80	10				4	80	4
		30	2	160	5			35	2	160	3
			4	80	10				4	80	4

6.3 Structure installation engineering of reinforced concrete single-story industrial building

Prefabricated reinforced concrete structure is often adopted in single-story industrial building.

The main load-bearing members are precast members, such as pillar, crown block beam, roof frame, skylight frame and roof plate.

According to the size and weight of components and the capacity of transportation, the larger prefabricated components are generally made on site, while the middle and smaller ones are mostly made in prefabrication plant.

The structural installation engineering is the leading type of engineering of the construction of single-story industrial building.

6.3.1 Preparation for structural installation

The preparation works before the structure installation includes: clearing the site, paving roads; and transportation, stacking, assembling, strengthening, checking, snapping lines and numbering of the structural components; preparation of the foundation, etc.

1. Transportation and stacking

1) Transportation

The components made in factory orin site shall be transported to the lifting location before hoisting, varieties vehicles could be selected. No any tipped, deformed or destroyed are allowed during transportation.

2) There for

When there is no specific design requirement for the strength of components, it shall not be lower than 75% of the standard design strength value of concrete; the support position of the component should be correct as well as the appropriate quantities, and the lifting point should meet the design requirements during loading and unloading; roads should be smooth with sufficient width and turning radius.

3) Stacking

Components should be stacked according to the location specified in the plan preventing from secondary carrying. The stacking of components shall comply with the following provisions:

(1) The site should be flat and solid with drainage measures.

(2) Components should be placed on the wood or bracket according to the designed bearing stress condition and should be stable.

(3) For overlapped stacked components, its lifting rings should be upward as well as signs outwards, the wood or brackets used for supporting components should be in the same vertical line; the stacking height of the overlapped components should be determined according to the bearing capacity of the components, mat, ground strength and the stacking stability.

(4) The components that relying on bracket should be stacked or lifted symmetrically, and its upper parts separated by wooden blocks.

2. Assembly and reinforcement of components

In order to transportation and avoiding damage to the members in the process of straightening, skylight frame and large roof frame can be made into the two half parts, which will be assembled into a whole after being transported to the site.

The assembly of component is divided into two kinds: flat assembly and vertical assembly. The former is generally applicable to small span members, such as skylight frame, as shown in Figure 6-9.

The latter is applicable to large-span roof truss with poor lateral rigidity. When assembling the members, they are in an upright state at the lifting position, which can reduce the process of moving and straightening, as shown in Figure 6-10.

For the skylight frame and roof frame with poor lateral rigidity, transverse rod should be temporarily used to prevent being out of shape and crack in assemble, welding, turn over, lifting process.

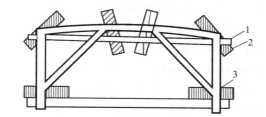

Figure 6-9 Flat assembly diagram of skylight frame
1—Pole; 2—Wood pad; 3—Skylight frame

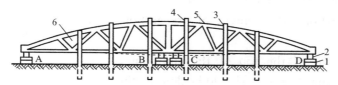

Figure 6-10 30~36m vertical assembly diagram of prestressed concrete roof frame
1—Brick bedding; 2—Square wood or reinforced concrete pad; 3—Triangular support;
4—8# lead wire; 5—Wooden wedge; 6—Roof block

3. Quality inspection of components

All components should be thoroughly inspected before lifting. The main contents of inspection are as follows:

1) Appearance of the components

Type, number, outside size (total length, section size, lateral bending), embedded parts, reserved holes, and whether there are cavities, honeycomb, pitted surface, cracks on the surface components.

2) Strength of the components

When there are no specific requirements from design, the general column should reach 75% of the concrete design strength, the large component (large cavity beam, roof frame) should reach 100%, and the pre-stressed concrete component tunnel grouting strength should not be less than 15MPa.

4. Setting out and number of the components

After the component is qualified in quality inspection, the positioning ink line can beset out on the component as the basis for positioning and adjusting during lifting.

(1) The positioning axis lines shall pop up on the three surfaces of the column body, which shall be consistent with the positioning axis lines on the top surface of the base cup. In addition, the lifting positioning line of crane beam and roof frame shall pop up on the middle part surface and the top surface of the column as shown in Figure 6-11.

(2) Geometric center line should be set out on the top surface of upper chord of the roof truss, and extends it to the lower part of both roof truss ends, then set out the lifting

positioning line of the skylight frame and roof plate from the middle of the roof truss to both ends.

(3) The lifting alignment line should be set out on top and both end side surfaces of the crane beam. While setting out on components, number should be written in the obvious parts of the components based on the design drawing. For Indicating the direction of the components in the case of its upper and down side, left and right sides are not easy to be identified in lifting.

5. Preparation of foundation

The foundation of prefabricated concrete column is usually cup foundation, the content of preparation work mainly includes:

Line setting out at the top of the cup foundation: longitudinal and transverse lines should be done as the basis of column alignment and correction.

Leveling: for ensuringthe corbel elevation accuracy, it is necessary to adjust the elevation of the bottom of the cup foundation before hoisting. Before adjustment, (1) measure

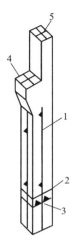

Figure 6-11   diagram or column setting out line
1—Position axis line; 2—Base elevation line; 3—Top elevation line; 4—Position line of crane beam; 5—Center line at top of the column

the actual elevation of the bottom of the cup, the midpoint for smaller column and four corner points for larger column; (2) according to the designed elevation of the top of the corbel, determine the calculation length between the bottom of the column and the top of the corbel; (3) measure the actual distance from the bottom of the column to the top of the corbel; (4) adjusted value of the cup bottom elevation is calculated.

Then fill with cement mortar or fine stone concrete to the required elevation. After the bottom elevation of the cup is adjusted, it should be protected to prevent sundries from falling into it.

6.3.2 Component installation process of the components

For any components in the prefabricated reinforced concrete single-storey industrial workshop, the installation process includes: tying, lift, alignment, temporary fixation, correction and final fixation.

1. Installation of column

1) Tying

Slings, clasp and transverse beam are the tools that mainly used in tying. In order to avoid the slings wearing away the surface of the column, cloth sack or wooden boards should be padded between the slings and the member surfaces. Active clasp should be adopted for making decoupling easier in the high altitude.

The number and position of tying spots shall be determined according to the shape, section, length, reinforcement of the column and the crane performance. One tying spot can be adopted for medium and small column (≤130kN), and the tying spot should be under the column corbel; two tying spots should be adopted for heavy and slenderness columns.

For ensuring the column rotation and standing upright after lifting, the resultant action line of the two slings should be higher than the gravity center of the column.

Commonly used tying methods include oblique tying and straight tying.

(1) Oblique tying

When the shorter side bending strength meets the requirements, this method can be adopted.

The column is lying on the longer side and lifted directly from the bottom mold. The column is slightly inclined after lifting, as shown in Figure 6-12.

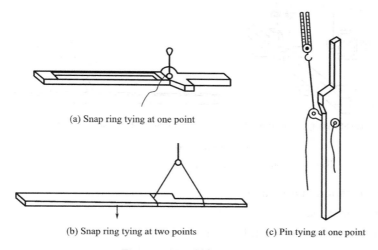

(a) Snap ring tying at one point

(b) Snap ring tying at two points   (c) Pin tying at one point

Figure 6-12   Oblique tying

By this method, the column does not need to be turned over, the length of the lifting boom and the lifting height can be smaller, but the column is inclined after lifting from the ground, which is not convenient for positioning.

(2) Straight tying

When the bending strength of the column cannot meet the requirement, the column should be turned over to improve the bending strength of the column section. After the column is lifted, the column body is in a vertical state. The slings are at both sides of the column and connected to the hook through the transverse beam (Figure 6-13).

The characteristic of this method is that the column is vertical after lifting, easy to position; the hook is on the top of the column, which requires a larger lifting height, so the required lifting boom is longer than the oblique tying method.

2) Lifting of the column

Lifting of the columncan be divided into two methods according to their movement

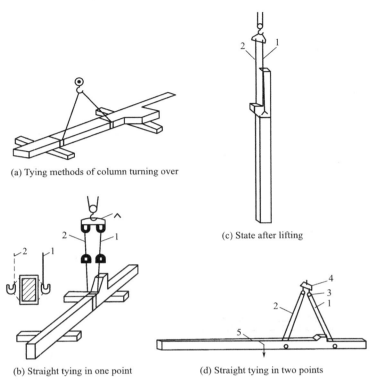

(a) Tying methods of column turning over
(b) Straight tying in one point
(c) State after lifting
(d) Straight tying in two points

Figure 6-13　Straight tying

1—1 sling；2—2 sling；3—Pully；4—Transverse beam；5—Gravity center

characteristics: rotation and sliding.

(1) Rotation methods

The lifting crane rotations while raising its hooks at the same time, till the column change into vertical state by means of rotation the column around its bottom, then lifting it off the ground, put the column into the entrance of the foundation after a bit of further rotating.

The column suffers little vibration during lifting, have higher efficiency. However, the better mobile performance is needed because the crane rotation and the hook rising at the same time.

This method should be adopted by self-propelled derrick crane, as shown in Figure 6-14.

(2) Sliding methods

When hoisting the column by sliding method, the crane only needs raising hook. The column foot slides along the ground, and the column body is in an upright state at the binding point. Then the column is lifted off the ground a bit, after the crane boom rotates slightly to put the column into the foundation entrance.

The column is suffered more vibration, protect procedures should be taken to protect the column feet.

However, the sliding method has a low requirement on the maneuvering performance

Chapter 6 Structure Installation Engineering

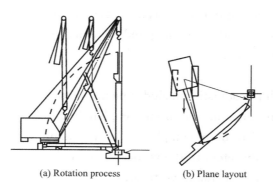

(a) Rotation process   (b) Plane layout

Figure 6-14 Rotation methods in column lifing

of the crane, and it only needs raising its hook while lifting, thus, it is often adopted for the derrick cranes, as shown in Figure 6-15.

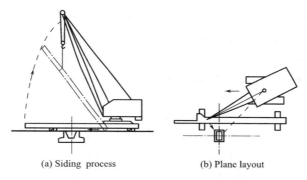

(a) Siding process   (b) Plane layout

Figure 6-15 Sliding methods in column lifting

3) Alignment and temporary fixation of the column

Column alignment starts at 30~50 mm from the cup bottom of the foundation bottom, instead of inserting to it immediately. By using 8 wood or steel wedges (wood or steel), 2 for each side of the column, make the lifting alignment line of the column align with the positioning axis on the cup foundation, and keep the column vertical by means of crane operation or crowbar adjusting to the bottom of the column, as shown in Figure 6-16. After alignment, loosen the hook to let the column sink into the bottom of the cup foundation, re-check the lifting aligning axis lines, tight the wedges symmetrically to temporarily fixing the column, unhook and remove the binding rigging.

For higher column, the ratio of cup foundation depth to column length less than 1/20, or larger

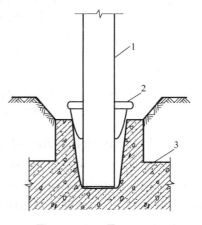

Figure 6-16 Temporary fixation of the column
1—Column; 2—Wedge;
3—Cup foundation

corbel, the wedges at the column foots cannot guarantee the stability of the temporary fixed column, in this case, other aligning and adjusting methods can be adopted such as cable ropes or diagonal bracings.

4) Adjusting column

The content of column adjusting includes plane position, elevation and verticality.

Since the elevation adjusting has been completed in leveling progress and the plane position adjusting has also been completed in the alignment process. Therefore, the adjusting of column mainly refers to the verticality adjusting.

The control method of column verticality is to use two theodolites to check the verticality of column lifting alignment axis on two adjacent sides of the column.

Allowable deviation: 5mm when column height $H<5m$; when column height $H=5\sim10m$, it is 10mm; when the column height $H>10m$, it is $(1/1000)H$ and not greater than 20mm.

Adjusting methods of the column verticality: when the vertical deviation of the column is smaller, it can be used to tighten or relax the wedge or steel drill to correct; when the deviation is larger, it can be done by means of screw jack inclined bracing or flat bracing, or steel pipe jack inclined bracing, etc., as shown in Figure 6-17.

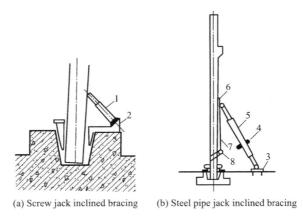

(a) Screw jack inclined bracing   (b) Steel pipe jack inclined bracing

Figure 6-17   Verticality adjusting

1—Screw jack; 2—Base of the jack; 3—Base plate; 4—Turning handle; 5—Steel pipe;
6—Head friction plate; 7—Steel wire ropes; 8—Clasp

5) Final fixation of the columns

Final fixation should be done immediately after the column adjusting finished. The fixing methods is to pour fine stone concrete into the gap between the column foot and the cup foundation, and its strength grade should be two grades higher than that of component concrete.

The pouring of fine stone concrete is divided into two steps. Step1: Pour fine stone concrete to the bottom of the wedge. Step2: After the strength of the first step casting fine stone concrete reaches 25% of the design strength, pull out the wedges and fill the cup with fine stone concrete.

2. Installation of crane beam

Installation of crane beam shall be carried out after the concrete in second step reaches 75% of the designed strength.

1) Tying, lifting, alignment and temporary fixation of the crane beam

The binding point of the crane beam shall be set symmetrically at both ends of the beam, make sure that the vertical line passing through the hook center and the gravity center of the crane beam, and the crane beam shall remain horizontal after lifting.

A sliding ropeshould be arranged at both ends of the crane beam to prevent it from rotating, avoiding collide with the column.

The hook should be lowered slowly when aligning, and the alignment line at the beam end and the lifting positioning line at the corbel should be aligned, as shown in Figure 6-18.

Generally, the cranebeam has a good self-stability and does not need to be temporarily fixed after alignment. However, when the height-to-width ratio of the crane beam is greater than 4, the crane beam can be temporarily fixed on the column with iron wire to prevent its dumping.

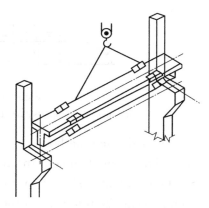

Figure 6-18 Installation of crane beam

2) Adjusting and final fixation of the crane beam

Adjusting contents includes elevation, plane position and perpendicularity.

Elevation adjusting has beencompleted basically while leveling the foundation.

For small and medium cranebeams, adjusting of plane position and perpendicularity should be carried out after the plant structure is adjusted and fixed. This is because the installation of roof truss, support and other components may cause changes in crane beam position, affecting the accuracy of crane beam position.

For heavy crane beams, it is difficult to correct them after decoupling, so they can beadjusted while lifting. However, after the roof truss and other components are fixed, they need to be re-check again.

The verticality of the crane beam shall be checked with a plumb, and when the deviation exceeds the allowable value (5mm) specified in the code, the inclined shim between the two ends of the beam and the corbel surface of the column shall be adjusted.

The adjusting of the crane beam plane position is mainly to check whether the straightness of the longitudinal axis of the crane beam meets the requirements. Common methods include through-line method and axis-shifting method.

(1) Through-line method

It is also called drawing steel wire method.

According to the positioning axis, the installation axis position of the crane beam is determined on the ground at both ends of the workshop by driving into wooden piles. The gauge of the two crane beams is checked with a steel ruler to see whether it meets the requirements. Then use theodolite to adjust the position of the four crane beams at both ends of the workshop. Finally, a supports with a height of about 200mm are set on the crane beam at both ends of the column, then, the steel wires are pulled through piles. According steel wire lines, the center line of the crane beam is adjusted with a crowbar, as shown in Figure 6-19.

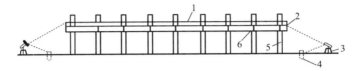

Figure 6-19 Through-line method

1—Through line; 2—Supports; 3—Theodolite; 4—Wooden piles; 5—Column; 6—Crane beam

(2) Axis-shifting method

Axis-shifting method is in the column on both sides.

Sit the theodolite aside of the column lines, project the column alignment lines onto the columns near the top of its corbel around the crane beam surface, and make a mark, as shown in Figure 6-20.

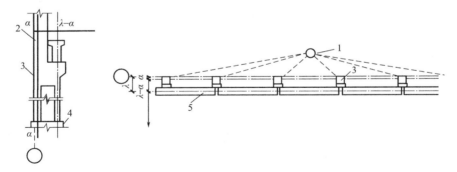

Figure 6-20 Axis-shifting method

1—Theodolite; 2—Mark; 3—Column; 4—Foundation; 5—Crane beam

If the distance between the mark line and the column positioning axis is a, then the distance between the mark line and the crane beam positioning axis is $\lambda - a$, wherein λ is the distance between the column positioning axis of the column and the crane beam positioning axis.

Adjusting crane beam center lines one by one, and check whether the gauge between the two crane beams meets the requirements.

This method is suitable especially in the case of more crane beams on the same axis. After being adjusted, it shall be finally fixed by electric welding onto the columns immediately, and the gap between the crane beam and column shall be filled with fine stone concrete.

3. Installation of the roof truss

1) Standing and placement of the roof truss

The roof truss of single-storey industrial workshop is generally poured in the construction sitewith lying flat. Therefore, before lifting them, it is necessary to let it standing upright from flat its horizontal state. Then lift it to the position specified in the design. This construction process is called the standing and placement of the roof truss.

2) Binding of the roof truss

The binding points of the roof truss should be placed at the upper chord node symmetrically. and the resultant action point binding sling (binding center) should be higher than the gravity center of the roof truss so as to not be overturned and rotated.

In order to prevent roof truss from bearing more lateral pressure, the horizontal angle of binding sling should not be less than 60°while binding, it should not be less than 45° while lifting.

In order to reduce the lifting height and lateral pressure of roof truss, the transverse beam can be used in lifting process.

Generally, two-points binding can be adopted if the span of roof truss is less than 18 m; two-point lashing is adopted; when the span of roof truss is larger than 18m, use two slings to bind it at four points; when the span is greater than 30m, the use of transverse beam should be considered in order to reduce the lifting height

For the roof truss with poor rigidity, such as triangular composite roof truss, since the lower chord cannot bear the pressure, the transverse beam should also be used for binding, as shown in Figure 6-21.

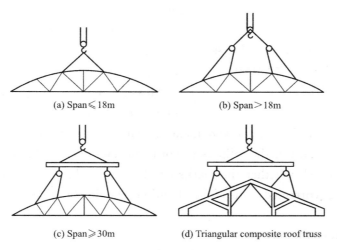

(a) Span≤18m (b) Span>18m
(c) Span≥30m (d) Triangular composite roof truss

Figure 6-21 Binding of the roof truss

3) Lifting, alignment and temporary fixation

Lifting the roof truss off away from the ground by 500mm, and then lift it to the top of the column by 300mm, lowering the roof truss onto the top of the column slowly for alignment.

Alignment processes are subject to the building axes which are projected onto the top of the column by theodolite before. After alignment finished, temporary fixation should be done immediately, followed by decoupling the crane hooks.

The temporary fixing method of the first roof truss is to use four cables to fasten it from both sides. If the wind column has been installed, the roof frame can be connected to the wind column. Calibrators should be adopted for second or subsequent trusses to temporary fix themselves to the previous ones. Each roof truss requires at least two calibrators.

4) Adjusting and final fixation

The content of roof truss adjusting is to check and correct its verticality.

(1) Theodolite inspection

Install three calipers on the roof truss, put one in the middle of the upper chord of the roof truss, put the other two at both ends of the roof truss. The calipers are perpendicular to the plane of the roof truss. Take 500mm from the geometric center line of the upper chord of the roof truss and mark it on the caliper. Then set a theodolite on the ground 500mm away from the center line of the roof truss and check whether the marks on the three calipers are on the same vertical plane, as shown in Figure 6-22.

(2) Plumb checking

The caliper is set in the same way as the theodolite.

Measure 300mm away from the geometric center line of the upper chord of the roof truss along the direction of the caliper and make a mark on the three calipers. Then draw a through line at the mark of the calipers at both ends and hang a plumb at the mark of the central caliper to check whether the marks on the three calipers are on the same vertical plane. After being corrected, the final fixation should be done by electric welding immediately.

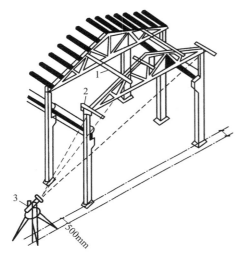

Figure 6-22 Adjusting of the roof truss
1—Temporary bracing; 2—Caliper;
3—Theodolite

4. Installation of the roof slabs

Generally, the roof board is embedded with hanging rings, and the hanging rings are hoisted by hanging cables with hooks. The installation sequence of roof boards should be symmetrical from both sides of the cornice to the roof ridge in order to avoid unbalancing under loads. After the roof boards are aligned, fix them by electric welding immediately.

5. Installation of skylight frame

The hoisting of the skylight frame should be carried out after the hoisting of roof slabs on both sides of the skylight frame is completed, and the hoisting method is basically the same as that of the roof truss.

6.3.3 Structural installation scheme

1. Structural hoisting methods

The structural hoisting methods of single-storey industrial factory buildings include the components hoisting and the comprehensive hoisting.

1) Components hoisting method

The crane lifts only one or a few components at its every running. Usually the installation is completed in three times: the first operation is to install the columns, adjusting and final fixing one by one; the second operation is to install crane beams, connecting beams and inter-column support, etc; the third operation is to install the roof truss, skylight frames and roof plate in terms of building units.

Component hoisting method hoists the same kind of components every time, the lifting crane can be selected according to the weight and height of the components. Also, there is no need to change the rigging frequently in the hoisting process, which makes it easier to operate skillfully. Therefore, the hoisting speed is fast, which can fully make used of the working performance of the crane; in addition, supplies of the components, the plane layout and adjusting are easier to be organized. Accordingly, the components hoisting methods is adopted often at present in single-story industry workshop installation.

However, it cannot provide the working space for subsequent projects because of its long running route and many stops while in progress.

2) Comprehensive hoisting method

The crane runs only once but installs all the components in one building units.

Firstly, the crane lifts 4~6 columns, flowing its adjusting and final fixation; then it lifts the crane beam, connection beams, roof truss and roof skylight frames at this building units.

The crane has short running route and less stopping points, therefor it can provide the working space for the subsequent project as earlier as possible.

However, because it holds all kinds of components at one time, the rigging is changed frequently and the operation are vary, which affects the improvement of production efficiency and the performance of the crane cannot be fully developed.

In addition, the supply and the plane layout of the components are complex, adjusting and final fixation process are in a hurry which goes against to the construction organization. Therefore, generally, this method is not adopted. It can be selected only in the case of that the derrick crane is adopted which is difficult to move.

2. Selection of the crane

1) Crane type selection

The crane type should bedetermined according to the structural type, the weight of the structure component, the installation height of plant, the lifting methods and the condition of the existing lifting equipment, consideration of its rationality feasibility and economical comprehensively.

For small and medium-sized plants, self-propelled derrick crane is commonly used, the most commonly used is the crawling crane, and home-made mast crane can be selected as well as in the case of lacking of the lifting equipment above.

For heavy workshops which have long span, heavy components and higher installation height, the workshop equipment installation and the structures members lifting operation are carrying out simultaneously, therefore, larger self-propelled derrick crane or the combination of heavy tower crane together with other hoisting machinery should be selected.

2) Crane model selection

The crane typeshould be determined according to the size, weight and installation height of the components. The three working parameters of the selected crane, namely load, lifting height and lifting radius, must meet the requirements of lifting components.

(1) Lifting load Q

The lifting load of the crane must be greater than or equal to the sum of the weight of the installed components and the weight of the rigging, i. e. as shown in Formula (6-8).

$$Q \geqslant Q_1 + Q_2 \tag{6-8}$$

There in:

Q ——Lifting load (kN);

Q_1 ——Components weight (kN);

Q_2 ——Rigging weight (kN).

(2) Lifting height

The lifting height of the crane must meet the installation requirements of the components, as shown in Figure 6-23 and Formula (6-9).

$$H \geqslant h_1 + h_2 + h_3 + h_4 \tag{6-9}$$

There in:

H ——Lifting height (distance from the stop surface to the hook center, m).

h_1 ——Distance from the top of the lower support surface (m).

h_2 ——Installation clearance, as the cases may be (m).

h_3 ——Distance from the component bottom to the binding point (m).

h_4 ——Height of the riggings (distance from the binding point to the hook center, m).

(3) Lifting radius

The determination of lifting radius is generally divided into two cases.

Chapter 6 Structure Installation Engineering

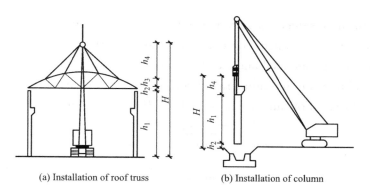

(a) Installation of roof truss (b) Installation of column

Figure 6-23 Diagram of the lifting height calculation

Case 1: when the crane can be operated at an unrestricted area near to the lifting position, it doesn't need to check the lifting radius R. According to the calculated lifting weight Q and lifting height H, check the performance curve or performance table of the crane to select the crane model and lifting boom length L; and the lifting radius R under the corresponding lifting weight and lifting height can be found, which can be used as the basis to determine the crane running route and stop point position.

Case 2: when the crane cannot be directly operated to the near of the lifting position, the minimum lifting radius R shall be determined according to the actual situation. Refer to crane performance curve or performance table according to lifting weight Q, lifting height H and lifting radius R, select crane type and lifting boom length L.

(4) The minimum length of the lifting boom L

When the crane boom runs over the installed structure components, such as lifting roof boards onto the installed roof truss, the minimum boom length should be determined for avoiding colliding.

The methods for determining the minimum boom length of a crane are calculating method and graphic method. The calculating method geometrical relationship is shown in Figure 6-24.

The minimum length of the lifting boom can be calculated according to the Formula (6-10).

$$L \geqslant l_1 + l_2 = \frac{h}{\sin\alpha} + \frac{f+g}{\cos\alpha} \tag{6-10}$$

There in:

L ——Length of the crane boom (m).

h ——Vertical distance from the lower hinge of the crane boom to the support surface ($h = h_1 - E$, m).

l_1 ——Vertical distance from the stopping surface to the support surface (m).

l_2 ——Vertical distance from the stopping surface to the lower hinge of the crane boom (m).

f ——Horizontal offsetting distance between hook center and the component installed (m).

g ——Horizontal safety operation distance between crane boom axis line and the components installed, normally not less than 1m.

α ——Horizontal contained angle of the crane boom axis.

In order to obtain the minimumcrane boom length, differential calculus is required as in Formula (6-11) and Formula (6-12), and set $dl/d\alpha$ equal to 0.

$$\frac{dl}{d\alpha} = \frac{-h\cos\alpha}{\sin\alpha} + \frac{f+g}{\cos^2\alpha} = 0 \tag{6-11}$$

$$\alpha = \text{arctg}\sqrt[3]{\frac{h}{f+g}} \tag{6-12}$$

Substitute the value of α into Formula (6-10), calculate the minimum crane boom length L_{min}.

Lifting radius will be calculated by the minimum lifting length L_{min} and horizontal contained angle, as shown in Formula (6-13).

$$R = F + L_{min}\cos\alpha \tag{6-13}$$

There in:

F ——Horizontal distance between the lower hinge of the crane boom and the turning center of the crane (m).

The other symbols are the same as above.

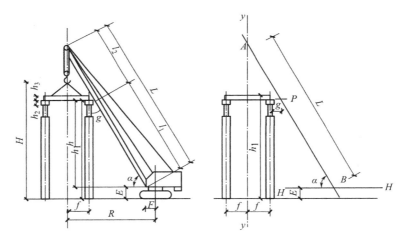

Figure 6-24 Calculation diagram for the minimum boom length of a crane when lifting roof slabs

3. The plane arrangement of the prefabricated components

Field plane arrangement of the prefabricated members is a very important work in single-story industrial workshop installation engineering. The reasonable arrangement of the components can exempt the secondary transportation in the working site and ensure lifting machinery efficiency.

Plane arrangement of the componentsis related to the lifting methods, lifting machinery performance, component production methods and so on. The main requirements are as follows:

(1) Trying to put the lifting components inside the span unit of the building, or put them aside of the span but convenience whenever there are difficulties to do that.

(2) The component arrangement should meet the requirements of lifting technology. Try to arrange components within the lifting radius to reduce the crane traveling distance and the number of lifting boom ups and downs.

(3) The component arrangement should meet the requirements of lifting sequence, and pay attention to the component directions during installation, ensure construction schedule and safety and avoid turning around in the air.

(4) A certain distance (generally no less than 1m) shall be kept between members so as to perform formwork and concrete casting. In addition, the operation place of drawing pipe and piercing bar should be considered too.

(5) Each component should try to cover the minimum working site area to ensure the smooth operation of lifting machinery, transport vehicles, and ensure that lifting machinery does not collide with components when turning.

(6) All components should be placed on a solid foundation to prevent the new filling up foundation from sinking, and ensure components quality.

Questions

1. What are the common used types of winch? How many kinds of anchorage methods are there for winch?

2. What are the specifications of steel wire ropes commonly used in structural installation? How do you calculate the allowable tension?

3. Briefly describe the types and applications of the derrick cranes.

4. How many types does the self-propelled derrick have? What are their characteristics?

5. What are the main technical parameters of crawling crane and what are the relationships between them? How to calculate the stability?

6. What are the types of tower cranes? Describe the applications respectively. How to choose tower crane according to its technical parameters?

7. What preparations works should be done before the installation of single-storey industrial workshop structure?

8. How many binding methods they have in column lifting process? How to do alignment, temporary fixation, adjusting and final fixation?

9. Briefly describe the method of verticality adjusting of the crane beam, how to do its final fixation?

10. How to determine the binding point when the roof truss is in place and lifting? How to do temporary fixation, adjusting and final fixation?

11. What are the two lifting methods in reinforced concrete single-story industrial workshop building structure? What are their characteristics? How to choose lifting machinery?

Chapter 7   Waterproof Engineering

Mastery of contents: Construction technology and methods of roof waterproof construction.

Familiarity of contents: Construction technology and methods of underground waterproof construction, interior waterproof construction.

Understanding of contents: Varieties of roof waterproof.

Key points: Construction technology and key points of roof waterproof construction on roof, as well as exterior wall of the building basement.

Difficult points: Construction key points of exterior wall of the building basement.

Waterproofing engineering is important for guaranteeing structure free from water erosion, its waterproofing quality has close relationship with the design, materials, construction, and directly affects the using function of civil engineering structure.

Waterproof engineering can be divided into structure self-waterproof and water proof layer by its construction methods. The structure self-waterproof mainly use its compactness of the structural materials itself and some structural measures (such as setting slog, embedding water stop belt, etc.) to make it water proofing. Waterproof layer, made of additional waterproof materials at the face or back joints of structure and members, such as coiled waterproof, coating waterproof and rigid waterproof layer, etc. to make it water proofing.

According to its performance, waterproof engineering can also be divided into: flexible waterproof (such as coiled sheet waterproof, coating waterproof, etc.), rigid waterproof (such as structure self-waterproof, cement mortar waterproof layer).

According its structural position, waterproof engineering can be divided into: roof waterproof, underground waterproof and interior waterproof.

Waterproof engineering construction process requirements much strict and details, for ensuring its construction quality, its construction schedule should avoid the rainy season and wintertime.

7.1   Roof waterproof engineering

Building roof waterproofing engineering is an important content in building engineering, its quality is not only related to the servicing life of the building, but also directly affects the normal runs of manufacturing and life.

Chapter 7 Waterproof Engineering

The flowing compulsory provisions must be enforced strictly according to the national code of Technical Code for Roof Engineering GB 50345—2012.

(1) Roof waterproof engineering takes different fortification levels according to building characteristics, importance, functional requirements and reasonable service life of waterproof layer.

Building roof with special requirements for waterproofing, special waterproof design should be carried out.

Waterproof grade and fortification requirements of building roof shall be in accordance with those specified in Table 7-1.

Waterproof grade and fortification requirements of building roof　　Table 7-1

Waterproof grade	Building category	Fortification requirement
Grade I	Important and high rise building	Two course fortification
Grade II	Normal building	One course fortification

(2) Waterproof grade and structure details shall be in accordance with those specified in Table 7-2.

Waterproof grade and structure details of coiled sheet and coating waterproof roof　　Table 7-2

Waterproof grade	Construction details of waterproof
Grade I	Coiled sheet waterproof layer and coiled sheet waterproof layer; coiled sheet waterproof layer and coating waterproof layer; composite waterproof layer
Grade II	Coiled sheet waterproof layer, coating waterproof layer, composite waterproof layer

Note: When waterproof grade is I, waterproof used as single course, should be in accordance with the technical regulations of single course coiled sheet waterproof.

(3) Minimum thickness in each coiled sheet waterproof layer shall be in accordance with those specified in Table 7-3.

Minimum thickness in each coiled sheet waterproof layer　　Table 7-3

Waterproof grade	High polymer waterproof sheet	Polyester base, fiberglass base and polyethylene base	Self-adhesive polyester base	Self-adhesive no-base
Grade I	1.2	3.0	2.0	1.5
Grade II	1.5	4.0	3.0	2.0

(4) Minimum thickness in each coating waterproof layer shall be in accordance with those specified in Table 7-4.

(5) Minimum thickness of composite waterproof layer shall be in accordance with those specified in Table 7-5.

Minimum thickness in each coating waterproof layer (mm)   Table 7-4

Waterproof grade	Synthetic polymer coating waterproof layer	Polymer cement waterproof coating	Polymer modified bitumen waterproof coating
Grade I	1.5	1.5	2.0
Grade II	2.0	2.0	3.0

Minimum thickness of composite waterproof layer (mm)   Table 7-5

Waterproof grade	Synthetic polymer waterproof coiled sheet + Synthetic polymer waterproof coating	Self-adhesive polymer modified bitumen waterproof coiled sheet + Synthetic polymer waterproof coating	Polymer modified bitumen waterproof coiled sheet + Polymer modified bitumen waterproof coating	Polypropylene coiled sheet + Polymer sement waterproof sementitious materials
Grade I	1.2+1.5	1.5+1.5	3.0+2.0	(0.7+1.3)×2
Grade II	1.0+1.0	1.2+1.0	3.0+1.2	0.7+1.3

(6) Waterproof grade and structure details of tile roofing shall be in accordance with those specified in Table 7-6.

Waterproof grade and structure details of tile roofing   Table 7-6

Waterproof grade	Construction details of waterproof
Grade I	Tile + waterproof layer
Grade II	Tile + waterproof cushion

Note: Thickness of waterproof should be in accordance with those specified grade II in Table 7-3 or Table 7-4.

(7) Waterproof grade and structure details of metal sheet roofing shall be in accordance with those specified in Table 7-7.

Waterproof grade and structure details of metal sheet roofing   Table 7-7

Waterproof grade	Structure details
Grade I	Molded metal sheet + Waterproof cushion
Grade II	Molded metal sheet + Double skin metal faced insulating panels

Note: 1. When waterproof is grade I, the thickness of substrate board of profiled aluminum alloy sheet should not be less than 0.9 mm; the thickness of substrate board of profiled steel sheet should not be less than 0.6 mm.
2. When waterproof is grade I, 360 degree lock and seam connection method should be adopted for profiled metal sheet.
3. In grade I waterproof roofing structure details, when applied only as profiled metal sheet, it should be accordance with those specified in *Technical Code for Application of Profiled Metal Sheets* GB 50896—2013.

(8) The following safety regulations must be complied with in construction of roofing engineering:

① Constructions are strictly forbidden under the situation of rain, snow and the wind more than grade 5.

② Safety guard rails and nets must be set up in accordance with the regulations near the edges and openings, such as surroundings of roof and reserved holes.

③ Anti-slippery measures must be taken when the roof slope angle is more than 30 degree.

④ Construction personnel must wear anti-slippery boots, construction operators must fasten safety belt and lock safety hooks when there is no reliable safety measures in some special situation.

7.1.1 Construction details of coiled sheet waterproof roof

Coil waterproof roof refers to applying adhesive materials to stick flexible coiled sheet to roof base, forming a whole impermeable layer, therefore, fulfilling its waterproof function. The general structure of coil waterproof roof is as shown in Figure 7-1.

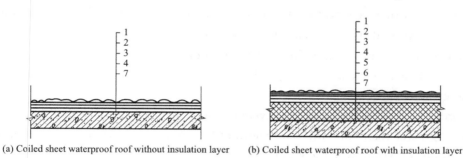

(a) Coiled sheet waterproof roof without insulation layer   (b) Coiled sheet waterproof roof with insulation layer

Figure 7-1 Construction details of coiled sheet waterproof roof
1—Protective layer; 2—Coiled sheet waterproof layer; 3—Binding course; 4—Leveling layer;
5—Insulation layer; 6—Vapor barrier; 7—Structure layer

1. Materials commonly used in coiled sheet waterproof roof

1) Coiled sheet

There are three main types that commonly used, that are bitumen coiled sheet, polymer modified bitumen coiled sheet, synthetic high molecular coiled sheet.

2) Adhesive

The adhesive used for pasting can be divided into three kinds: base and coil sheet pasting agent, lapping adhesive between coiled sheets, adhesive tape for bonding seal.

It also can be divided into bitumen adhesive and synthetic high molecular adhesive according to its constituent materials.

3) Base treatment agent

Base treatment agent is to strength the bonding between waterproof layer and base layer, it brush onto the base layer before construction of waterproof layer. It might be cold primer oil, or base coat which is suitable for polymer modified bitumen coiled sheet, synthetic high molecular coiled sheet.

2. Construction technology and method of coiled sheet waterproof roof

The general construction technological process of coiled sheet waterproof roof is shown as following:

Cleaning and repairing the base layer (leveling layer) surface →spraying or brushing base treatment agent → addition layers enhancing at joints → positioning, layout and trial laying → end part treatment, sealing at joint → clean, check and adjust → protective layer construction.

1) Construction of leveling layer

Leveling layer is the base of laying waterproof coiled sheet, which should have sufficient strength and stiffness. 1 : 2.5 or 1 : 3 cement mortar, C15 fine aggregate concrete or 1 : 8 bituminous mortar with thickness of 15~35mm can be adopted as the materials of the leveling layer.

The leveling layer shall be smeared evenly with cement mortar and polished two times, the concave and convex corners connecting with the structure protruding from roofs as well as the corners connecting the base layer with vertical wall, drainage ditch and trench ridge edge shall be smeared into smooth arch shape, with the radius at 50 mm in general. The area around drain gullies for internal drainage shall be concave.

The leveling layer shall be firm, flat and free of looseness, sand streak, pitting, sags and crests.

In order to avoid cracking of the leveling layer, separation joints shall be provided at spacing no greater than 6m with seam width of 20~30mm, insert sealing materials or lay 100 mm wide coil strap over the seam.

2) Spraying of base treatment agent

Before spraying or brushing base treatment agent, the quality and dryness of leveling layer should be checked and make sure it is clean. Construction can start only after those requirements are met.

The parts of joints, periphery and corner should be sprayed or brushed first, then the large area spraying or brushing follows.

While in construction, thickness should be uniform. Blank, spots or bubbles are not allowed. Laying coiled sheet should start timely after the base treatment agent layer dries.

3) Laying of coiled sheet waterproof

(1) Laying direction

Laying direction should be determined according to roof slope and whether it bears vibration.

Parallel to the roof ridge when roof slope equal or less than 3%.

Parallel or perpendicular to the roof ridge when roof slope is 3%~15%.

In case of roof slope is greater than 15% or bearing vibration, bitumen coiled sheet should be perpendicular to the roof ridge, while polymer modified bitumen coiled sheet and synthetic high molecular coiled sheet might be parallel or perpendicular to the roof ridge.

In case of roof slope greater than 25%, the coiled sheet waterproof should be perpendicular to the roof ridge, and the fixing measures should be taken to prevent coiled sheet waterproof sliding, the fixing points should also be sealed.

The upper and lower coiled sheet waterproof shall not be laid vertically with each other.

(2) Laying sequence

Joints, additional layers and drainage parts should be paved first, then, large areas follows.

When parallel to the roof ridge paving methods is adopted, the paving should start from cornice to the ridge; while perpendicular to the roof ridge is adopted, the paving should start from ridge to the cornice. In multi-span roof, higher and far parts should be paved first, then the lower and near parts.

(3) The pasting methods between coiled sheet and the base layer

Empty paving, point pasting, strip pasting and full pasting methods can be adopted according to the relative stipulations.

(4) Overlapping of coiled sheet

While paving, coiled sheet should be lapped, and the overlapping joints parallel to the roof ridge should be lapped in the direction of water stream flows; as for overlapping joints perpendicular to the roof ridge, it should be lapped in the direction of annual prevailing wind.

No joint is allowed at the ridge, the coiled sheet must be lapped across the roof ridge, the overlapping joints shall be staggered each other between adjacent sheets of same layer, as well as that of in different layers.

The width of overlapping joints should meet the requirement specified.

4) Additional reinforcement layer

An additional layer of coiled sheet or coating should be applied to some special parts of the roof, such as, flashing, cornice, coiled sheet end, gutter, water droop mouth, deformation joins.

5) Protective layer

After the waterproof layer is laid and checked to be qualified, the protective layer should be constructed immediately to avoid damage to the just finished waterproof layer of the roof.

Protective layers can be selected according to characteristics of waterproof layer.

Protective layer of mung bean sand, mica or vermiculite can be adopted onto bitumen coiled sheet water proof roof.

Protective layer of light-colored reflection coating can be adopted onto waterproof roof of polymer modified bitumen coiled sheet and synthetic high molecular coiled sheet.

Prefabricated block materials protective layer, cement mortar protective layer and fine aggregate concrete protective layer, can be used for all kinds of coiled sheet waterproof roof.

7.1.2 Construction of coating waterproof roof

Brushing waterproof paint onto base layer first, then an membrane with a certain thickness and elasticity is formed after its curing and hardening, so as to achieve the purpose of waterproof.

It has the advantages such as, light weight, good temperature adaptability, simple operation, fast construction speed. Cold construction is mostly adopted to improve working conditions, easy to be repaired and lower cost.

However, it is difficult to keep uniform membrane thickness in construction.

1. Coating waterproof materials

Coating waterproof materials are classified into three categories according to the main substance that forming membrane: bitumen-based waterproof paint, polymer modified bitumen waterproof paint and synthetic high molecular waterproof paint.

The construction process of coating waterproof roof is as follows:

Repairing and cleaning the surface of the base layer→ spaying base treatment agent (bottom coating) → adding reinforcement treatment in specials parts → spraying and reinforce materials laying →clean, inspect and repair → construction of protective layer.

2. Construction technology and method of coating waterproof materials

1) Brushing base treatment agent

There are three types of base treatment agent: solvent-based waterproof paint, water-emulsion waterproof paint and polymer modified bitumen waterproof paint.

Brush thin layer with forces and try to brush base treatment agent into base layer surface wool stoma evenly and completely. Only after base treatment agent dried, can coating waterproof construction start.

2) Brushing waterproof materials

Some joint parts should be applied enhancing treatment before construction, such as drain gully, gutter, eaves ditch, flashing, pipe roots protruding from roofs, an additional layer with a matrix reinforcement materials should be applied to those areas described above, then, large area coating follows.

In multi-span roof, higher and far parts should paved first, then the lower and near parts.

The compactness of coating layer is the key to ensure the waterproof quality of roof.

3) Laying matrix reinforcement materials

Matrix reinforcement material might be fiber nonwoven fabric, glass fiber mesh and other materials used for reinforcement in the waterproof coating.

Laying direction should be determined according to roof slope. Parallel to the roof ridge when roof slope less than 15%; perpendicular to the roof ridge when roof slope is equal or greater than 15%; the sequence should be from the lowest part of the roof upward to the higher parts.

The upper and lower matrix reinforcement materials shall not be laid vertically with each other and their joints should be staggered both in same layer and layers between.

The matrix reinforcement materials can be laid after the second or the third waterproof material brushing.

4) Closed head processing

Multi-brushing methods or sealing materials methods can be adopted at the closed head parts for ensuring its construction quality.

5) Protective layer

Fine sand, mica or vermiculite, light-colored reflection materials, prefabricated block materials, cement mortar and fine aggregate concrete can be selected as protective layer.

7.1.3 Construction of rigid waterproof roof

Rigid waterproof roof is the roof that uses rigid waterproof materials as waterproof layer, such as ordinary fine aggregate concrete, compensation shrinkage concrete, fiber concrete and prestressed concrete waterproof roof. The following only introduce construction of ordinary fine aggregate concrete waterproof roof.

Comparing with coiled sheet and coating waterproof, its advantages and disadvantages are described as following: its materials are easy to get, lower price, good durability, construction and maintenance are more convenient; however, due to its high apparent density and low tensile strength, it is easy to crack under the influence of temperature deformation, dry and wet deformation and structural displacement of concrete.

The structure of rigid waterproof layer of fine aggregate concrete is as shown in Figure 7-2. The thickness should not be less than 40 mm. Embedded pipeline, lines or parts are strictly forbidden inside the waterproof layer.

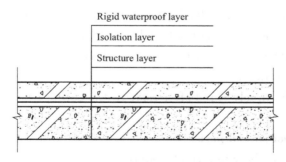

Figure 7-2 Structure waterproof layers

1. Isolation layer settings

The isolation layer should be set between the rigid waterproof layer and the structure layer to avoid the cracking of rigid waterproof layer caused by structural deformation.

The following materials or process are commonly used as isolation layers, such as paper strip mixed lime mortar, clay mortar, lime mortar, cement mortar, or dry laying coiled sheet or cement mortar plastic membrane over cement mortar leveling layer.

2. Dividing joints

The rigid waterproof layer should be equipped with dividing joints to avoid cracking of the rigid waterproof due to temperature difference, concrete dry shrinkage, load and vibration, foundation settlement and other deformation.

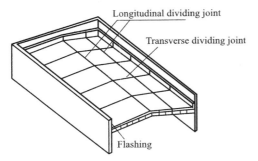

Figure 7-3   Dividing joints

The dividing joints should be located at the supporting end of the roof plate, the turning corner of the roof, the intersection of waterproof layer and the extruding structure parts; and should be aligned with the plate joint as shown in Figure 7-3.

The vertical and horizontal spacing of the dividing joints should not be more than 6m, or "one pattern one room", and its area should not exceed 60 m^2. The seam width of the joints should be 5~30 mm, the seam should be filled with sealing materials and the upper part should be provided with a protective layer.

3. Steel mesh sheet settings

The waterproof layer shall be equipped with two-way steel mesh sheets with diameters of 4 ~ 6mm and spacing of 100 ~ 200mm.

The steel mesh sheet shall be disconnected at the dividing joint, and its concrete cover thickness shall be no less than 10mm, its position shall be at the upper part of the concrete layer during construction.

4. Construction of fine aggregate concrete

Ordinary Portland cement or Portland cement can be adopted in fine aggregate concrete construction, while Portland pozzolanic cement is not allowed, when using Portland slag cement, measures should be taken to reduce the seepage.

The concrete within each partition point shall be casted at one time and no construction joints is allowed.

Concrete should be timely cured after casted, the curing time should not be less than 14 d.

5. Treatment at roof special parts

At intersection part between rigid waterproof layer and those of gable, parapet, extruding roof structure, a gap should be left with a width of 30 mm and filled with sealing materials.

Coiled sheet or coating layer should be laid at the flashing part.

Intersection parts between pipe roots protruding from roofs and the rigid waterproof should leave gaps and fill them with sealing materials, add additional coiled sheet or coating layer.

Cement mortar can be selected for slope making at the area of gutter, eaves gutter. Fine aggregate concrete should be selected whenever the slope leveling thickness is greater than 20 mm.

7.2 Underground waterproof engineering

The flowing compulsory provisions must be enforced strictly according to the national code of *Technical Code for Waterproofing of Underground Works* GB 50108—2008.

(1) As for the upstream-face main structure of underground works, it shall adopt waterproof concrete and take other waterproof measures in accordance with the requirement of waterproof grade.

(2) If there is segregation after the transport of waterproof concrete mixture, it must be mixed for a second time. When the slump fails to meet the requirement of construction, it shall be added with cement slurry of the original water-binder ratio or mixed with water reducer of the same variety but no water shall be added.

(3) The construction of construction joint shall be in accordance with the following requirements:

① Before the concreting of horizontal construction joint, it shall remove the surface laitance and sundries and lay cement paste or paint concrete interface treating agent and cementitious capillary crystalline waterproofing (CCCW) coating. Then it shall lay 1 : 1 cement mortar of 30~50mm and cast concrete in time.

② Before the concreting of vertical construction joint, it shall clean up the surface first, then paint concrete interface treating agent or cementitious capillary crystalline waterproofing (CCCW) coating and shall cast concrete in time.

(4) The waterproofing of underground works shall be divided into four grades, each of which shall be in accordance with those specified in Table 7-8.

The waterproof standard of underground works   Table 7-8

Waterproof rating	Waterproof standard
Grade 1	No water seepage and no wet stains on structural surface
Grade 2	No water leakage and with a few of wet stains on structural surface. Industrial and civil buildings: The total area of wet stains shall be less than or equal to 1/1000 of the total waterproof area(including roof, wall and floor); there shall be less than or equal to 2 wet stains in any 100m² waterproofing area and the maximum area of each wet stain shall be less than or equal to 0.1 m². Other underground works: The total area of wet stains shall be less than or equal to 2/1000 of the total waterproof area; there shall be less than or equal to 3 wet stains in any 100m² waterproofing area and the maximum area of each wet stain shall be less than or equal to 0.2m²; as for tunnel works, it is also required that the average seepage volume shall be less than or equal to 0.05L/(m²·d) and the seepage volume of any 100m² waterproofing area shall be less than or equal to 0.15L/(m²·d)

续表

Waterproof rating	Waterproof standard
Grade 3	With a few of leakage points but no filiform flow and silt leakage. There shall be less than or equal to 7 leakage points or wet stains in any 100 m^2 waterproofing water. The maximum leakage volume for each leakage point shall be less than or equal to 2.5L/d and the maximum area of each wet stain shall be less than or equal to 0.3 m^2
Grade 4	With leakage points but no filiform flow and silt leakage. The average leakage volume of the whole works shall be less than or equal to 2L/(m$^2 \cdot$ d) and the average leakage volume of any 100 m^2 waterproofing area shall be less than or equal to L/(m$^2 \cdot$ d)

(5) As for the different waterproof grades of underground works, its application scope shall be determined according to the importance of the project and the in-use requirement for waterproofing, which are set out in Table 7-9.

The application scope of different waterproof grades　　　　Table 7-9

Waterproof rating	Application scope
Grade 1	The place for long-term personnel stop; the position for storing materials which will go bad or lose effectiveness due to a few of wet stains and the position which will have a strong impact normal equipment operation and safe engineering operation; vital war preparation engineering and subway station
Grade 2	The place for frequent personnel activity; the position for storing materials which will not go bad or lose effectiveness due to a few of wet stains and the position which will have little impact normal equipment operation and safe engineering operation; important war preparation engineering
Grade 3	The place for temporary personnel activity; common war preparation engineering
Grade 4	The project without strict requirement for leakage water

The waterproof scheme of underground construction projects can be roughly divided into three categories:

① Waterproof concrete structure scheme, that is, to improve the compactness and impermeability of the concrete structure itself for waterproof.

② Additional waterproof layer scheme, that is to set waterproof layer on the structure surface to achieve waterproof purpose.

③ The scheme of prevention and drainage scheme, that is to rain away the groundwater, so as to assist the waterproof structure to meet the requirements.

7.2.1　Construction of concrete itself waterproof

Self-waterproof concrete is adopted in engineering structure so that combine the structure load-bearing, enveloping and waterproofing functions into one, it has better compactness and impermeability, and can be divided into ordinary waterproof concrete and admixture waterproof concrete. Its advantages are simple construction, short construction period, reliable waterproof, better durability, lower cost, etc. therefore, it is widely used in underground works.

1. Construction of formwork

Formwork used in waterproof concrete engineering should be flat surface, small water absorption, firmly jointing without leakage, fastness and stable.

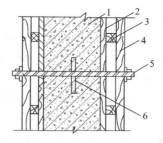

Figure 7-4   Diagram of bolts plus waterstop
1—Waterproof structure; 2—Formwork;
3—Waler; 4—Stiff back; 5—Pulling bolts;
6—Water stop ring

In order to prevent water from infiltrating along bolt, certain measures should be taken when puling bolts are adopted in formwork. Their structural details are as shown in Figure 7-4~Figure 7-6.

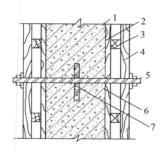

Figure 7-5   Diagram of blots plus stop cap
1—Waterproof structure; 2—Formwork; 3—Waler;
4—Stiff back; 5—Pulling bolts; 6—Water stop ring;
7—Stop cap

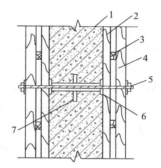

Figure 7-6   Diagram of embedded sleeve supporing
1—Waterproof structure; 2—Formwork; 3—Waler;
4—Stiff back; 5—Pulling bolts; 6—Wooden mat;
7—Embedded sleeve

2. Construction of waterproof concrete

Waterproof concrete must use mechanical mixing and mixing time should not be less than 120 seconds or should be determined according to the technical requirement of admixture.

Measures should be taken to prevent concrete segregation, loss of slump and air content during the transportation, as well as leakage of slurry.

In casting, the free falling height should not exceed 1.5m, otherwise, downspouting, tumbling barrel, should be adopted. Concrete should be casted in layers, each layer thickness should not exceed 300~400 mm, the interval time of adjacent layers concrete casting should not exceed 2 hours, it should be shortened appropriately in summer.

High frequency machine shall be adopted in waterproof concrete compacting and the vibration time should be 20~30 seconds. The vibrating can be stopped after the concrete surface has no obvious settlement, cement slurry appears and no air bubble appears.

The curing of waterproof concrete has a great influence on its impermeability, especially at the early wet curing phase.

Generally, after the concrete reaches final setting time (4~6 hours after casting), it should be covered and watered for curing and the curing time should not be less than 14 days.

Making holes in the completed waterproofing concrete is strictly forbidden.

3. Construction joint of waterproof concrete

Waterproof concrete should be casting continuously as far as possible without or less construction joints. Whenever there are construction joints, the following regulations shall be complied with:

(1) Roof and bottom plate are not suitable for setting construction joints.

(2) Horizontal construction joints between the wall and the floor shall be left on the wall not less than 300 mm above the surface of the floor.

(3) Horizontal construction joints between the wall and the roof shall be left 150~300mm below the roof.

(4) When there are holes in the wall, the distance between construction joints and the edges of the holes should not be less than 300mm.

Vertical construction joint should avoid the parts which has abundant ground water and fissure water and try to combine with deformation joint (post-cast concrete strip) as possible.

Waterproof measures must be taken at construction joints, the structure can be selected according to Figure 7-7.

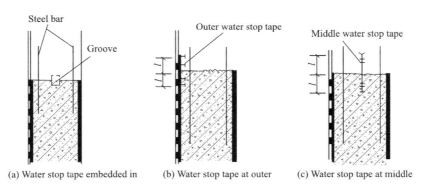

Figure 7-7 Setting of waterstop tape

4. Settings of post-cast strap

When structure area is relative larger, in order to avoid structure harmful cracks due to excessive temperature and shrinkage stress, a post-cast-strap can be set up to temporarily divide the structure into several segments. Post-cast-strap can also be a substitution of the settlement joints at where the settlement joint required to be set up.

The width of post-cast strap is generally 700～1000mm with the spacing of 30～60m. Details of post-cast strap is shown in Figure 7-8.

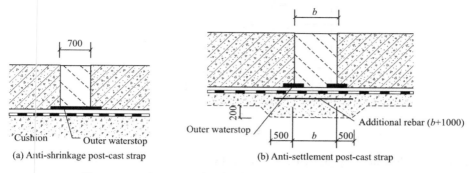

Figure 7-8　Structure details of post-cast strap (unit: mm)

Before casting of post-cast strap, the whole concrete surface must be treated according to the requirements of construction joints.

The casting time of post-cast strap: for anti-shrinkage type, it shall be casted after 30～40 days of main structure concrete casting and the both sides concrete basically stops shrinking; for anti-settlement type, it shall be casted after the completion of the main structure casting and the both sides concrete settlement basically complete.

The concrete should adopt micro-expansion or non-shrinkage cement, or ordinary cement with corresponding additives, however, the strength should be one grade higher than that of main structure, and the wet curing time should not be less than 15 days.

7.2.2　Construction of additional waterproof layer

The additional waterproof layer scheme is a method of making a waterproof layer on the water face of the structure. The additional waterproof layer can be coiled sheet waterproof layer, coating waterproof layer or cement mortar waterproof layer, etc. All of these different type of waterproof layers can be selected according to different project object, waterproof requirement and construction condition.

1. Construction of coiled sheet waterproof layer

Coiled sheet waterproof layer is generally set in the outer side of the underground structure (facing the water), known as external waterproof. It has good water resistance and ductility, can adapt to structural vibration and small deformation, and can resist the corrosive medium.

The coiled sheet of underground waterproof project should choose high polymer modified bitumen waterproof sheet or synthetic high molecular waterproof sheet, laying methods is the same as of roof waterproof engineering.

There are two kinds of setting methods for the externally-protected waterproof layers, details are shown in Figure 7-9 and Figure 7-10.

Construction of internally-painted method is relatively simple, the waterproof layer

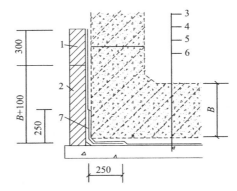

Figure 7-9　The externally-protected and externally-laying structure of coiled sheet waterproof layer (unit: mm)

1—Temporary protective wall; 2—Permanent protective wall; 3—Fine aggregate concrete protective layer;
4—Coiled sheet waterproof layer; 5—Cement mortar leveling layer; 6—Concrete cushion;
7—Reinforcement layer of coiled sheet waterproof layer

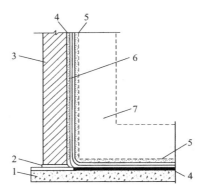

Figure 7-10　The externally-protected and internally-laying structure of coiled sheet waterproof layer

1—Concrete cushion; 2—Dry laying coiled sheet; 3—Permanent protective wall; 4—Cement mortar leveling layer; 5—Protective layer; 6—Coiled sheet waterproof layer; 7—Concrete structure

of bottom slab and wall can be laid at one time and there is no overlapping gap. However, the uneven settlement of the structure has a great influence on the waterproof layer, this might lead to leakage phenomenon which is difficult to be repaired. Therefore, the internally-painted method might be selected only there are limitation of construction conditions.

2. Construction of coating waterproof layer

The construction of coating waterproof layer has great flexibility, no matter base plane with complex shape, or the node with narrow small area, all the parts where could be brushed to can take the method of coating waterproof, so it is widely used in underground engineering.

The coating waterproof layer of underground engineering can be categorized into internal waterproof layer, external waterproof layer and combination of both.

The construction methods and requirements of coating waterproof layer in underground engineering are basically the same as roof waterproof engineering.

3. Construction of cement mortar plaster waterproof layer

Cement mortar plaster is a rigid waterproof layer.

By means of that a certain thickness cement mortar and pure cement slurry plastered on the bottom and side of structure, impermeability and waterproof effect can be realized by capillary channels clog each other in different mortar layers. However, this type of waterproof layer has a poor resistance to deformation, so it is not suitable for projects affected by vibration load and those have uneven settlement.

Cement mortar waterproof layers should be constructed continuously, no construction joints allowed.

When construction joints are required, it should meet the related requirements:

(1) Ladder slop joints should be adopted at plane area, as shown in Figure 7-11.

(2) Joint of ground and wall can be set on ground or wall, but all joints need to be away from the concave or convex corner 200 mm. Joint at corner is shown in Figure 7-12.

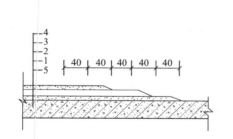
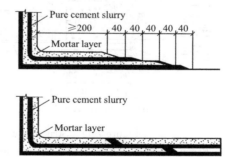

Figure 7-11 Joint at plane area (unit: mm)
1, 3—Pure cement slurry layer; 2, 4—Mortar layer; 5—Structure base

Figure 7-12 Joint at corner (unit: mm)

7.2.3 Construction of prevention and drainage method

The combined method of prevention and drainage is not only waterproof, but also to drain the underground water through the drainage system so as to decrease pressure of the underground water, and reduce the infiltration of water, so as to assist the underground waterproof to achieve the requirement.

Permeation and drainage, blind drain methods can be adopted at construction of prevention and drainage in underground engineering. The structure details are as shown in Figure 7-13 and Figure 7-14.

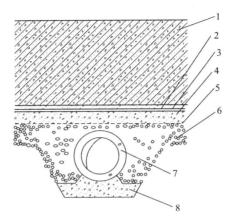

Figure 7-13 Structure details of permeation and drainage

1—Structure bottom slab; 2—Fine aggregate concrete; 3—Waterproof layer; 4—Concrete cushion;
5—Slurry-stop layer; 6—Coarse sand filter layer; 7—Collector pipe; 8—Base of collector pipe

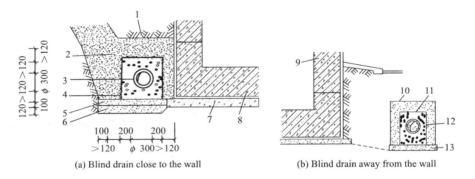

(a) Blind drain close to the wall   (b) Blind drain away from the wall

Figure 7-14 Structure details of blind drain (unit: mm)

1—Rammed earth; 2, 10—Medium sand inverted filter layer; 3, 12—Collector pipe; 4, 11—Pebble inverted filter layer; 5, 13—Layer of cement, sand and brick rubble; 6—Compacted layer of brick bubble; 7—Concrete cushion; 8, 9—Main structure

7.3 Interior waterproof engineering

Interior waterproof project refers to waterproof engineering of various water rooms, such as toilet, bathroom, kitchen, those water room of laboratory and industrial buildings.

The flowing compulsory provisions must be enforced strictly according to the national code of *Technical Code for Interior Waterproof of Residential Building* JGJ 298—2013.

(1) Solvent-based waterproof paint shall not be used in interior waterproof project of residential buildings.

(2) The floor of toilet, bathroom, ground should set waterproof layer, wall surface and ceiling should set moisture-proof layer, the measures must be taken at the door

opening to prevent water overflowing. Extension range of waterproofing at ground, floor entrance is shown in Figure 7-15.

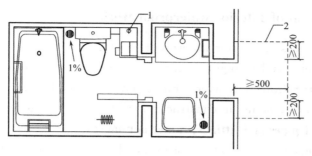

Figure 7-15 Schematic diagram of extension range of waterproofing at ground, floor entrance (unit: mm)
1—Pipe and sleeves that goes across floor; 2—Waterproofing extension range at door opening

(3) Drainage upright pipe should not run through the lower layer habitable room, when floor drain set in the kitchen, its drainage branch pipe should not pass through the lower layer habitable room.

(4) No leakage is allowed at waterproof layer.

Methods of inspection: water close test shall be conducted after the completion of waterproof layer. The height of water storage on the floor and ground shall not be less than 20 mm, and the water storage time shall not be less than 24 hours. The independent water container should be full of water for not less than 24 hours. Quantity of inspection: each room or container should be inspected one by one.

7.3.1 Selection of waterproof materials

Comparing with roof waterproof, underground waterproof, the characteristics of the interior waterproof are listed as followings:

(1) Due to none or less influence of natural climate, small temperature difference deformations and small resistance to water pressure, therefore, the requirements of temperature and thickness of waterproof materials are small.

(2) Interior plane shape is more complex, the construction space is relativity small and narrow.

(3) There are more pipes passing through the floor or wall and the more concave and convex corners make it difficult to the construction of waterproof materials.

(4) Waterproof material is direct or indirect contact with peoples, thus, waterproof materials should satisfy the requirements of non-toxic, flame resistance, pro-environment, construction and use safety.

Considering the characteristics above, the coating waterproof methods which is convenient in construction, non-seam, should be adopted.

Commonly used waterproof materials are: high elastic polyurethane waterproofing paint, elasto-plastic chloroprene latex bitumen waterproof paint, polymer cement waterproof paint, polymer emulsion waterproof paint, additional matrix reinforcing materials can be added if necessary.

7.3.2 Construction of interior waterproof works

The general process of interior waterproof construction:

Installation of pipe and fittings, water appliances→Leveling layer→Waterproof layer →Closed water test→Protective layer→Finish layer→Closed water test→Acceptance of construction quality. Key points in construction are listed as followings.

1. Installation of pipes and fittings, water appliances

Pipes and fittings (such as pipes, sleeves, floor drains, etc) that passing through the floor or walls must be firmly installed.

After positioning the pipe and fittings, the gap between the pipe and surrounding structure should be sealed with 1 : 3 cement mortar. When the gap is larger than 20 mm, the fine aggregate concrete mixed with expansion agent can be casted tightly, and bottom formwork should be hanged below for concrete casting.

A groove shall be left between the root of pipe and the cement mortar or concrete, with a depth of 10 mm and width of 20 mm. Fill the groove with sealing paste and scrape upward 30~50 mm height.

The installation of water appliances should be stable, the installation position should be accurate, periphery of the water appliance must be sealed with sealing materials. The waterproof construction where the pipe goes across through floor is as shown in Figure 7-16, waterproof construction details of floor drain is as shown in Figure 7-17.

2. Construction of leveling layer

The leveling layer is usually 1 : 2.5 or 1 : 3 cement mortar with a thickness of 20 mm.

The leveling layer should be flat, smooth and solid, free of looseness, sand streak, dusting and mortar ash dropping phenomenon. The slope of the leveling layer should be 1%~2%, keep the pipe root slightly above around ground, form a pit around floor drain area a little below than the surrounding ground. At turning corners, the leveling layer should be made into small evenly rounded corners with a radius not less than 10 mm.

3. Construction of waterproof layer

When the leveling layer is basically dry and the moisture content is not more than 9%, the waterproof layer construction can be carried out.

Dust and debris on the surface of the leveling layer should be thoroughly cleaned before construction.

Water penetration happens easier at concave and convex corner between floor and wall, root of the pipe going across through the floor and the floor drain, therefore, additional reinforcement treatment must be carried firstly at these areas by means of adding matrix reinforcement materials and adding brushing waterproof paint.

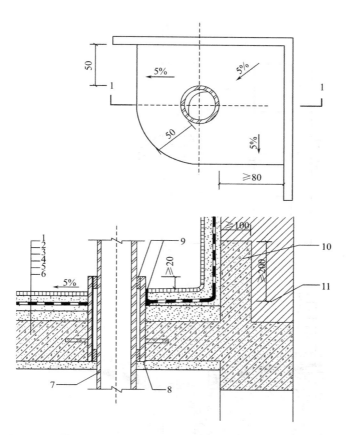

Figure 7-16 Waterproof construction where the pipe goes across through floor (unit: mm)
1—Surface layer of ground or floor; 2—Binding course; 3—Waterproof layer; 4—Leveling layer;
5—Cushion or slope making layer; 6—Reinforced concrete floor; 7—Upright drainage pipe;
8—Sleeve; 9—Sealant; 10—Grade C20 fine aggregate concrete upside edge;
11—Elevation of finished decoration layer

For the pipes that go across through the floor, it should be brushed upward and beyond the top of the pipe sleeve.

At the intersection parts between floor and wall, it should be brushed (or laid) onto the wall, the height is 200~300mm above ground surface. Waterproof construction at intersection area of wall and ground or floor is as shown in Figure 7-18.

For the walls of toilets with shower facilities, the height of waterproof layer should not be less than 1.8 m, and the floor waterproof construction should be carried out after that of wall.

In order to increase the bond between coating layer and cement mortar protective layer, a few clean sands can be spread onto the last course ahead of its hardening.

4. Closed water test

According to national standard and specification, 24 hours clos water test should be carried out after the construction of waterproof layer is completed and shaded drying, the

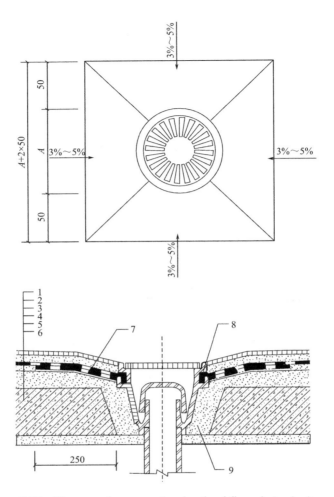

Figure 7-17 Waterproof construction details of floor drain (unit: mm)
1—Surface layer of ground or floor; 2—Binding course; 3—Waterproof layer; 4—Leveling layer; 5—Cushion or slope making layer; 6—Reinforced concrete floor; 7—Waterproof layer or additional layer; 8—Sealant; 9—Filled with Grade C20 fine aggregate concrete mixed with polymer substance

depth of water shall be 20~30 mm higher than slope-making layer highest point. The construction of protective layer and decorative layer can only be carried out after confirming that the waterproof layer has no leakage. After the completion of the construction of the equipment and decorate layer, the second close water test shall be carried out. Only when the final no leakage and smooth drainage are qualified, can the formal acceptance be conducted.

5. Construction of protective layer

After the close water test is qualified and the waterproof layer is completely cured and hardened, 1:2 cement mortar protective layer with thickness of 15~25mm can be laid, the protective layer should be applied moisture curing.

Chapter 7 Waterproof Engineering

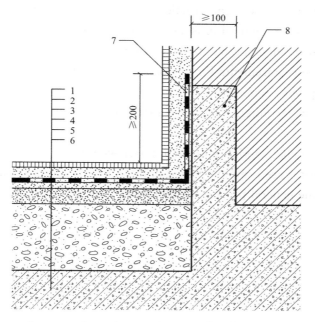

Figure 7-18 Waterproof construction at intersection area of wall and ground or floor (unit: mm)
1—Surface layer of ground or floor; 2—Binding course; 3—Waterproof layer; 4—Leveling layer;
5—Cushion or slope making layer; 6—Reinforced concrete floor; 7—Upside height of waterproof;
8—Grade C20 fine aggregate concrete upside edge

6. Construction of decorate layer

Floor tiles or other decoration materials can be paved onto the cement mortar protective layer, the 107 glue should be added into cement mortar used in laying of decorate layer, the cement mortar should be filled with compactness, no empty bulge and uneven phenomenon.

Attention should be paid to the drainage slope and slope direction in the room during construction, at 50 mm around floor drain, the drainage slope can be appropriately increased.

Questions

1. Describe the construction details of coiled sheet waterproof roof.

2. What are the materials commonly used in coiled sheet waterproof roof?

3. Write out the general construction technological process of coiled sheet waterproof roof.

4. Why should the leveling layer set the separation joints?

5. How to determine the direction of waterproof coiled sheet?

6. What the laying sequence of waterproof coiled sheet?

7. How many types of protective layers? Which kind of coiled sheet are they suitable for respectively?

8. Write out the contents of construction process of coating waterproof roof.

9. What materials can be used in isolation layer of rigid waterproof roof?

10. Why should the dividing points be set in rigid waterproof roof and how to set?

11. What are the requirements for formwork that used in concrete itself waterproof of underground waterproof engineering?

12. What are the requirements for construction joints in underground waterproof engineering?

13. What are the characteristics of the interior waterproof?

14. What materials are commonly used in interior waterproof engineering?

15. Write out the general process of interior waterproof construction.

Chapter 8   Principle of Flow Construction

Mastery of contents: The concept and feature of flow construction, the content and determination of flow construction parameters, various methods and characteristics of flow construction organization.

Familiarity of contents: Expression form of flow construction.

Key points: Features of flow construction, concept of flow construction parameters, various methods and characteristics of flow construction organization.

Difficult points: Methods of flow construction organization, analysis of different methods in relation to flow construction.

8.1   Overview flow construction

Flow construction is one of the scientific methods to effectively organize construction engineering. It is similar to the assembly line production of other industrial product. However, due to the different characteristics of construction products and production, the concept, characteristics and effects are different from those of other industrial products. The latter is that the workpiece and product flow from one process to another, the producer is fixed, while in flow construction, the professional construction team move from one construction section to another, producer is movable.

8.1.1   Concept of flow construction

In civil engineering construction, sequential, parallel and flow construction can be adopted as construction organization methods. Different organization mode, its technology economy benefit is different. Through the analysis and comparison in Table 8-1, the basic concept and superiority of flow construction are illustrated.

Example 8-1:

There are 4 identical buildings named Ⅰ, Ⅱ, Ⅲ and Ⅳ, each foundation divisional work is divided into four construction sections, the construction process are earth work, concrete cushion, masonry work and back fill, and workers for each construction process are 10, 15, 20 and 12 respectively. Construction duration for each process at each section is 5 days.

1. Sequential construction

The sequential construction organization method is to divide the entire project into several construction divisions, after the previous division is totally completed, the next division starts. It is the most basic and primitive construction organizing way, its characteristics are listed as bellow:

(1) Working spaces is un-efficient usage cause of its pause, project duration is longest.

(2) Professional team can't work continuously.

(3) The resource supply amount is relatively small per unit time, which is benefit to the supply of resources, but the supply is not continues and balanced.

(4) Organization and management in construction site are simple.

When the project scale is relatively small, the construction space is limited and the project duration is longer, the sequential construction is suitable, also, it is the common organization way.

2. Parallel construction

Theparallel construction organization method is to organize several professional team constructing at different working space but start and finish at exactly the same time. its characteristics are listed as bellow:

(1) Working spaces is efficiently usage, project duration is shortest.

(2) Professional team can't work continuously; it is suitable for organizing integrated professional team construction.

(3) The resource supply amount applied in construction per unit time increase multiply, temporary on-site facilities have also been increased.

(4) Organization and management in construction site are complex, management costs increase.

Parallel construction is generally applicable to the construction which has urgent time limitation, large scale, tasks in stages and batches, as well as under a reliable resources supplying.

3. Flow construction

The flow construction method is to divide the whole project construction into several construction process in term of technology, several construction sections on plane and several layers on floors; establishing corresponding professional team according to the construction process, with the number and material and tools unchanged; applying the established professional team to the construction sections on plane and layers on floors sequentially and continuously under the construction sequence specified, ensuring the adjacent professional team can execute construction in simultaneous and overlapping way as possible as they can, complete the construction task within the specified duration.

Its characteristics are listed as bellow:

(1) Working spaces is efficiently usage, project duration is relative shorter.

(2) Professional team can work continuously, it is conducive to specialized production, improving operation technology and engineering quality, as well as labor productivity.

Chapter 8　Principle of Flow Construction

(3) The resource supply amount applied in construction per unit time is more balanced, which is conducive to the organization and supply of resources.

(4) The construction site is easy to organize and manage, which creates favorable conditions for civilized construction and scientific management.

Flow construction method is widely applied and used at present.

The organizations for the three mode of organizing construction are listed by means of Gantt chart as shown in Figure 8-1. It includes two parts: construction schedule and labor curve which described the relationship between construction process, time and labor forces consumption.

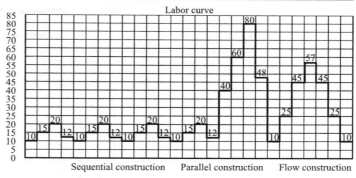

Figure 8-1　Comparison of construction organization methods

Comparison of the three modes of organizing construction is listed in Table 8-1.

Comparison of the three modes of organizing construction　　Table 8-1

	Sequential construction	Parallel construction	Flow construction
Working area	Un-efficient usage	efficiently usage	Reasonable usage
Duration	Longest	Shortest	Medium
Professional team continuous	—	—	Yes
Resources	Small amount and dispersed	Large amount and concentrated	Medium, Equilibrium
Efficiency	No	No	Yes

8.1.2 Expression of flow construction

The expressions of flow construction mainly include line chart and network chart. The line chart can be further categorized into horizontal chart and vertical chart as shown in Figure 8-2.

1. Horizontal chart

The horizontal axis represents the duration of flow construction, i.e. the construction schedule; the vertical axis represents the construction process or professional team; horizontal line segments and number in circle express the number of construction sections and the construction sequence.

2. Vertical chart

The horizontal axis represents the duration of flow construction, i.e. the construction schedule; the vertical axis represents the construction sections; oblique line segments represent the time and sequence in which a construction process or professional team puts work into each construction section.

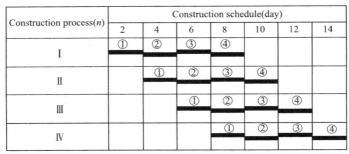

Figure 8-2 Expressions of line chart

8.1.3 Categories of flow construction

Flow construction organization might be categorized into 4 types according to its different organizational goals:

(1) Sub-divisional work flow construction.

(2) Divisional work flow construction.

(3) Unit project flow construction.
(4) Group engineering flow construction.

8.2 Parameters of flow construction

The parameters used to express the state of flow construction, include technology, time and space parameters.

8.2.1 Technology parameter

The parameters related to technology characteristics include construction process and intensity of flow construction.

1. Construction process

In the project construction, the scope of the construction process can be larger or smaller, so the construction process might be sub-division, division, unit-project or sub-project. Generally, it can be categorized into the following categories according to different properties of construction process: prefabrication, transportation, masonry and installation.

Whenever those that occupy the construction space of the project and affect the total construction period, they must be considered in the construction schedule.

The number of construction process is expressed in "n", which is one of the basic parameter of flow construction.

The following issues should be paid more attention while determining the number of construction process included in the construction schedule.

(1) Only those occupy the construction space and have a direct impact on the construction period can be included in the construction schedule.

(2) The number of construction process should be appropriate. Generally, for a controlling schedule, the division of the project can be rough, while for an implementation schedule, it should be divided more detailed and usually giving the names of the sub-divisions.

(3) The leading construction process should be found and determined in order to grasp the key links.

(4) Some interpenetration or smaller additional construction process can be combined into the dominant construction process.

(5) Water, heating, electricity, sanitation and equipment installation projects are usually carried out by other construction processes, so the construction schedule just only reflect their coordination relationships among them.

2. Intensity of flow construction

The amount of work completed in a certain construction process per unit time is called the flow construction intensity of the construction process, it expressed by "V" as shown in

Formula (8-1) and Formula (8-2).

$$V = \sum_{j=1}^{x} R \cdot S \qquad (8\text{-}1)$$

There in:

V——Intensity of mechanical in a construction process.

R——The number of construction machines in construction process.

S——Output quota of construction machines in construction process.

x——Variety of machines in construction process.

$$V = R \cdot S \qquad (8\text{-}2)$$

There in:

V——Intensity of labor in a construction process.

R——The number of labor in construction process.

S——The average output quota of labor in construction process.

8.2.2 Space parameter

The parameters related to space arrangement state include working surface, construction section and construction layer.

1. Working surface

The activity space that professional workers must have in operation is called the working surface, which is determined according to the production quota and the requirement of safe construction technical regulations.

2. Construction section

Division of construction section is the basis of flow construction organization, and the construction layer is often divided into a number of sections on the plane, known as construction section and expressed by "m" which is one of the basic parameters of flow construction.

1) Principles of division of construction section:

Construction section number should be appropriate, it should comply with the follow principles.

(1) The labor quantity of the same professional team in each construction section should be roughly equal, the difference should not be more than 10%~15%.

(2) For ensuring the integrity of the structure, the dividing line of construction section should be set at the building joints as possible; if it has to stay on the wall, it should be in the opening of door and window in order to decrease and deal with the racking easily.

(3) In order to improving productive efficiency of workers (or machinery), construction section should not only meet the requirements of professional team to working space, but also make the amount of labors (or machinery) in which construction section can meet the requirements of labor optimization combination.

Chapter 8　Principle of Flow Construction

(4) In order to meet the requirements of reasonable flow construction organization, number of sections should more than or equal that of construction processes ($m \geq n$).

(5) For multi-storey buildings, the construction section should be divided on plane along horizontal and on layer along the vertical direction. For ensuring that professional team can carry out rhythmic, balanced and continuous flow construction between the construction sections and layer, the position of dividing line and number of sections of upper and lower layers should be aligned and consistent with.

2) Relationships between number of construction sections (m) and that of construction processes (n).

(1) In the case of section number less than that of construction processes ($m < n$), as shown in Figure 8-3.

Construction layer	Construction process	Construction schedule(d)									
		5	10	15	20	25	30	35	40	45	50
Floor 1	Formwork	①	②								
	Steel installation		①	②							
	Concrete placing			①	②						
Floor 2	Formwork					①	②				
	Steel installation						①	②			
	Concrete placing							①	②		

Figure 8-3　Flow construction with number of sections less than that of construction processes

The construction sections are fully used, but the construction processes cannot work continuously. Even though, it could be organized as group flow construction for several buildings in order to make up the stopping of construction processes, it is not suitable for single building.

(2) In the case of section number more than that of construction processes, ($m > n$), as shown in Figure 8-4.

The construction processes are able to work continuously, but there is unused working space at construction sections. This certain free time at construction can be used to make up for the time needed for technical, organizational stopping and material preparation.

(3) In the case of section number equal to that of construction processes ($m = n$), as shown in Figure 8-5.

The construction processes can work continuously and the construction sections are fully used, it is the ideal flow construction scheme.

Therefore, in order to ensure the continuous construction of construction processes,

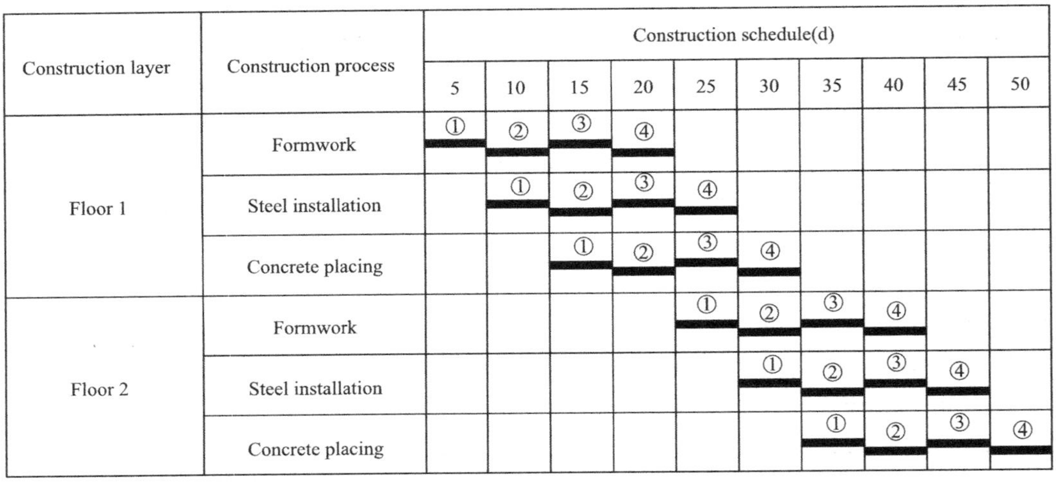

Construction layer	Construction process	Construction schedule(d)									
		5	10	15	20	25	30	35	40	45	50
Floor 1	Formwork	①	②	③	④						
	Steel installation		①	②	③	④					
	Concrete placing			①	②	③	④				
Floor 2	Formwork						①	②	③	④	
	Steel installation							①	②	③	④
	Concrete placing								①	②	③	④

Figure 8-4 Flow construction with number of sections more than that of construction processes

Construction layer	Construction process	Construction schedule(d)									
		5	10	15	20	25	30	35	40	45	50
Floor 1	Formwork	①	②	③							
	Steel installation		①	②	③						
	Concrete placing			①	②	③					
Floor 2	Formwork				①	②	③				
	Steel installation					①	②	③			
	Concrete placing						①	②	③		

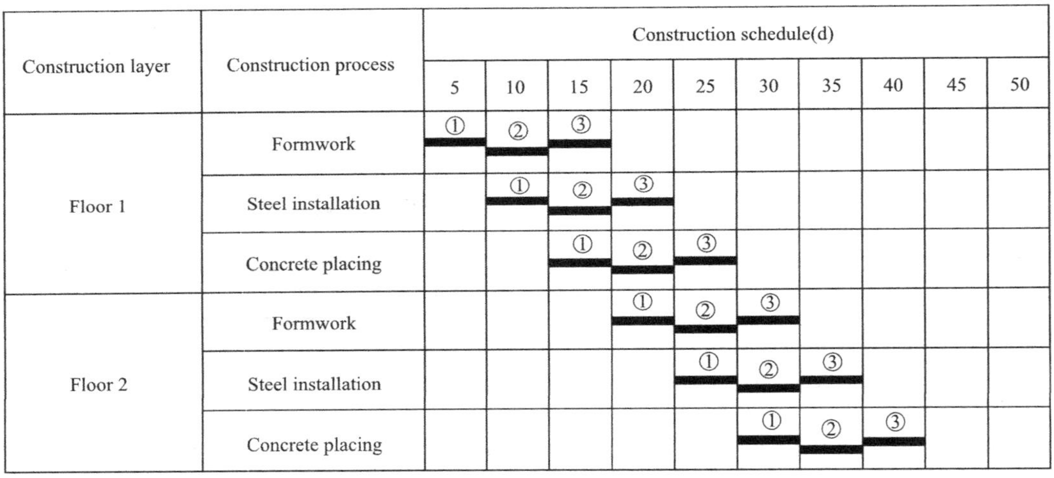

Figure 8-5 Flow construction with number of sections equal to that of construction processes

the requirements of $m \geqslant n$ must be complied with.

3. Construction layer

The project can be divided into several construction sections (or operation layers) in vertical direction, which is called construction layer denoted by r.

The division of construction layer should be determined according to the details of the project structure, and generally, a structure layer could be construction layer.

8.2.3 Time parameter

The parameters used to express the state of flow construction in time arrangement are called time parameters which include flow construction rhythm, flow construction spacing, technical pause, organizational pause and overlapping time.

Chapter 8　Principle of Flow Construction

1. Flow construction rhythm

Flow construction rhythm is the time that a professional team takes on a construction section, denoted by "t", it is one of the important parameters in flow construction organization.

The value of Flow construction rhythm reflects the speed and the supply of resources in flow construction and the value can be determined by the following methods.

1) Quota method

$$P = Q \cdot H \tag{8-3}$$

$$t = \frac{P}{R \cdot N} \tag{8-4}$$

There in:

P——The amount of labor or machinery required within a construction section.

Q——The engineering quantity within a construction process.

H——Time quota (the amount of labor or machinery per unit engineering quantity).

R——Number of professional team or number of machinery.

N——Number of shift of professional team.

2) Inverse scheduling

For some projects that must be completed within the specified date, reverse scheduling methods is usually adopted.

According to the reverse scheduling, determine the working duration of a construction process, then, calculate the flow construction rhythm of the construction process on sections, as shown in Formula (8-5).

$$t = \frac{D}{m} \tag{8-5}$$

There in:

D——The working duration of a construction process.

m——The number of construction section of a construction process.

3) Empirical estimation method

Generally, the shortest, longest and normal time of the flow construction are estimated at first according to previous construction experience, then, calculate the expected time by Formula (8-6), and take it as the flow construction rhythm of a professional team at construction section.

$$t = \frac{a + 4c + b}{6} \tag{8-6}$$

There in:

a——The shortest estimated time (/best optimism time) of a construction process at a construction section.

b——The longest estimated time (/most pessimism time) of a construction process at a construction section.

c——The normal estimated time of a construction process at a construction section.

2. Flow construction spacing

Flow construction spacing is the time interval between two adjacent construction processes entering the first construction section, denoted by "K" which is one of important parameter in flow construction.

When the construction section is determined, the flow construction spacing directly affects the construction period same as flow construction rhythm. The large "K" is, the longer the construction period will be; on the contrary, it is shorter.

The number of "K" depends on the number of construction processes participation in flow construction, the number of "K" equals to $(n-1)$ or (n_1-1), where n, n_1 is the number of construction process or professional teams respectively.

1) Principles of construction spacing

(1) The process sequence shall be ensured, and the constraints on adjacent professional working team shall be satisfied, the maximum and reasonable lapping in construction should be achieved.

(2) Ensure continuous operation of professional teams as far as possible, properly handle the technical or organizational pause and avoid stopping or work slowdown.

(3) The flow construction spacing should be at least one or half shift's duration, and keep relation to the flow construction rhythm in order to organize the flow constructionconveniently.

2) Method in determining construction spacing

There are many ways to calculate the flow construction spacing, which should be determined according to the characteristics of flow construction rhythm.

3. Technical pause, organizational pause and overlapping time

(1) Technical pause is the intermission time determined by the engineering materials or the technological properties of the construction process, such as the curing time of cast-in-situ concrete members, the drying and hardening time of plastering layer and paint layer, etc.

(2) Organizational pause is the interval time cause by construction organization, such as the inspection and acceptance of underground pipes before backfilling, the time required for the transfer of construction machinery, leveling and laying-out before masonry, and any other time spent for preparing of construction.

The two types of time above can be denoted by "P".

(3) Overlapping time:

Generally, the next construction process can only start after the former is completely finished, but, in order to shorten the construction period, the two processes could be overlapped in construction section if there is enough operation space, so both of the two adjacent construction processes are constructed simultaneously on the same construction section.

The duration of its simultaneous construction at the same construction section is called overlapping time between two adjacent construction process or professional teams, denoted by "O"

8.3 Organization of flow construction

The division (professional) flow construction is the foundation in the civil engineering construction. According to the characteristics of engineering project construction and different flow construction parameters, professional flow construction organization can be divided into three types: fixed rhythm, multiple rhythms and no rhythm flow construction.

Basic principles in organization of flow construction:

(1) Construction processes (professional teams) sequence shall be complied with. That means only after the previous process (professional team) completed or mainly completed, can the next process start.

(2) Construction process (professional team) conflict must be avoided. That means any construction process (professional team) cannot be carried out at different construction sections at same time.

(3) Generally, only after the first construction section of the previous layer completed, can the first construction section of the next layer start.

8.3.1 Fixed rhythm Flow Construction

The rhythms of any construction process (or professional team) are constant at any construction section, therefore, it is also known asisorhythmic or all equal rhythm.

Fixed rhythm Flow Construction is generally applicable to the construction of houses or some structures with small scale, simple structures, and few construction processes. It is often used to organize the division engineering.

1. Characteristics

(1) The rhythms are equal for each other, if there is a number n construction process, then:

$$t_1 = t_2 = \cdots\cdots = t_{n-1} = t_n = t \text{ (Constant)}$$

(2) The spacing is equal for each other, and equal to the rhythm, that is:

$$K_{1,2} = K_{2,3} = \cdots\cdots = K_{n-1,n} = K_n = K = t \text{ (Constant)}$$

(3) Each professional team can work continuously on each constructionsection, there is no free time in construction sections.

(4) The number of a professional teamis equal to that of construction processes, that is $n_1 = n$.

2. Flow Construction sections

(1) Without interlayer relationships

The sections number "m" can be calculated by Formula (8-7) with $P_2 = 0$.

(2) With interlayer relationships

If the sum pauses in the same layer is $\sum P_1$, the sum of overlapping in the same layer is $\sum Z_1$, the pauses between adjacent construction layers are P_2, and $\sum P_1$, $\sum Z_1$ and P_2 are equal in every layer respectively, then the minimum number of construction sections "m" to ensure continuous construction for each professional team is determined by Formula (8-7).

$$m = n + \frac{(\sum P_1 - \sum Z_1) + P_2}{t} \tag{8-7}$$

There in:

$\sum P_1$ ——Sum of pauses in the same layer.

$\sum Z_1$ ——Sum of overlapping in the same layer.

P_2 ——Pauses between adjacent construction layers.

If $\sum P_1$, $\sum Z_1$ and P_2 are not equal from inside layer or between adjacent layers respectively, take the greatest value.

The rest of the signs are the same as before.

3. Flow Construction period

(1) Without interlayer relationship.

The Flow Construction period "T" can be calculated by Formula (8-8) with $r = 1$.

(2) With interlayer relationship, as shown in Formula (8-8).

$$T = (m \cdot r + n - 1) \cdot t + (\sum P_1 - \sum Z_1) \tag{8-8}$$

There in:

r ——Number of construction layers (structure floors).

Note: Z_2 is no longer exist in Formula (8-8), because it has been included in $m \cdot r \cdot t$.

The rest of the signs are the same as before.

Example 8-2:

Division engineering is divided into four construction processes A, B, C, and D, choose 4 as the number of construction sections according to the principles of construction section division, the rhythm is 5 days for all construction processes at every construction section.

Answer:

According to the given conditions, fixed rhythm Flow Construction should be adopted.

(1) Construction spacing

$$K = K_{A,B} = K_{B,C} = K_{C,D} = t = 5 \text{ (days)}$$

(2) Construction section determination

$$m = n + \frac{(\sum P_1 - \sum Z_1) + P_2}{t} = 4 + \frac{(0 - 0) + 0}{2} = 4$$

(3) Construction period calculation
$$T = (m+n-1) \cdot t + (\sum P_1 - \sum Z_1) = (4+4-1) \times 3 + 0 - 0 = 35 \text{(days)}$$

(4) Construction schedule

The construction schedule is drawn as in Figure 8-6.

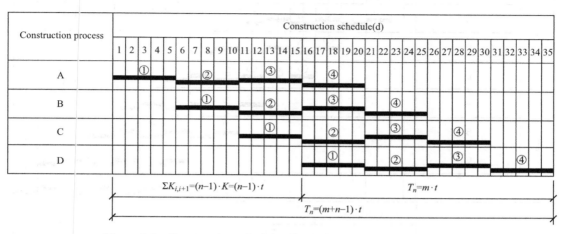

Figure 8-6 Construction schedule (fixed rhythm with a single layer)

Example 8-3:

Division engineering is divided into four construction processes A, B, C, and D, there are 2 construction layers (floors), and the rhythm is 2 days for all construction processes at every construction section. There are 2 days of technical pause between processes B and C and 2 days of technical pause between interlayers. To ensure the construction process operation continuously, try to determine the number of construction sections, calculate the construction period and draw the construction schedule.

Answer:

According to the given conditions, fixed rhythm Flow Construction should be adopted.

(1) Spacing of Flow Construction
$$K = K_{A,B} = K_{B,C} = K_{C,D} = t = 2 \text{ (days)}$$

(2) Construction section determination
$$m = n + \frac{(\sum P_1 - \sum Z_1) + P_2}{t} = 4 + \frac{(2-0)+2}{2} = 6$$

(3) Construction period calculation
$$T = (m \cdot r + n - 1) \cdot t + (\sum P_1 - \sum Z_1) = (6 \times 2 + 4 - 1) + (2 - 0) = 32 \text{(days)}$$

(4) Construction schedule

The construction schedule is drawn as in Figure 8-7.

8.3.2 Multiple rhythm flow construction

Multiple rhythms can also be called different rhythms. It means even the construction

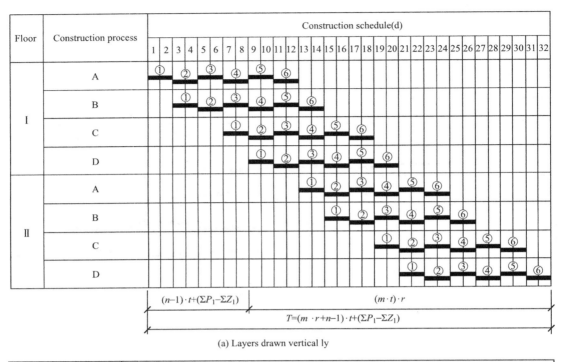

(a) Layers drawn vertically

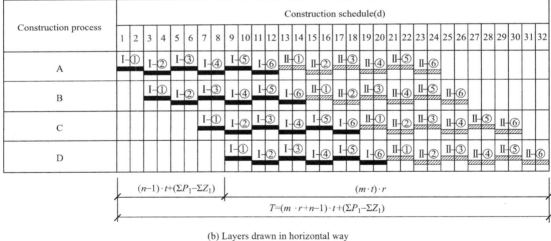

(b) Layers drawn in horizontal way

Figure 8-7 Construction schedule (fixed rhythm with multiple layers)

process has the same rhythm in all of its construction sections, the different construction process has different rhythm, but there is a maximum divisor between the different rhythms.

Usually, in order to speed up the flow construction progress, the professional team of each construction process can be organized in multiples number of the maximum divisor.

1. Characteristics

(1) Same construction process has same rhythms at all of its construction sections,

construction sections' rhythms in different construction sections has a maximum divisor.

(2) Spacing of construction flow is equal to each other and equals to the maximum divisor.

(3) The professional team can work continuously, and, there is no pause at construction sections.

(4) Number of professional team is more than that of construction processes, that is $n_1 > n$.

2. Spacing of flow construction

Take the maximum divisor as the spacing of professional team which is the minimum value of the construction process's rhythm as shown in Formula (8-9).

$$K = Maximum\ divisor = t_{min} \qquad (8\text{-}9)$$

3. Number of professional teams

Number of professional team can be determined by Formula (8-10)

$$b_i = \frac{t_i}{K}, n_1 = \sum b_i \qquad (8\text{-}10)$$

There in:

b_i ——The number of professional teams needed in a construction process.

n_1 ——Total number of professional teams.

4. Number of construction sections

(1) When there are no interlayer relationships, it can be determined according to the basic requirements of the construction sections, generally, $m = n_1$.

(2) When there are interlayer relationships, the minimum number of construction sections of each layer can be determined by Formula (8-11), just change n to n_1 comparing with the previous Formula (8-7).

$$m = n_1 + \frac{\sum P_1 + P_2}{K} \qquad (8\text{-}11)$$

There in:

$\sum P_1$ ——Sum of technical and organizational pause in same layer.

P_2 ——The technology and organizational pauses between adjacent construction layers, if it is not equal, take the greater.

5. Construction period

Construction period can be determined by Formula (8-12).

$$T = (m \cdot r + n_1 - 1) \cdot K + \sum P_1 - \sum Q_1 \qquad (8\text{-}12)$$

Note: P_2 is no longer exist in Formula (8-12), because it is included in $m \cdot r \cdot K$. Comparing with Formula (8-8), it changes n to n_1.

Example 8-4:

A division engineering is consisted of construction processes A, B, and C, the rhythm are 2, 6 and 4 days respectively. Try to organize the flow construction and drawing the construction schedule.

Answer:

According to the given conditions, multiple rhythm flow construction should be adopted.

(1) Determining spacing of flow construction from Formula (8-9).
$$K = Maximum\ divisor = t_{min} = 2 (days)$$

(2) Determining number of professional teams from Formula (8-10).
$$b_A = \frac{t_A}{K} = \frac{2}{2} = 1; b_B = \frac{t_B}{K} = \frac{6}{2} = 3; b_C = \frac{t_C}{K} = \frac{4}{2} = 2$$
$$n_1 = \sum b_i = (1 + 3 + 2) = 6$$

(3) Determining number of construction sections.
$$m = n_1 = 6$$

(4) Calculation of construction period from.
$$T = (m + n_1 - 1) \cdot K + \sum P_1 - \sum O_1 = (6 + 6 - 1) \times 2 + 0 - 0 = 22 (days)$$

(5) Construction schedule is as shown in Figure 8-8.

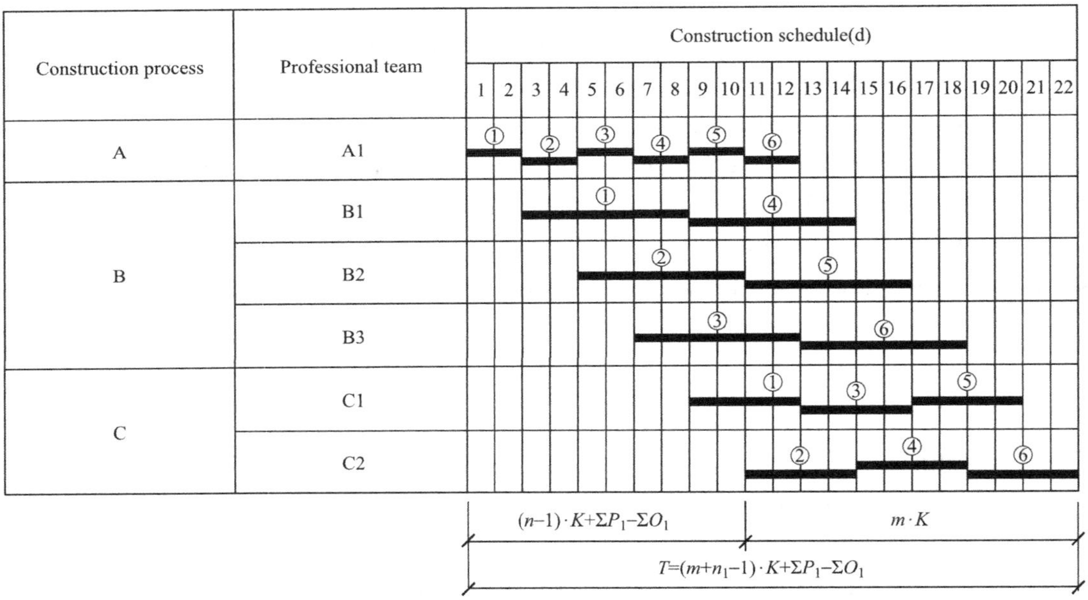

Construction process	Professional team	Construction schedule(d)
A	A1	
B	B1	
	B2	
	B3	
C	C1	
	C2	

Figure 8-8 Construction schedule (multiple rhythm)

8.3.3 No rhythm flow construction

In projects, working quantity of each construction process is usually different from each other at each construction sections, as well as the labor quantity cause of their different production efficiency of each professional team, as a result, most of flow construction rhythm are not equal to each other, making it difficult to organize rhythm flow construction.

In this condition, based on the requirements of construction order, by means of certain calculation methods, the flow construction spacing between adjacent construction process shall be determined at first, ensuring continues operation of each construction process shall be complied with in organization of flow construction.

No rhythm flow construction is common form in organization of flow construction.

1. Characteristics

(1) The rhythm of each construction process are not all the same.

(2) Construction spacing of different construction processes is not completely equal.

(3) Construction process or professional team can work continuously, but there might be pause at construction sections.

(4) Number of professional teamis equal to that of construction processes, that is $n_1 = n$.

2. Spacing of no rhythm flow construction

The essence of no rhythm flow construction is that of profession team working continuously, flow construction spacing is determined by certain calculation methods in order to ensure its working continuously, as well as no interfering with each other in a construction section, or connecting closely between previous and next professional team.

The key to organize rhythm flow construction is to calculate the flow construction spacing. Paterkovsky method is adopted in this chapter for those calculations.

Steps of Paterkovsky method:

(1) Adding up the rhythms for each construction section.

(2) Minus in stagger way.

(3) Take the maximum value as the construction spacing of the two adjacent processes.

3. Construction period

Different from the other flow construction which has a certain time constraint, no rhythm flow construction is flexible and free in arrangement, which is applicable to various construction organizations with different structural properties and scales, and has a wide range of practical applications. However, when organizing multi-layer structure construction, construction process or professional teams cannot work continuously, so the calculation period is only applicable to the case without inter-layer relationship.

Construction period can be calculated by Formula (8-13).

$$T = \sum K_{i,j+1} + \sum P_1 - \sum O_1 \qquad (8-13)$$

Example 8-5:

A division engineering is consisted of construction processes A, B, C, D and E, there are 4 construction sections in the construction layer. Rhythms of each process at each construction section are shown in Table 8-2. There are 2 days technical pause after process B completed, and 1 day organization pause after D completed.

Try to work out the construction scheme.

Rhythms of each process at each construction section   Table 8-2

Process \ Duration on section	1	2	3	4	5
A	3	1	2	4	3
B	2	3	1	2	4
C	2	5	3	3	2
D	4	3	1	3	3
E	2	1	3	1	2

Answer:

According to the given conditions, no rhythm flow construction should be adopted.

(1) Determining spacing of flow construction by Paterkovsky method

Adding up the rhythms for each construction section, minus in stagger way, take the maximum value as the construction spacing of the two adjacent processes.

```
A   3   4   6   10  13
B       2   5   6   8   12          K_{A,B} = max{3,2,1,4,5,-12} = 5
C           2   7   10  13  15      K_{B,C} = max{2,3,-1,-2,-1,-15} = 3
D               4   7   8   11  14  K_{C,D} = max{2,3,3,5,4,-14} = 5
E                   2   3   6   7   9   K_{D,E} = max{4,5,5,5,7,-9} = 7
```

$K_{A,B} = \max\{3,2,1,4,5,-12\} = 5$

$K_{B,C} = \max\{2,3,-1,-2,-1,-15\} = 3$

$K_{C,D} = \max\{2,3,3,5,4,-14\} = 5$

$K_{D,E} = \max\{4,5,5,5,7,-9\} = 7$

(2) Calculate the time of the project

$$T_C = \sum K_{ij} + T_n + (t_+ - t_-) = (5+3+5+7) + 9 + (2+1) - 0 = 32 \text{ (days)}$$

(3) Drawing the construction schedule as shown Figure 8-9.

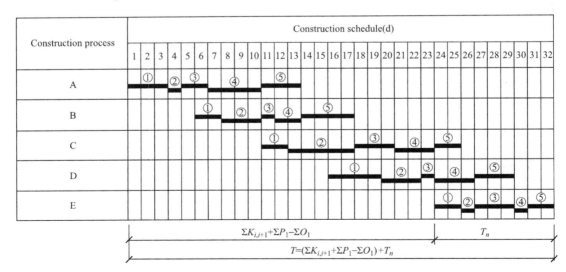

Figure 8-9 Construction schedule (no rhythm)

Chapter 8 Principle of Flow Construction

Questions

1. Write out the characteristics of sequential construction and parallel construction organization?
2. What are the characteristics in flow construction organization?
3. How many types can flow construction be divided according to its different organizational goals?
4. What are the technology parameters in flow construction organization?
5. What contents should be paid more attention while determining the number of construction process?
6. Write out the concept of flow construction intensity.
7. What are the space parameters in flow construction organization?
8. Briefly describe the principles of construction section division.
9. Briefly analyze the relationships between number of construction sections (m) and number of construction processes (n).
10. What are the time parameters in flow construction organization?
11. What's flow construction spacing?
12. Write out the principles of construction spacing.
13. What are the basic principles in organization of flow constructions?
14. Write out the characteristics of fixed rhythm flow construction.
15. Write out the characteristics of multiple rhythm of flow construction.
16. Briefly introduce the determination methods of professional teams and construction sections in multiple rhythm of flow construction organization.
17. Write out the characteristics of no rhythm flow construction.
18. What's the key to organize no rhythm flow construction?
19. Write out the steps of Paterkovsky method.
20. Write out the construction period formula of no rhythm flow construction.

Exercise

Duration time of each process at each section is shown as follows. Calculate the spacing and determine the time of project, drawing the construction schedule.

Table 8-3

Process	Duration on section	1	2	3	4	5
	A	3	1	2	4	3
	B	2	3	1	2	4
	C	2	5	3	3	2
	D	4	3	5	3	3

Table 8-4

n \ m	2	4	6	8	10	12	14	16	18	20	22	24	26	28	30
A															
B															
C															
D															

Chapter 9 Network Plan Technology

Mastery of contents: Drawing of activity on arrow network diagram and its concept and calculation of time parameters, drawing of activity on node network diagram and its calculation of time parameters, drawing of time scale network (activity on arrow) and its determination of time parameters, determination methods of critical path and critical works in network plan, optimizing methods in construction time.

Understanding of contents: Optimizing methods of construction time-cost tradeoff.

Knowing: Concept of the various time distance of the overlapping network plan, calculation of time parameter, concept of optimization of time-resource tradeoff.

Key points: Drawing and time parameters calculation of activity on arrow network diagram, activity on node network diagram, time scale network diagram, determination methods of critical work and critical path of network diagram, optimizing methods in construction time.

Difficult points: Drawing and time parameters calculation of activity on arrow network diagram, drawing of time scale network and meaning of waveform line, time parameters calculation of overlapping network plan, determination of critical work and critical path, optimizing methods of construction time and construction time-cost tradeoff.

9.1 Basic concepts of network plan

9.1.1 Implication of network plan technology

Network plan refers to the network diagram to present the process plan of the project.

Network diagram is a directed, ordered net shaped diagram, also known as process flow diagram or arrow diagram, which uses arrow lines and nodes to express the sequence and time of each works.

Network plan technology is a scientific planning management technology. With the help of network plan, it ensures the realization of predetermining goals by arrangement and controlling to the work scheduled.

9.1.2 The principle of network plan

The principle of network plan is same as that of overall planning methods. Basic principles are:

Applying network diagram to express the sequence and relationship of various tasks in a plan (or an engineering project).

Find out the critical work and critical path in the plan by calculating the time parameters of the network plan.

Seek for the best scheme through continuous improvement of network plan, so as to efficiently control and supervise the plan in implementation process, ensure the reasonable use of human forces, material and financial resources, and achieve the maximum economic benefits with the minimum consumption.

This method has been recognized by countries around the world, widely used in the planning and management of all walks of life.

9.1.3 Characteristics of network plan

In engineering construction, there are two ways to express the progress plan, horizontal chart and network plan.

Even though the two methods can both be used to express the engineering progress plan, they have their own characteristics. Horizontal chart method has been discussed in chapter 8 (principle of flow construction), comparing with that of horizontal chart, the network plan characteristics are as followings:

(1) Reflect the mutual restriction and dependence of each other comprehensively and clearly.

(2) Calculating various time parameters.

(3) Find out the critical works and critical paths among various intricate plans, so that the plan managers could focus on the main contradiction, ensure the scheduled and avoid blind construction.

(4) Be able to choose the best solution from many reasonable schemes.

(5) Be able to reflect the flexibility of works. When a certain work goes ahead or delay comparing its plan, its impactions to the works and project period could be foreseen, as well as quickly adjusted according to the changed situation, therefore, ensure the effective control and supervision of the plan from the beginning to end.

(6) Utilize time reserves for various works reflected in the network plan to adjust the human forces, material resources in order to achieve the purpose of reducing project cost.

(7) Network plan can be compiled and adjusted by means of computer software. It can select the best schemes among variety by calculation, optimization and adjustment to plan on different objectives (such as duration, cost, resources, etc), provide the construction information for the construction organizer, which is beneficial to strength the construction management.

However, network plan technology is more difficult to calculate labor resource consumption than horizontal chart plane.

9.1.4 Categories of network plan

The network plan is a major advance in management science. This technique is based on the basic characteristics of all project, that all work must be done in well-defined steps. For example, for completing a foundation the various steps are (1) layout (2) digging (3) placing side boards and (4) concreting. The net-work technique exploits this characteristic by representing the step of the project objective graphically in the form of a network or arrow diagram.

The network plan can be called by various names such as PERT, CPM, UNETICS, LESS, TOPS and SCANS. However, these and other systems have emerged from the following two major network systems: PERT and CPM.

(1) PERT stands for "Program Evaluation and Review Technique". PERT system is preferred for those projects or operations which are of non-repetitive nature or for those projects in which precise time determination for various activities can't be made. In such projects management cannot be guided by the past experience. They are referred to as once through operation or projects.

(2) CPM stands for "Critical Path Methods". In CPM, networks, the whole project consists of a number of clearly recognizable jobs or operations called activities. Activities are usually the operations which take time and resources to carry out. CPM networks are generally used for repetitive type projects or for those projects for which fairly accurate estimate of time for completion of each activity can be made.

Among many types of network planes, the CPM is commonly used in construction engineering. It can be further divided into activity on line network plan, activity on node network plan, activity on line network plan with time scaled and activity on node network plan with lapping. The first three network plans will be discussed in this chapter.

9.2 Activity on line network plan (double codes network diagram)

9.2.1 Content and basic symbols of activity on line diagram

1. Content

Any project is consist of many of construction work (activities, courses, processes).

Express a work with an arrow line, write the work name above of it and work time below of it, a circle at the end of the arrow line indicates the beginning of the work while a circle at the head of the arrow line indicates the end of the work. The circles have different numbers inside and the numbers of the two circles represent the work as shown in Figure 9-1.

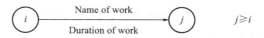

Figure 9-1 Expression of works in activity on line diagram

As shown in Figure 9-2, activity on line network diagram is consisted of arrow lines, nodes (circles) and paths, while dummy arrow line expresses only logical relationship between works.

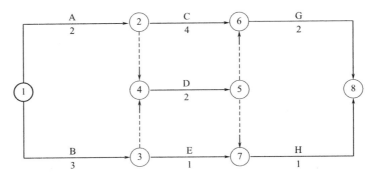

Figure 9-2 Expression of works in activity on line diagram

2. Basic symbols

1) Arrow lines

Arrow line can also be called arrow rod.

Each work must have a unique arrow line and the corresponding pair of numbers at arrow end and arrow head, where the numbers are non-repetition.

The work could be unit project, division, sub-division or construction process.

Generally, works consume not only time, but also the resources such as labor force, material, construction machinery and tools; a few of work consume time only, such as concrete curing and plaster drying, these works can be expressed as a "solid arrow line".

In activity on line network diagram, there exist the "dummy arrow line" which is used for expressing the constraints and dependencies among works, it takes none time nor resources.

Under the condition of no time coordinate constraint, the length of arrow line is independent of the reflected duration; the direction of the arrow indicates the progress of the work; the drawing of arrowheads must use double-sided arrowheads which can be drawn as straight line, oblique line or polyline.

There are two types of interrelationships between one work and another work in a network diagram, direct and indirect relationships.

The work that has direct relationship can be divided into predecessor work, successor work and parallel work.

A successor work is the one which follows another work, and work that occurs before another work is called predecessor work to that work, parallel works have the time overlapping in their processes.

For example, Formwork1 is the predecessor work of the Formwork2 and Reinforcement, on the contrary, Formwork2 and Reinforcement are successor works of the work Formwork1, Formwork2 and Reinforcement are parallel works in the network diagram as shown in Figure 9-3.

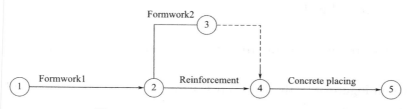

Figure 9-3 Function of dummy arrow line

Dummy work (3-4) must be added for distinguishing the formwork (2-3) and reinforcement work (2-4).

Dummy work has the functions of connecting, breaking and distinguishing, for example, there are 4 works of A, B, C and D, C and D starts after the A finished, D starts after the B finished, as shown in Figure 9-4, the dummy work in this diagram connected work A and D, and break down the path from B to C, that's its connecting and breaking functions.

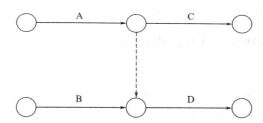

Figure 9-4 Function of dummy arrow line

2) Node

Node can be called events, expressed by circle, it is an instant that consumes none of time nor of that resources, it is just a connection between works of after and before. As shown in Figure 9-5, there are 3 types of nodes.

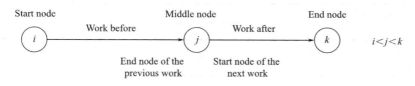

Figure 9-5 Nodes in activity on line network

The first node is the "start node", it means the start of a construction or a task; the last node is the "end node", it means the finish of construction or a task; any other nodes are "middle nodes", it means both of the finish of the previous works and the start of the next works.

Every node has the unique number and never repeated in network diagram, the numbers discontinues are allowed, but always ensure that the value at the arrowhead number larger than that of at the arrow tail.

3) Path and critical path

(1) Path

Path refers to the whole paths from the start node to the end node in the network diagram.

As shown in Figure 9-2, there are 5 paths from the start node to the end node totally in the network diagram.

(2) Critical path

Among many paths, the paths which have the longest working duration is called the critical paths, where the works on the critical paths are called critical work, and the nodes on the critical paths are called critical nodes.

There is at least one critical path in the network diagram, and the critical path can be expressed in double arrow lines, bold arrow lines or color arrow lines in order to show its different from the non-critical path.

The non-critical works at the non-critical paths have float time, that means these type of works can be delay but have no impacts on overall planned project duration.

9.2.2 Drawing of activity on line diagram

1. Basic principle

(1) Logical relationship between works must be properly expressed.

Network diagram are made of logical relationships which is the objective sequence relationship between works, it includes technical and organization logical relationships.

In order to express the logical relationships properly, three questions must be solved:

What are the predecessor works of this work; what works must be finished before this work starts; what works are parallel with this work. Bases on these question, the network diagram can be drawn out. The common logical relationship are shown in Table 9-1.

Expression of logical relationship in network diagram　　　　Table 9-1

No	Logical relationship between works	Expression in the network diagram	Explanation
1	Work A and B are executed in sequence	○—A→○—B→○	Work B depends on work A, work A constrains work B
2	Works A, B and C start in the mean	○ with branches A, B, C to three nodes	Works A, B and C are called parallel works for each other
3	Works A, B and C end in the meantime	Three nodes A, B, C converging into ○	Works A, B and C are called parallel works for each other

Chapter 9 Network Plan Technology

Continued

No	Logical relationship between works	Expression in the network diagram	Explanation
4	There are 3 works A, B and C, only work A finished, can work B and C start		Work A constrains the start of works B and C, works B and C called parallel works for each other
5	Work C can only start after works A and C finished		Works A and B constrain the start of work C together, works A and B are called parallel works
6	Only after the finish of works A and B, can work C and D start		Works C and D are constrained by works A and B, it is expressed by the event j
7	There are 4 works of A, B, C and D, work C starts after work A finish; after both work A and B finish, work D starts		There is logical connection line between works A and D, it is the only way to express the constraint relationships between them
8	There are 5 works of A, B, C, D and E, after both works A and B finish, work D can start; after both work B and C finish, work E can start		Dummy work $i\text{-}j$ establishes the constraint relationship between works B and D; dummy work $i\text{-}k$ establishes the constraint relationship between works B and E. ($i<j, i<k$)
9	There are 5 works of A, B, C, D and E, after work A finish, works C and D can start; after work B finish, works D and E can start		Dummy work $i\text{-}j$ reflects the constraint of work A to work D; dummy work $j\text{-}k$ reflects the constraint of work B to works D and E ($i<k, i<k$)
10	There are 5 works of A, B, C, D and E, after works A, B and C finish, work D can start; after work B and C finish, work E can start		Dummy work $i\text{-}j$ reflects the constraint of works B and C to D
11	There are 6 works of A, B, C, D, E and F, after work A finish, works C and E can start; after works A, B and D finish, work E can start; after work D finish, work G can start		The two dummy works express the constraint of works A and D to work E
12	There are two works A and B, organize a flow construction in 3 sections; after work A_1 finish, works B_1 and A_2 can start; after work A_2 finish, work B_2 starts; works A_3 and B_2 are after work B_1; work B_3 starts after both works A_3 and B_2 finished		Every construction process has a professional team, and every professional team goes to the construction sections sequentially. The logical relationship of different construction works are expressed by logical and overlapping relationships

(2) In network diagrams, arrow tail node without predecessor works is not allowed except the start node of the network diagram. In Figure 9-6 (a), there are two nodes "1" and "3" which have no predecessor works, it will cause the logical relationship confusions, the work 3-5 start time can't be determined and that's not allowed in network diagram. Adjust it to the Figure 9-6(b) by adding a dummy work, and keep the exist logical relationship no changed, then the problem solved.

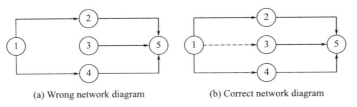

(a) Wrong network diagram   (b) Correct network diagram

Figure 9-6 The necessary predecessor work in intermediate nodes

(3) In network diagrams, arrow head node without successor works is not allowed except that end node of the network diagram. In Figure 9-7(a) there are two nodes "5" and "7" from which have no successor works starting, it causes the logical relationship confusions, that is the work 4-5 can't clearly express the logical relationship for its successor works, it is also not allowed in the network diagram. Figure 9-7(b) is correct after adjusting by adding a dummy work 5-7, it gives clear logical relationship, after work 4-5 finished, work 5-7 starts (though work 5-7 is a dummy work).

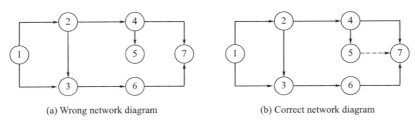

(a) Wrong network diagram   (b) Correct network diagram

Figure 9-7 The necessary successor work at intermediate nodes

(4) Circulation loop is forbidden in network diagram. In Figure 9-8(a), the path 1-2-3-1 is a circulation loop, it makes the network diagram express a wrong logical relationship because of the process sequence is contradictory. Figure 9-8(b) has been adjusted to be correct.

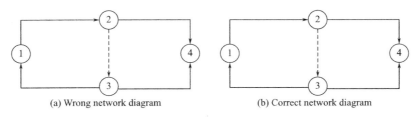

(a) Wrong network diagram   (b) Correct network diagram

Figure 9-8 The necessary predecessor work at intermediate nodes

(5) Sameserial number works are forbidden in the network diagram.

Every work in network diagram have and only have one unique serial number. In Figure 9-9 (a), both the works A and B have the same serial number of 1-2, it is not clear whether A or B is mentioned when saying work 1-2. In this case, an additional node and a dummy arrow line will be adopted for solving this type of problems. Figure 9-9 (b) is correct after adjusting base on the rules above.

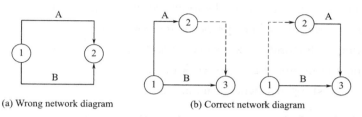

Figure 9-9 Serial number of work

(6) The work without start node is forbidden in network diagram. The leading out line in Figure 9-10 (a) shows that work B starts a certain time after work A starts, but doesn't give the exact time when the work B start, it is wrong network diagram for its lack of starting node. Figure 9-10 (b) is adjusted to be correct by adding a starting node at its arrow tail, at the same time, work A breaks into two works A1 and A2.

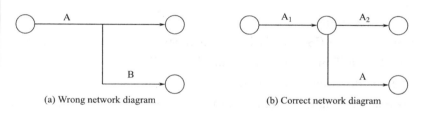

Figure 9-10 Express of the leading out arrow line

(7) Try to avoid "cross arrow lines" in network diagram, if it can't be avoided, "bridge method" or "directional method" might be adopted as shown in Figure 9-11.

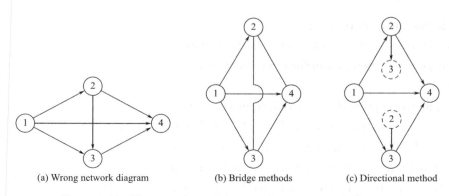

Figure 9-11 Express of crossing arrow lines in network diagram

2. Drawing methods and procedures

Before drawing network diagram, construction scheme must be determined, as well as construction sequences, and make list of working items and their logical relationships of each other.

Drawing procedures:

(1) The works without predecessor works should be drawn firstly.

(2) The successor works should be drawn after their predecessor work.

(3) Using dummy arrow lines properly and correctly.

(4) Checking, adjusting work sequences and logical relationship.

(5) Settling and numbering to network diagram.

Drawing can be stopped whenever no any successor work exists.

9.2.3 Calculation of time parameters in activity on line network diagram

After marking the duration of each work in the network diagram, time parameters can be calculated to provide a clear concept of time for the optimization, adjustment and execution of network plans.

The content of time parameters includes:

(1) Base on nodes of arrow line: Earliest (T_E) and latest time (T_L) at every nodes.

(2) Base on arrow line:

Earliest Start Time (EST), Earliest Finish Start Time (EFT), Latest Start Time (LST), Latest Finish Time (LFT), Total Float (TF), Free Float (FF), Duration of critical path (Time duration by calculation T_c).

There are many calculation methods in time parameters of network diagram, such as analysis calculation method, graphic calculation method, table calculation method, matrix calculation method and computer aided method.

Graphic calculation method will be discussed in this section by the helps of calculation formulas.

1. Duration of activity (t^{ij})

Time is the most essential and basic variable network diagram of planning and control. In general, two approaches may be used for the assessment of duration for activity completion.

(1) The first approach is the deterministic approach in which a single estimate of time gives reasonably accurate results (refer to 8.2.3-1). In this case, the activity time estimate are based on deterministic approach in which we may assume that we know enough about each job or operation, so that a single time estimate of their duration is sufficiently accurate to give reasonable results. No un-certainties are taken into consideration. The time of completion of any activity i-j is denoted by symbol t^{ij}.

(2) The second approach is the probabilistic approach in which one may only be able

to state limits within which it is virtually certain that the activity duration will lie. Between these limits, we must guess what is the probability of executing the activity.

Thus, to take the uncertainties into account, there are kinds of time estimates: the optimistic time estimate; the pessimistic time estimate; the most likely time estimate; expected time.

① The optimistic time estimate (t_O)

This is the shortest possible time in which an activity can be completed, under ideal conditions. This particular time estimate represents the time in which we could complete the activity or job if everything went along perfectly, with no problems or adverse conditions. Better than normal conditions are assumed to prevail.

② The pessimistic time estimate (t_P)

It is the best guess of the maximum time that would be required to complete the activity. This particular time estimate represents the time it might take us to complete a particular activity if everything went wrong and abnormal situations prevailed.

③ The most likely time estimate (t_L)

The most likely time or most probable time is the time that, in the mind of the estimator, represents the time the activity would most often require if normal conditions prevail. This time estimate lies between the optimistic and pessimistic time estimates. This time estimate reflects a situation where conditions are normal, things are as usual and there is nothing exciting.

④ Expected time (t_E)

The three time estimates t_O, t_P and t_L must be combined into one single time the average time taken for the completion of the activity. This average time or single workable time is commonly called expected time (t_E) and can be determined by the Formula (9-1).

$$t^{ij} = t_E = \frac{t_O + 4t_L + t_P}{6} \qquad (9-1)$$

2. Event time (Node time)

1) Earliest event time (T_E)

The earliest event time or earliest occurrence time is the earliest time at which an event can occur. It is the time by which all the activities discharging into the event under consideration are completed. For an activity path T_E for an event j can be completed from Formula (9-2).

$$T_E^j = T_E^i + t^{ij} \qquad (9-2)$$

There in:

T_E^i——Earliest occurrence time for the tail event.

T_E^j——Earliest occurrence time for head event.

ij——Activity under consideration.

t^{ij}——Time of completion of activity ij.

The above relation using forward pass is fully illustrated in Figure 9-12.

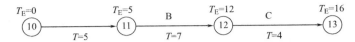

Figure 9-12 Calculation of T_E

In a network, there can be several activity paths leading to an event. However, no event can be considered to have reached until all activities leading to the event are completed. Hence T_E for that event will be greatest of the ones obtained from different paths. Hence the earliest occurrence time T_E^j for any event j is found from the expression in Formula (9-3).

$$T_E^j = (T_E^i + t^{ij})_{max} \tag{9-3}$$

2) Latest allowable occurrence time (T_L)

The latest allowable time (T_L) or the latest event time is the latest time by which an event must occur to keep the project on schedule. If the scheduled completion time (T_S) of the project is given, the latest even time of the end event will be equal to T_S. If the T_S is not specified, then T_S of end event can be taken equal to T_E.

The latest event time for an activity is computed by starting from head event and using the backward pass. In general, for an activity path, T_L^i for a predecessor event i can be determined from the known value of T_L^j of successor event, by the Formula (9-4).

$$T_L^i = T_L^j - t^{ij} \tag{9-4}$$

For a network, where several activity paths originate or radiate from the event under consideration; T_L for the event will be different from different path. However, minimum value of T_L will be the latest even time for that event as shown in Formula (9-5).

$$T_L^i = (T_L^j - t^{ij})_{min} \tag{9-5}$$

Combined tabular computations for T_E and T_L.

As stated earlier, the earliest even time (T_E) is computed using forward pass [Formula (9-2) and Formula (9-3)] while the latest event time (T_L) is computed using backward pass [Formula (9-4) and Formula (9-5)]. A combined tabular form, for the network shown in Figure 9-13, is illustrated in Table 9-2.

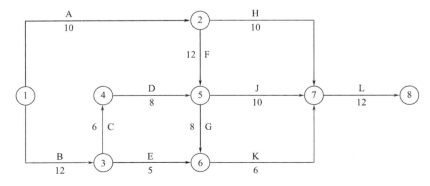

Figure 9-13 T_E and T_L calculation

Chapter 9 Network Plan Technology

Column 1 of Table 9-2 gives the event number, starting with the initial event and proceeding in the direction of increasing numbers of the event. Column 2 gives the predecessor events while column 6 gives the successor events to the events of column 1. These columns are completed first, using the network. An event under consideration (column 1) may have one or more than one predecessor events (column 2), and one or more than one successor events (column 6). A horizontal line is drawn after entering all predecessor events and successor events to every event of column1.

Calculations table of T_E and T_L Table 9-2

Event No. (1)	Predecessor event(i) (2)	t^{ij} (3)	T_E^j (4)	T_E (5)	Successor event(j) (6)	t^{ij} (7)	T_L^i (8)	T_L (9)
1	—	—	0	0	2 3	10 12	14−10=4 12−12=0	0
2	1	10	10	10	5 7	12 10	26−12=14 28	14
3	1	12	12	12	4 6	6 5	18−6=12 40−5=35	12
4	3	6	12+6=18	18	5	8	26−8=18	18
5	2 4	12 8	10+12=22 8+18=26	26	6 7	8 10	34−8=26 40−10=30	26
6	3 5	5 8	12+2=17 26+8=34	34	7	6	40−6=34	34
7	2 5 6	10 10 6	10+10=20 26+10=36 34+6=40	40	8	12	52−12=40	40
8	7	12	40+12=52	52	—	—	52	52

Then, computations are done for the earliest event time (T_E) in column 3, 4 and 5. Column 3 is for the activity time t^{ij} where j is the event under consideration (column 1) and i is the predecessor event (column 2), T_E^j is computed from the relation. $T_E^j = T_E^i + t^{ij}$.

Where there are more than one predecessor events, several values of T_E^j are obtained, which entered in column 4. The maximum value T_E^j is underlined. This underlined value is the appropriate value of the earliest event time for the event under consideration (column 1), and is entered as T_E in column 5. For the computation of T_E, we thus use forward pass starting from initial event and proceeding in the downward direction (↓) in the Table 9-2.

Then we compute the latest event (occurrence) time of the same events under consideration (column 1), in columns 7, 8 and 9. Column 7 is the activity time t^{ij} where i is the event under consideration and j is the successor event (column 6). T_L^i is computed from the relation $T_L^i = T_L^j - t^{ij}$.

Computations are done by backward pass, starting with the end event and proceeding upwards (↑) in the Table 9-2. If T_S is not given, T_L of the last event is taken equal to its T_E. Where there are more than one successor events, several values of T_L^i are obtained, which are entered in column 8. The minimum value of T_L^i is underlined. This underlined value is the appropriate value of the latest event time for the event under consideration, and is entered as T_L in column 8.

3. Activity time

1) Earliest Start Time (EST)

Earliest start time of an activity is the earliest time by which it can commence. This is naturally equal to the earliest event time associated with the tail of the activity arrow. Thus, for an activity i-j, there is Formula (9-6).

$$EST = T_E^i \tag{9-6}$$

2) Earliest Finish Time (EFT)

If an activity proceeds from its early start time and takes the estimated duration for completion, then it will have an early finish. Hence EFT for an activity is defined as the earliest time by which it can be finished. This is evidently equal to the earliest start time plus estimated duration of the activity. EFT = earliest start time + activity duration, as shown in Formula (9-7).

$$EFT = T_E^i + t^{ij} \tag{9-7}$$

3) Latest Finish Time (LFT)

The latest finish time for an activity is the latest time by which an activity can be finished without delaying the completion of the project. Naturally, the latest finish time for an activity will be equal to the latest allowable occurrence time for the event at the head of the arrow. Hence, LFT = Latest event time at the head of activity arrow, as shown in Formula (9-8).

$$LFT = T_L^j \tag{9-8}$$

4) Latest Start Time (LST)

Latest start time for an activity is the latest time by which an activity can be started without delaying the completion of the project. For "no delay" condition to be fulfilled it should be naturally equal to the Latest Finish Time (LFT) minus the activity duration, LST = LFT − activity duration, as shown in Formula (9-9).

$$LST = T_L^j - t^{ij} \tag{9-9}$$

5) Total Float (TF)

In certain activities, it will be found that there is a difference between maximum time available and actual time required to perform the activity. This difference is known as the total float. In other words, it is the excess of the maximum available time over the activity time. They are shown as in Formula (9-10) and Formula (9-11).

$$TF = LST - EST = LFT - EFT \tag{9-10}$$

$$TF = (T_L^j - T_E^i) - t^{ij} \qquad (9\text{-}11)$$

6) Free Float (FF)

Free Float is the free time that can be used by an activity without delaying any succeeding activity, it can be calculated by Formula (9-12) and Formula (9-13).

$$FF = (T_E^j - T_E^i) - t^{ij} \qquad (9\text{-}12)$$
$$FF = T_E^j - EFT^i \qquad (9\text{-}13)$$

Example 9-1:

Activity on line network diagram is shown in Figure 9-14 (activities and their durations). Calculate its time of project by the node time method.

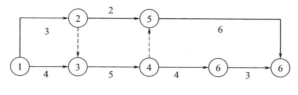

Figure 9-14 Network diagram in the example

Solution:

(1) Take $T_E^1 = 0$
(2) $T_E^j = T_E^i + t^{ij}$, or $T_E^j = \max\{T_E^i + t^{ij}\}$
(3) $T_E^n = T_c$
(4) Calculation process (Figure 9-15)

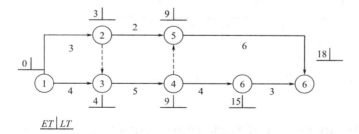

Figure 9-15 Calculation process

(5) Time of project is 18 (days).

Example 9-2:

An activity on line network diagram is shown in Figure 9-16. Calculate the time parameter, determine the time of the project and the critical path.

Solution:

(1) Calculate the time parameter, as shown in Figure 9-17.
(2) The time of the project is 45 (days).
(3) The critical path is: 1→2→4→8→11→12.

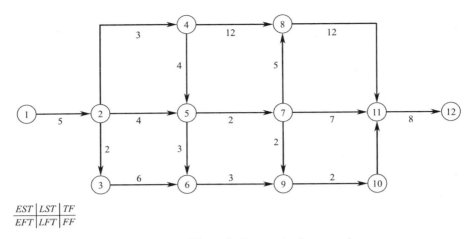

Figure 9-16　Network diagram in the example

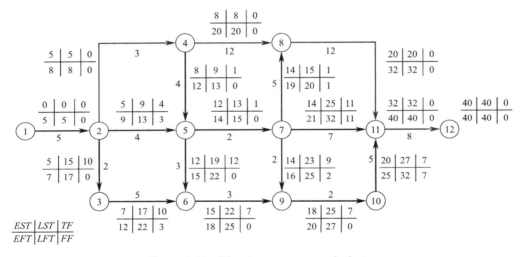

Figure 9-17　The time parameter calculation

9.3　Activity on node network plan (Single code network diagram)

9.3.1　Content and basic symbols of activity on line network diagram

1. Content

Activity on node diagram is also composed of nodes, arrow lines and paths, but they have different meanings comparing with the activity on nodes diagram.

In activity on node diagram, nodes express the activities contents and duration time, while the arrow lines express logical relationships between activities, so there is no any dummy arrow lines in activity on node network.

It can be called node network diagram because it expresses activities at nodes.

2. Basic symbols

1) Node

Node is the main symbols in the network diagram, it can be expressed as a circle or rectangular with the activity contents inside, as shown in Figure 9-18.

Number
Activity
Duration

Number		Resource
Activity		Duration
ES_i	LS_i	TF
EF_i	LF_i	FF

Figure 9-18 Node expression of activity on node network diagram

2) Arrow line

In activity on node network diagram, arrow lines consume none of time and resources.

The arrow line head points to the forward direction project progress, and the arrow tail indicates the predecessor work as shown in Figure 9-19. Where, A is the predecessor work of B and C, B and C are parallel works, D is the successor work of B and C.

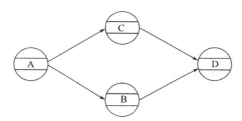

Figure 9-19 Expression of logical relationship in activity on node network diagram

3) Number

One work can only have an unique serial number, repeated serial numbers are forbidden in network diagram. Numbering method is same as that of activity on line network discussed in 9.2.1-2 of this Chapter.

9.3.2 Drawing of activity on node network diagram

1. Expression way of logical relationship in activity on node network diagram

In activity on node network diagram, logical relationship among different works depends on objective order from technical or organizational requirement, it is easier to be expressed that of in activity on line network diagram.

Common expression ways for logical relationships among works are show in Table 9-3.

Expression of logical relationship in network diagram Table 9-3

No.	Logical relationship between works	Expression in the network diagram
1	Works A and B are executed in sequence	Ⓐ → Ⓑ
2	Work D can only start after works B and C finished	Ⓑ → Ⓓ, Ⓒ → Ⓓ
3	After A finished, work C can start; work D can start after both works A and B finished	Ⓐ → Ⓒ; Ⓐ → Ⓓ; Ⓑ → Ⓓ
4	After A finished, work C can start; after B finished, work E can start; after both works A and B finished, work D can start	Ⓐ → Ⓒ; Ⓐ → Ⓓ; Ⓑ → Ⓓ; Ⓑ → Ⓔ

2. Basic principle

The basic principles are similar comparing with that of activity on line network diagram.

Dummy nodes might be needed if there are many works start at the same time or finished at the same time, in order to satisfy the requirements of "one start node" and "one end node".

3. Drawing methods and procedures

Its drawing methods and procedures are similar with that of activity on line network diagram.

Drawing procedures:

(1) The works without predecessor works should be drawn firstly.

(2) The successor works should be drawn after their predecessor work.

(3) Using dummy nodes properly and correctly.

(4) Checking, adjusting work sequences and logical relationship.

(5) Settling and numbering to network diagram.

Drawing can be stopped whenever no any successor work exists.

Example 9-3:

Draw activity on node network diagram according to the given logical relationship in Table 9-4.

Logical relationship table Table 9-4

Name of work	Predecessor work	Successor work
A	—	B、C、E
B	A	D、E

Chapter 9 Network Plan Technology

Continued

Name of work	Predecessor work	Successor work
C	A	G
D	B	F,G
E	A,B	F
F	D,E	G
G	D,F,C	—

Solution:
(1) The works without predecessor work A is drawn at first.
(2) The successor works are drawn after their predecessor works.
(3) Using dummy nodes at the start node and end node.
(4) Checking, adjusting work sequences and logical relationship.
(5) Settling and numbering to network diagram.
Process is shown in Figure 9-20.

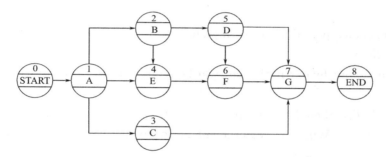

Figure 9-20 Network diagram in the example

9.3.3 Calculation of time parameters in activity on node network diagram

They are similar comparing from that of activity on line network diagram, but there are different in expressions style and parameter symbols.

1. Earliest Start Time (ES)

Take start node of network plane:
$$ES_0 = 0 \tag{9-14}$$

For any other works:
$$ES_j = \max\{ES_i + D_i\} = \max\{EF_i\} \tag{9-15}$$

There in:
ES_i——Earliest occurrence time predecessor work of work j.
D_i——Duration time of work i.

2. Earliest Finish Time (EF)
$$EF_i = ES_i + D_i \tag{9-16}$$

335

3. Calculated period of network plan (T_c)

$$T_c = EF_n \text{ or } T_c = ES_n + D_n \tag{9-17}$$

There in:

EF_n——Earliest finish time at end node (n).

ES_n——Earliest start time at end node (n).

D_n——Duration of end node (n).

4. Calculation of LAG and float

1) LAG

LAG means difference of successor work's Earliest Start Time (ES_j) and predecessor work's Earliest Finish Time (EF_i)

$$LAG_{ij} = ES_j - EF_i \tag{9-18}$$

2) Free float time

$$FF_l = \min\{LAG_{ij}\} \tag{9-19}$$

For final work of the network plan,

$$FF_n = T_p - T_c \tag{9-20}$$

There in:

T_p——Scheduled period of network plan.

3) Total float

(1) Calculate one by one in the backward directions from the end work of the network plan.

If $T_p = T_c = T_r$, then $TF_n = 0$; if $T_p \neq T_c$, then $TF_n = T_p - T_c$.

Wherein, T_r——Which is requirement period of network plane from contractor or superior.

(2) For any other works, $FF_n = \min\{LAG_{ij} + TF_j\}$ (where work j is the successor work of the work i).

5. Calculation of latest start time

$$LS_i = ES_i + TF_i \tag{9-21}$$

6. Calculation of latest finish time

$$LF_i = EF_i + TF_i \text{ or } LF_i = LS_i + D_i \tag{9-22}$$

Example 9-4:

The given activity on line network diagram is shown in Figure 9-21. Try to transform it into activity on node network diagram and calculate the time parameters.

Solution:

As an example, pick one time parameter to do calculate for every type of time parameters.

(1) The earliest start time: take $ES_0 = 0$, then $ES_1 = ES_0 + D_0 = 0$ (day)

(2) The earliest finish time: $EF_1 = ES_1 + D_1 = 0 + 3 = 3$ (days)

(3) Calculated period: $T_c = EF_n = 12$ (days)

(4) LAG calculation: $LAG_{1-2} = ES_2 - EF_1 = 3 - 3 = 0$ (day)

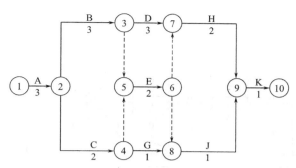

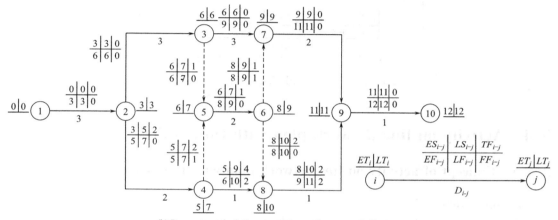

(b) Parameter calculation at activity on line network diagram

Figure 9-21 Network diagram in the example

$$LAG_{1-3} = ES_3 - EF_1 = 3 - 3 = 0 \text{ (day)}$$
$$LAG_{8-9} = ES_9 - EF_8 = 11 - 9 = 2 \text{ (days)}$$

(5) Free float: $EF_1 = \min \{LAG_{1-2}, LAG_{1-3}\} = \min \{0, 0\} = 0$ (day)

(6) Total float: as for, $T_c = T_p = T_r$, then $TF_9 = 0$ (day)

$$TF_8 = \min \{LAG_{8-9} + TF_9\} = \min \{2+0\} = 2 \text{ (days)}$$
$$TF_1 = \min \{LAG_{1-2} + TF_2, LAG_{1-3} + TF_3\}$$
$$= \min \{0+0, 0+2\}$$
$$= 0 \text{ (day)}$$

(7) Latest start time: $LS_1 = ES_1 + TF_1 = 0 + 0 = 0$ (day)

(8) Latest finish time: $LF_1 = EF_1 + TF_1 = 3 + 0 = 3$ (days)

or $LF_1 = LS_1 + D_1 = 0 + 3 = 3$ (days)

All of the time parameters are listed in Figure 9-22.

7. Critical path

In activity on node network plan, paths without any LAG are critical path, and works on critical path are called critical work which has the smallest total float.

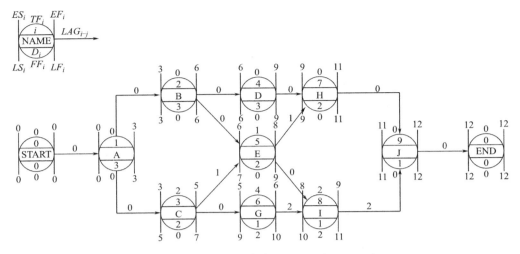

Figure 9-22 Network diagram in the example

9.4 Activity on line network plan with time scaled

9.4.1 Concept of activity on line network plan with time scaled

1. Implication

It represents working time by time scale coordinate, an intuitive expression way for activity on line network plan.

2. Basic symbol

Time scaled network plan is an activity on line network plane but drawn on the time scale table, the time unit could be hour, day, week, month, season or year, etc.

The real works are expressed by real arrow lines, dummy works are expressed by dummy arrow lines.

Except the wave lines, the horizontal projective lengths are work duration, as shown in Figure 9-23.

3. Characteristic of network plan with time scared

Comparing with the normal activity on line network plan, time scale network plan has characteristic as followings:

(1) Time parameters are obviously and clearly, it has both advantages of Gantt chart and that of activity on line network plan without time scaled.

(2) Drawing and modifying is relatively difficult because its arrow line length is controlled by time scale coordinate, and it might need to re-draw if the work duration changed.

(3) Some of time parameters have been shown in the network diagram after finishing the drawings, so it can save time in time parameter's calculation.

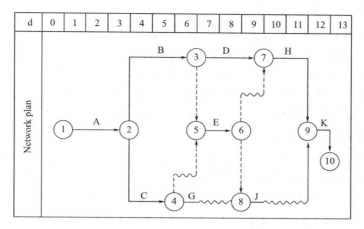

Figure 9-23　Time scale network plan

9.4.2　Drawing activity on line network plan with time scaled

1. Basic principles

(1) Time durations are expressed by the horizontal length of arrow lines length at time table, their time values corresponded each other.

(2) Dummy work must be expressed by dummy arrow line, or expressed by wave lines whenever there is Time Different (Free Float).

(3) Working time in scaled network plan should be drawn according to Earliest Start Time (EST).

(4) Activity on line network plan must be done before the activity on line network with time scale starts to compile.

2. Drawing procedures

Drawing procedures are discussed by the given activity online network in Figure 9-24.

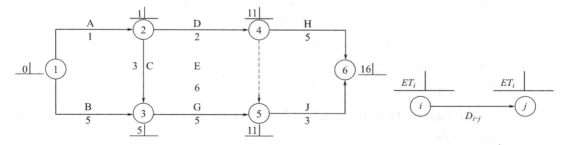

Figure 9-24　The given activity on line network diagram

1) Direct drawing

(1) Time table.

(2) Locate the start event 1 onto the initial time scale mark.

(3) Outward works from event 1 should be drawn according its work duration time,

339

refer to 1-2 and 1-3 in Figure 9-24.

(4) Arrow head must be located at the latest finish event, refers to event 3, 4, 5 and 6 in Figure 9-24.

(5) Any other successor work duration time should be drawn according to the event Latest Time (LT), the insufficient parts should be completed with wave lines, refer to 2-3, 2-4, 3-5 and 5-6 in Figure 9-25. For any dummy arrow lines (dummy works), take same rules as mentioned above.

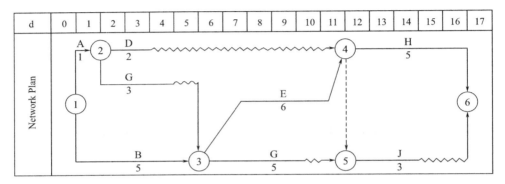

Figure 9-25 Drawing procedures of time scale network plan

(6) The drawing direction should be from left to right, till the location of end event. The layout should be similar with the activity on line network plane.

(7) Checking, modifying and numbering of the network diagram.

2) Indirect drawing

(1) Calculate the Earliest Time (ET) of all events in network diagram.

(2) Draw time scaled table.

(3) Locate the Earliest Time (ET) of all events in time scaled table, take layout the same with that of the activity on line network diagram.

(4) Drawing work duration time by means of Earliest Start Time (EST) methods, any insufficient parts should be completed with wave lines.

(5) The drawing direction should be from left to right, till the location of end event. The layout should be similar with the activity on line network plane.

(6) Checking, modifying and numbering of the network diagram.

9.4.3 Calculation of time parameters in activity on node network diagram with time scaled

1. LAG determination

LAG is the horizontal projection length between the work and its successor works, except the work which take end event as its working end.

As shown in Figure 9-25, $LAG_{C,E}=LAG_{C,G}=1$; $LAG_{D,H}=LAG_{D,J}=8$; $LAG_{G,J}=1$, other LAG are zero except whose that take end event as its working end.

The projective length for which take the end event as its working end, is the time difference between the work's Earliest Finish Time (EST) and the project plan period.

The projective length of wave lines on dummy arrow line, expresses the time difference of its predecessor and successor works.

2. Critical path and project period

1) Calculated project period

Project period is the event time difference between start event and end event, $T_c = 16 - 0 = 16$ (days).

2) Critical path

The paths which have no any wave lines from end event to start event, are critical paths. That means the LAG are zero at critical paths, because there are no any wave lines, on the premise of $T_c = T_p = T_r$, for any works on critical paths, $TF = 0$, $FF = 0$.

In Figure 9-25, critical path is ①→③→④→⑥.

3. Time parameters

A given activity on line network with time scaled is shown in Figure 9-26.

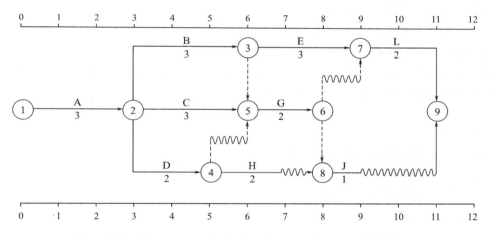

Figure 9-26 Parameters calculation of time scaled network diagram

1) EST and EFT

The time mark corresponding to the center of the left end node of the working arrow line is the EST of the work.

When there is no wave lines in working arrow line, the time mark corresponding to the center of the right end node is the EFT of the work.

When there is wave lines in working arrow line, the time mark corresponding to the center of the solid line part is the EFT of the work.

In Figure 9-26, $EST_{2,3} = 3$, $EST_{4,8} = 5$, $EST_{3,7} = EST_{5,6} = 6$, $EST_{8,9} = 8$; $EFT_{2,3} = 6$, $EFT_{4,8} = 7$, $EFT_{3,7} = 9$, $EFT_{5,6} = 8$, $EFT_{8,9} = 9$.

The maximum value of the $EFT_{i,n}$ is the calculate project period, that is: $T_c = \max \{EFT_{i,n}\}$

2) TF

The determination of TF should take the end event as start event, calculate along backward direction.

(1) $TF_{i,n}$

For the works which take the project plan end event as its finish event, the $TF_{i,n}$ equals to the difference of project planned period and the work's Earliest Finish Time ($EFT_{i,n}$), that is:

$$TF_{i,n} = T_p - EFT_{i,n} \tag{9-23}$$

There in:

$TF_{i,n}$——Total float for the works which take the project plan end event as its finish event.

T_p——Planned period of network plan.

$EFT_{i,n}$——Earliest finish time for the works which take the project plan end event as its finish event.

Suppose that T_p equals to 11, then:

$TF_{7,9} = T_p - EFT_{7,9} = 11 - 11 = 0$

$TF_{8,9} = T_p - EFT_{8,9} = 11 - 9 = 2$

(2) $TF_{i,j}$ of other work

Other work's $TF_{i,j}$ is equal to the minimum value of sum of successor work's TF, and LAG of this work and its successor work.

$$TF_{i,j} = \min\{T_{j,k} + LAG_{i,j-j,k}\} \tag{9-24}$$

There in:

$TF_{i,j}$——Total float of work i-j.

$TF_{j,k}$——Total float of work j-k, which is the successor work of the work i-j.

$LAG_{i,j-j,k}$——LAG of work i-j and its successor work j-k.

In Figure 9-26,

$TF_{3,7} = TF_{7,9} + LAG_{3,7-7,9} = 0 + 0 = 0$

$TF_{4,8} = TF_{8,9} + LAG_{4,8-8,9} = 2 + 1 = 3$

$TF_{6,7} = TF_{7,9} + LAG_{6,7-7,9} = 0 + 1 = 1$

$TF_{6,8} = TF_{8,9} + LAG_{6,8-8,9} = 2 + 0 = 2$

$TF_{5,5} = \min\{TF_{6,7} + LAG_{5,6-6,7}, TF_{6,8} + LAG_{5,6-6,8}\} = \min\{1+0, 0+2\} = 1$

3) FF

(1) $FF_{i,n}$

For the works which take the project plan end event as its finish event, the $FF_{i,n}$ equals to the difference of project planned period and the work's Earliest Finish Time ($EFT_{i,n}$), that is:

$$TF_{i,n} = T_p - EFT_{i,n} \tag{9-25}$$

There in

$FF_{i,n}$——Total float for the works which take the project plan end event as its finish event, other symbols has the same meanings as before.

In Figure 9-26, FF of Work L and J are:
$$FF_{7,9}=T_p-EFT_{7,9}=11-11=0$$
$$FF_{8,9}=T_p-EFT_{8,9}=11-9=2$$

Actually, FF must be equal to that of TF for the works which take the project plan end event as its finish event.

(2) $FF_{i,j}$ of other work

Other work's $FF_{i,j}$ is its projective length of wave line, or the dummy arrow line's minimum projective length if its successor work are dummy works.

$FF_{1,2}=FF_{2,3}=FF_{2,4}=FF_{2,5}=FF_{3,7}=0$

$FF_{4,8}=1$

$FF_{5,6}=\min\{0, 1\}=1$

4) LST and LFT

(1) Latest Start Time (LST) equals to the sum of Earliest Start Time (EST) and its Total Float (TF), that is:
$$LST_{i,j}=EST_{i,j}+TF_{i,j} \qquad (9\text{-}26)$$

LST of work E, G, H and J are:

$LST_{3,7}=EST_{3,7}+TF_{3,7}=6+0=0$

$LST_{5,6}=EST_{5,6}+TF_{5,6}=6+1=7$

$LST_{4,8}=EST_{4,8}+TF_{4,8}=5+3=8$

$LST_{8,9}=EST_{8,9}+TF_{8,9}=8+2=10$

(2) Latest Finish Time (LFT) equals to the Earliest Finish Time (EFT) and its Total Float (TF), that is:
$$LFT_{i,j}=EFT_{i,j}+TF_{i,j} \qquad (9\text{-}27)$$

LFT of work E, G, H and J are:

$LFT_{3,7}=EFT_{3,7}+TF_{3,7}=9+0=9$

$LFT_{5,6}=EFT_{5,6}+TF_{5,6}=8+1=9$

$LFT_{4,8}=EFT_{4,8}+TF_{4,8}=7+3=10$

$LFT_{8,9}=EFT_{8,9}+TF_{8,9}=9+2=11$

Questions

1. Explain the concept of network plan, network diagram and network plan technology.
2. Write out the principle of network plan.
3. Briefly introduce the characteristics of network plan.
4. Introduce briefly the basic symbols of activity on line diagram.
5. Write out the basic principle in drawing of activity on line diagram.
6. Briefly describe of the drawing procedures of activity on line diagram.
7. Write out the concept of activity on node network diagram.
8. Write out the drawing procedure of activity on line network diagram.

9. What's the concept of activity on line network plan with time scaled?

10. What are basic symbols in activity on line network plan with time scaled?

11. Briefly introduce the characteristic of network plan with time scared.

12. What are the basic principles in drawing of activity on line network plan with time scaled?

Exercise

An activity on line network diagram is shown as follows. Calculate the time parameter, and determine the critical path, draw the activity on line network with time scaled.

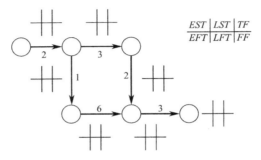

Figure 9-27 An activity on line network diagram

Table 9-5

Duration	1	2	3	4	5	6	7	8	9	10	11	12	13	14

Reference

[1] Code for design of concrete structures: GB 50010—2010 [S]. Beijing: China Architecture & Building Press, 2010.
[2] Technical code for composite steel-form: GB/T 50214—2001 [S]. Beijing: China Architecture & Building Press, 2010.
[3] Roy Chudley, Roger Greeno. Building construction handbook [M]. Routledge. 2016.
[4] Code for quality acceptance of concrete structure construction: GB 50204—2015 [S]. Beijing: China Architecture & Building Press, 2015.
[5] Code for construction of concrete structures: GB 50666—2011 [S]. Beijing: China Architecture & Building Press, 2011.
[6] Steel for the reinforcement of concrete—Part 1: Hot rolled plain bars: GB 1499.1—2017 [S]. Beijing: China Architecture & Building Press, 2017.
[7] Steel for the reinforcement of concrete—Part 2: Hot rolled ribbed bars: GB 1499.2—2018 [S]. Beijing: China Architecture & Building Press, 2018.
[8] Technical specification for concrete structures with cold-rolled ribbed steel wires and bars: JGJ 95—2011 [S]. Beijing: China Architecture & Building Press, 2012.
[9] Quenching and self-tempering ribbed bars for the reinforcement of concrete: GB/T 13014—2013 [S]. Beijing: China Architecture & Building Press, 2014.
[10] Screw-thread steel bars for theprestressing of concrete: GB/T 20065—2016 [S]. Beijing: China Architecture & Building Press, 2017.
[11] Technical code for roof engineering: GB 50345—2012 [S]. Beijing: China Architecture & Building Press, 2012.
[12] Technical code for waterproofing of underground works: GB 50108—2008 [S]. Beijing: China Architecture & Building Press, 2009.
[13] Technical code for interior waterproof of residential buildings: JGJ 298—2013 [S]. Beijing: China Architecture & Building Press, 2013.
[14] Double skin metal faced insulating panels for building: GB/T 23932—2009 [S]. Beijing: China Architecture & Building Press, 2009.
[15] Technical code for application of profiled metal sheets: GB 50896—2013 [S]. Beijing: China Architecture & Building Press, 2014.
[16] Code for construction organization plan of building engineering: GB/T 50502—2009 [S]. Beijing: China Architecture & Building Press, 2009.